Dynamics of Earth Science

Dynamics of Earth Science

Edited by **Russell Sands**

SYRAWOOD
PUBLISHING HOUSE

New York

Published by Syrawood Publishing House,
750 Third Avenue, 9th Floor,
New York, NY 10017, USA
www.syrawoodpublishinghouse.com

Dynamics of Earth Science
Edited by Russell Sands

International Standard Book Number: 978-1-68286-030-4 (Hardback)

Printed in the United States of America.

Contents

Permissions

List of Contributors

Preface

The discipline of earth science has undergone rapid development in the recent past. This book elucidates the concepts and innovative models around prospective developments with respect to the study of earth science. Some of the topics included in this text are upper mantle structure, lithosphere, sea-level indicators, etc. Comprising contributions of an eminent panel of internationally renowned scholars, this book is highly recommended for students pursuing geology, earth sciences and related fields of study. It will also prove to be a valuable reference material for research scholars and academicians.

All of the data presented henceforth, was collaborated in the wake of recent advancements in the field. The aim of this book is to present the diversified developments from across the globe in a comprehensible manner. The opinions expressed in each chapter belong solely to the contributing authors. Their interpretations of the topics are the integral part of this book, which I have carefully compiled for a better understanding of the readers.

At the end, I would like to thank all those who dedicated their time and efforts for the successful completion of this book. I also wish to convey my gratitude towards my friends and family who supported me at every step.

Editor

Upper mantle structure around the Trans-European Suture Zone obtained by teleseismic tomography

I. Janutyte[1,5,6], M. Majdanski[2], P. H. Voss[3], E. Kozlovskaya[4], and PASSEQ Working Group[7]

[1]NORSAR, Kjeller, Norway
[2]Institute of Geophysics Polish Academy of Sciences, Warsaw, Poland
[3]Geological Survey of Denmark and Greenland – GEUS, Copenhagen, Denmark
[4]Sodankylä Geophysical Observatory/Oulu Unit, University of Oulu, Oulu, Finland
[5]Vilnius University, Vilnius, Lithuania
[6]Lithuanian Geological Survey, Vilnius, Lithuania
[7]indicated in Acknowledgements

Correspondence to: I. Janutyte (ilma@norsar.no)

Abstract. The presented study aims to resolve the upper mantle structure around the Trans-European Suture Zone (TESZ), which is the major tectonic boundary in Europe. The data of 183 temporary and permanent seismic stations operated during the period of the PASsive Seismic Experiment (PASSEQ) 2006–2008 within the study area from Germany to Lithuania was used to compile the data set of manually picked 6008 top-quality arrivals of P waves from teleseismic earthquakes. We used the TELINV nonlinear teleseismic tomography algorithm to perform the inversions. As a result, we obtain a model of P wave velocity variations up to about $\pm 3\%$ with respect to the IASP91 velocity model in the upper mantle around the TESZ. The higher velocities to the east of the TESZ correspond to the older East European Craton (EEC), while the lower velocities to the west of the TESZ correspond to younger western Europe. We find that the seismic lithosphere–asthenosphere boundary (LAB) is more distinct beneath the Phanerozoic part of Europe than beneath the Precambrian part. To the west of the TESZ beneath the eastern part of the Bohemian Massif, the Sudetes Mountains and the Eger Rift, the negative anomalies are observed from a depth of at least 70 km, while under the Variscides the average depth of the seismic LAB is about 100 km. We do not observe the seismic LAB beneath the EEC, but beneath Lithuania we find the thickest lithosphere of about 300 km or more. Beneath the TESZ, the asthenosphere is at a depth of 150–180 km, which is an intermediate value between that of the EEC and western Europe. The results imply that the seismic LAB in the northern part of the TESZ is in the shape of a ramp dipping to the northeasterly direction. In the southern part of the TESZ, the LAB is shallower, most probably due to younger tectonic settings. In the northern part of the TESZ we do not recognize any clear contact between Phanerozoic and Proterozoic Europe, but further to the south we may refer to a sharp and steep contact on the eastern edge of the TESZ. Moreover, beneath Lithuania at depths of 120–150 km, we observe the lower velocity area following the boundary of the proposed paleosubduction zone.

1 Introduction

1.1 Tectonic settings

The Trans-European Suture Zone (TESZ) is the most fundamental lithospheric boundary in Europe (Pharao, 1999) that marks the transition between the old Proterozoic lithosphere of the East European Craton (EEC) and the younger Phanerozoic lithosphere of central and western Europe (Fig. 1a). The EEC, the Baltica segment to the east of the TESZ, comprises three paleocontinents: Sarmatia, Volgo-Uralia and Fennoscandia (Bogdanova et al., 2006), with significant sutures in between them. The territories in the northeastern part of the EEC consist of several Svecofennian crustal units (Fig. 1b), such as the Belarus–Podlasie Granulite Domain (BPG), the East Lithuanian Domain (EL) and the West Lithuanian Granulite Domain (WLG), which continue in a NE–SW direction into Poland and terminate at the TESZ (Bogdanova et al., 2006). The area in between the EL and the WLG is called the Middle Lithuanian Suture Zone (MLSZ), which was interpreted as a paleosubduction zone along which the EL subducted under the WLG about 1.83 Ga (Motuza, 2004, 2005; Motuza and Staškus, 2009).

To the west of the TESZ, the structure of the lithosphere is much more complex compared to the lithosphere of the EEC (e.g., Zielhuis and Nolet, 1994; Dadlez et al., 2005; Knapmeyer-Endrun et al., 2013a; Babuška and Plomerova, 2001) (Fig. 1a). The territories in central–western Europe consist of various continental fragments that were subsequently rifted off the northern margin of Gondwana and accreted to the southwestern margin of the Precambrian Baltica during a number of orogenic events (Nolet and Zielhuis, 1994; Pharaoh, 1999; Winchester and the PACE TMR Network Team, 2002; Banka et al., 2002). The TESZ contains two pronounced linear segments: the Sorgenfrei–Tornquist Zone (STZ) in the northwestern part of the TESZ between Sweden and Denmark, and Germany, and the Teisseyre–Tornquist Zone (TTZ) stretching from the Baltic Sea in the northwest to the Black Sea in the southeast. The territories around the TESZ formed during four major geological stages: (1) Caledonian collision tectonics, (2) Variscian orogeny, (3) Mesozoic rifting, and (4) Alpine orogenic events (Bogdanova et al., 2007; Thybo, 2000). During the Cambrian period, the terrains of Lysogory, Malopolska and Bruno-Silesian accreted to Baltica, forming southern Poland and the eastern edge of the Bohemian Massif (Belka et al., 2000). During the Caledonian orogeny, the Avalonian segment closing the Tornquist Ocean accreted to the eastern margin of Baltica (Pharaoh, 1999). The Variscan orogeny from the late Silurian to early Carboniferous resulted in a junction of three paleomicrocontinents: Saxothuringian, Moldanubian and Tepla-Barrandian, in the territory of Vogtland and northwestern Bohemia (Franke and Zelazniewicz, 2000). The Saxothuringian is juxtaposed with the Moldanubian in a broad contact indicating a paleosubduction of the Saxothuringian, possibly with a piece of the oceanic lithosphere beneath the Moldanubian (Plomerova et al., 1998). The "triple junction" resulted in the crust and lithosphere thinning as well as the tectono-sedimentary evolution of the Cheb Basin situated above the junction. The basin formed between the late Oligocene and Pliocene by reactivation of the Variscan junction of the three lithospheric blocks (Babuška et al., 2007). During the Cretaceous to Cenozoic periods, a number of terrains accreted to western Europe, resulting in the Alpine and Carpathian orogenies. During the middle to late Eocene, rifting processes took place in central Europe, followed by the quaternary volcanism (Wagner et al., 2002; Babuška et al., 2007) that was possibly related to the upper mantle reservoir (Babuška and Plomerova, 2001; Zhu et al.,

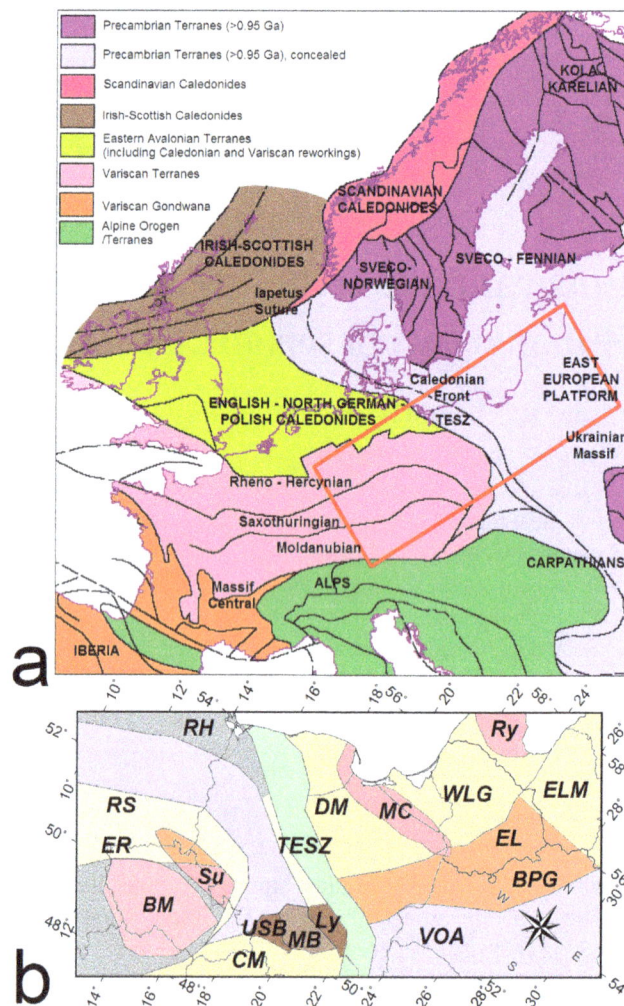

Figure 1. (a) Simplified tectonic sketch of Precambrian and Phanerozoic Europe (after Blundell et al., 1992). Study area indicated by red rectangle. (b) Tectonic sketch of the study area compiled from Skridlaite and Motuza (2001), Malinowski et al. (2008), Guterch et al. (1999), Bogdanova et al. (2001), and Gee and Stephenson (2006). Units: BM, Bohemian Massif; BPG, Belarus–Podlasie Granulite Belt; CM, Carpathian Mountains; DM, Dobrzyn Massif; EL, East Lithuanian Domain; ELM, East Latvian Massif; ER, Eger Rift; Ly, Lysogory; MB, Malopolska Block; MC, Mazury Complex; RH, Rheno-Herzynian Front; RS, Rheic Suture; Ry, Riga batholith; Su, Sudetes Mountains; TESZ, Trans-European Suture Zone; USB, Upper Silesian Coal Basin; VOA, Volyn–Orsha aulacogen; WLG, West Lithuanian Granulite Domain.

2012). The developed Tertiary Eger Rift continues 300 km in the ENE–WSW direction and follows the late Variscan mantle transition between the Saxothuringian and the Tepla-Barrandian.

1.2 Review of previous studies

Due to a long evolution and complex tectonic structure, the TESZ and the surrounding territories have always been a subject of great interest in geosciences. The structure of the crust and uppermost mantle around the TESZ has been studied intensively during the controlled-source seismic experiments – long-range deep seismic sounding (DSS) profiles (e.g., Guterch et al., 1999, 2004; Grad et al., 2002, 2006; EUROBRIDGE Seismic Working Group, 1999; Pharaoh and TESZ Project Core Group, 2000). The obtained results show large variations of average thickness of the continental crust: the Moho depth varies from 28–35 km beneath the Paleozoic platform (Guterch and Grad, 1996; Pharaoh et al., 1997; Guterch et al., 1999) to 40–50 km beneath the western part of the EEC adjoining the TESZ and even deeper farther to the northeast (Grad et al., 2006; Guterch et al., 2004). The projects provided sufficient information about the crustal structure around the area, which was used to compile some precise 3-D crustal models (e.g., Majdanski, 2012). Using data of the DSS projects, the EUROBRIDGE Working Group (1999), Czuba et al. (2001), Yliniemi et al. (2004), Grad et al. (2002) and Thybo et al. (2003) found some reflectors in the upper mantle just beneath the Moho going down to 75 km in Fennoscandia, which could be related to different crustal units. Similar subhorizontal lithospheric reflectors were observed beneath the TESZ (Grad et al., 2002; Guterch et al., 2004) and the Baltic Sea (Hansen and Balling, 2004). However, the depths of resolution of the DSS profiles are usually limited to about 50–80 km.

Compared to the crust, the structure of the lithosphere and the lithosphere–asthenosphere boundary (LAB) in the TESZ and its surroundings is poorly known. While it was found that the cratonic lithosphere extends much deeper than that of the younger continental regions (e.g., Plomerova et al., 2002; Eaton et al., 2009; Shomali et al., 2006; Gregersen et al., 2010), the studies revealed that the structure of the lithosphere and the LAB differs a lot on both sides of the TESZ (e.g., Zielhuis and Nolet, 1994; Majorowicz et al., 2003; Artemieva et al., 2006; Koulakov et al., 2009; Wilde-Piórko et al., 2010). Regarding different physical properties and geophysical techniques, the LAB has different practical definitions: (1) the seismic LAB defines the transition between the solid outer layer of the Earth, which is characterized by higher seismic velocity values, and its interior, which is characterized by lower seismic velocity values; (2) the thermal LAB defines the transition between the outer layer with dominating conductive heat transfer above the convective mantle that usually coincides with a depth of a constant isotherm of about 1300 °C (McKenzie, 1967); (3) the electrical LAB is a transition between the generally electrically resistive outer layer of the Earth and the conductive layer in the upper mantle.

The studies by Majorowicz et al. (2003) and Artemieva et al. (2006) based on global tomography and heat flow measurements indicate that beneath the EEC the thickness of the thermal lithosphere is about 180–200 km, while the thickness of the seismic lithosphere is more than 250 km. The results by Artemieva et al. (2006) were obtained using all available data resulting from the wide-angle studies by Vinnik and Ryaboy (1981), Garetskii et al. (1990), Grad and Tripolsky (1995), Kostyuchenko et al. (1999), the EUROBRIDGE Working Group and EUROBRIDGE'95 (2001), Grad et al. (2002), and Thybo et al. (2003), and the results of P and S wave tomography by Matzel and Grand (2004). These data are sparse compared to the study area, and the spatial resolution is questionable; however, the thick seismic lithosphere reported by Artemieva et al. (2006) was also found in the area during other studies. Koulakov et al. (2009) observed the positive P wave velocity anomaly beneath the EEC down to at least 300 km, which indicates even thicker lithosphere compared to Artemieva et al. (2006). Legendre et al. (2012) find no indications of a deep cratonic root below about 330 km for the EEC, while Geissler et al. (2010) do not observe any clear indications of deep seismic LAB beneath the EEC either.

In central–western and northern Europe, the TOR 1996–1997 passive seismic project, which was carried out across the STZ, provided a detailed model of the upper mantle and the LAB (Gregersen et al., 1999; Plomerova and Babuska, 2002; Shomali et al., 2006; Artlitt, 1999; Cotte et al., 2002). The results show that the average thickness of the seismic lithosphere is about 100 km in central Europe, which coincides with global tomography studies by Artemieva et al. (2006) and the studies of S receiver functions by Geissler et al. (2010). The results obtained from the TOR data indicate that beneath the TESZ the thickness of the seismic lithosphere is about 120 km, which is an intermediate value between that of the EEC and western Europe (Shomali et al., 2006; Wilde-Piórko et al., 2010), while the transition beneath the STZ is near-vertical, with only a weak tendency to the northeastern slope (Gregersen et al., 2010). Geissler et al. (2010) indicate the lithosphere thickness of about 115–130 km in the vicinity of the TESZ, while the LAB beneath the southwestern part of the Variscan Bohemian Massif is estimated at a depth of 115 km, and the thin lithosphere of only about 75 km is reported beneath some parts of the Pannonian Basin. Beneath the Bohemian Massif, an extensive low-velocity heterogeneity in the upper mantle is found (Koulakov et al., 2009; Karousova et al., 2013), while the high-resolution tomography studies indicate the most distinct low-velocity perturbations along the Eger Rift down to about 200 km (Karousova et al., 2013). Plomerova et al. (2007) interpret the broad low-velocity anomaly beneath the Eger Rift as an upwelling of the LAB. The authors also find different orientations of seismic anisotropy corresponding to the major tectonic units in the Bohemian Massif (i.e., Saxothuringian, Moldanubian and Tepla-Barrandian), while the studies of shear-wave splitting (e.g., Wüstefeld et al., 2010; Vecsey et al., 2013; Sroda et al., 2014) show that anisotropy in the

Bohemian Massif is higher compared to the anisotropy observed in the TESZ and even smaller, but still noticeable, for the EEC (Plomerova et al., 2008).

Jones et al. (2010) performed a comparison between the delineation of the LAB for Europe based on seismological and electromagnetic observations, and concluded that the LAB, as an impedance contrast from receiver functions, as a seismic anisotropy change and as an increase in conductivity from magnetotellurics, are consistent with the deeper LAB beneath the EEC and the shallower LAB beneath central Europe, which coincides with conclusions by Korja (2007), who made a review of previous studies of magnetotelluric imaging of the European lithosphere. Jones et al. (2010) found that the seismic and electric LABs beneath Phanerozoic Europe are at depths of about 90–100 km, while for the EEC, they differ, and the electric LAB is at a depth of about 250 km. The studies also show anomalously thick electrical LAB beneath the TESZ, whereas the seismic LAB should be much shallower. The authors imply that the difference could be caused by increased partial melting or by hydration beneath the TESZ.

An opportunity to enhance knowledge of the lithosphere structure and the LAB around the TESZ was implemented during the international PASsive Seismic Experiment (PASSEQ) 2006–2008 (Wilde-Piorko et al., 2008), which aimed to study the lithosphere and asthenosphere around the TESZ. The aim of this study is to obtain a model of the upper mantle and the seismic LAB on a regional scale in the territory around the TESZ (Fig. 1b) using data from the seismic stations operated in the region during the PASSEQ project and the method of teleseismic tomography.

2   Data set

The PASsive Seismic Experiment (PASSEQ) 2006–2008 (Wilde-Piorko et al., 2008) was carried out from June 2006 to July 2008 in the territory extending from Germany and the Czech Republic throughout Poland to Lithuania where 139 short-period and 49 broadband temporary seismic stations were deployed (Fig. 2). In this study, we use data of the PASSEQ project and some permanent seismic stations operated in the area during the period of the PASSEQ project. Although there were over 200 temporary seismic stations deployed in the region, due to some technical peculiarities, in total we used data of 183 seismic stations. From the seismological bulletins of the International Seismological Centre (ISC), we selected 101 teleseismic earthquakes (EQs) with a magnitude range of 5.5 to 7.2 and an epicentral distance of 30 to 92 degrees with respect to the point at the Lithuanian–Polish border at 23° E and 54° N (Table A1). The majority of the selected EQs are located to the east of the target area (i.e., Sumatra, Japan, Kamchatka and the Aleutian regions) due to naturally higher seismicity compared to the regions to the west of the study area; thus, the largest seismic gap of about

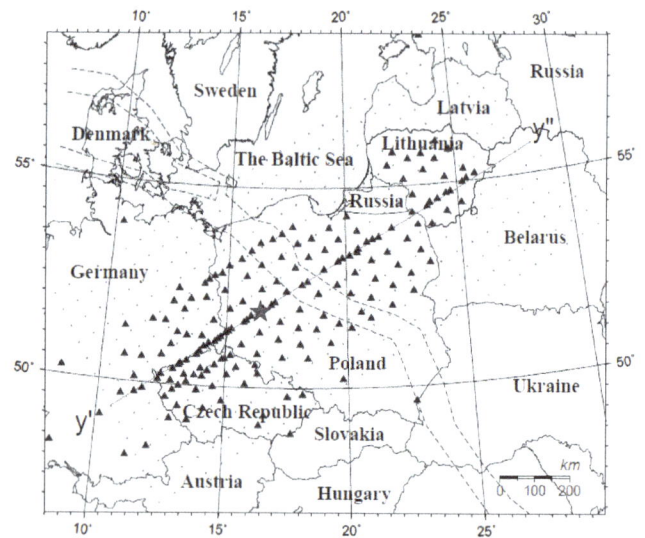

Figure 2. Seismic stations used in this study marked as triangles. Dots indicate nodes of the model grid. Star indicates origin of the local Cartesian coordinate system used. Dashed lines indicate the TESZ. Solid line $y'y''$ marks the main PASSEQ transect at $y = 0$ in the local Cartesian coordinate system.

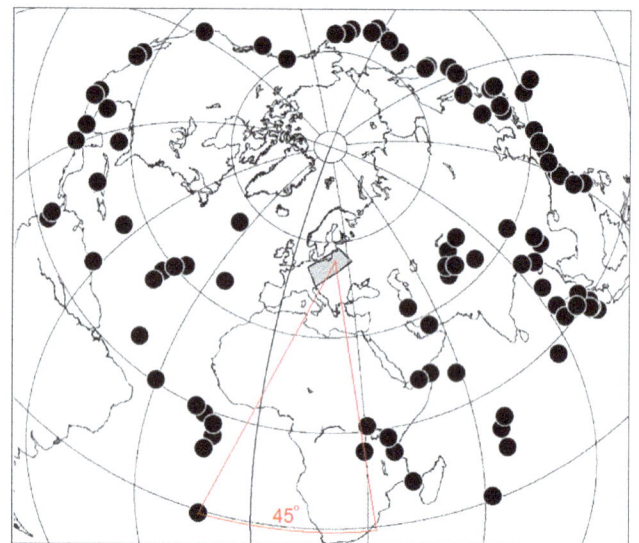

Figure 3. Map of epicenters of EQs (black circles) used in our study. Grey rectangle indicates the study area. Red lines show the largest seismic gap.

45 degrees is for the region of Africa and the southern part of the Atlantic Ocean (Fig. 3). Using the Seismic Handler Motif (SHM) (http://www.seismic-handler.org/) program package, we analyzed the data and compiled the data set of 6008 manually picked top-quality absolute P wave arrivals. The weighting factor of the picks was assigned according to the picking error, which was set to less than 0.2 s for the top-quality data. The picking error of the top-quality picks was

usually much smaller (< 0.1 s) because of good data quality; however, the large interval of the error was selected before the data analysis in order to ensure a reasonable number of the top-quality picks.

The calculation of theoretical travel times (TT) of the P wave arrivals was performed using the EQ location information from the ISC seismological bulletins and the Seismic Handler (SH) program, which applies the IASP91 velocity model. The TT residuals were calculated as follows:

$$T_{\text{picked}} - T_{\text{theoretical}} = \text{TR}, \tag{1}$$

where T_{picked} is the observed TT, $T_{\text{theoretical}}$ is the theoretical TT calculated with SH, and TR is the TT residual. It was noticed that the calculated values of the TT residuals are higher to the west and lower to the east from the TESZ, which might be related to different tectonic–geological settings in the area.

3 Teleseismic tomography inversion method

We used the TELINV nonlinear teleseismic tomography code (Weiland et al., 1995) to perform the inversions. In teleseismic tomography, the perturbations of the TT are used to estimate the size and magnitude of the velocity variations within the given volume. The TT residuals TR_{ij} (at the ith station for the jth event) include effects of origin time uncertainty, hypocenter location errors and velocity perturbations outside the study area. These effects are eliminated while subtracting some reference residual TR_j, and the relative residuals RTR_{ij}, which are used in the inversion, are calculated.

To invert the data set, the ACH inversion method by Aki et al. (1977), which later was developed by Evans and Achauer (1993), was used. According to Evans and Achauer (1993), the problem can be linearized through block parameterization, disregarding refraction by the slowness perturbations:

$$b = Gm, \tag{2}$$

where b is a vector derived from the relative TT residuals RTR_{ij}, m is a vector of perturbations of slowness, and G is a matrix derived from unperturbed TT of a ray ij in block k. To estimate m, the damped least squares can be used, and the basic inversion equation for the TELINV code can be written as

$$m^{\text{est}} = \left(G^T W_D G + \varepsilon^2 W_M\right)^{-1} G^T W_D b, \tag{3}$$

where m^{est} are estimated model parameters, W_D is a weighting matrix of the data, ε^2 is a damping factor, and W_M is the smoothing matrix of the model. The abilities of the ray geometry and model parameter grid to resolve the velocity perturbations can be estimated by a resolution matrix (Menke, 1984):

$$R = \left(G^T W_D G + \varepsilon^2 W_M\right)^{-1} G^T W_D G. \tag{4}$$

The code is an iterative process where each iteration involves a complete one-step inversion, including both ray tracing and model estimations. Iterations stop when the model ceases to change significantly and the root-mean-square (RMS) difference between predicted and observed TT residuals is comparable to data variance. The data is a relative measure, thus, one can estimate only relative perturbations to the used reference model.

The ray tracing is crucial in teleseismic tomography. A ray path is determined through a model, i.e., which nodes the ray crosses and how much time it spends at each node. An algorithm produces the theoretical TT that are used in computing the relative residual arrival time data. In our study, the 3-D ray tracing algorithm of Steck and Prothero (1991) was used. The procedure performs a simplex search for the fastest path of a planar wavefront to a point at the surface. In this procedure, the departure point of a ray from the plane wave is not fixed, but determined by the algorithm itself. It assumes that the ray bending and distortions are caused by heterogeneities along their paths (Weiland et al., 1995; Sandoval, 2002).

4 Model parameterization

Our study area is shown in Fig. 2. The model parameterization must be fine enough in order to capture the structure that can be resolved. Regarding the seismic signal frequency and spacing between the seismic stations, we set a spacing of 50 km between the grid nodes in horizontal directions. The 1-D IASP91 velocity model (Kennett and Engdahl, 1991) was used to parameterize the reference 3-D velocity model with 16 layers of constant velocities (i.e., all nodes in one layer were assigned the same values) down to 700 km. We set the inverted layers (between 70 and 350 km) every 30 km from 90 to 300 km and two more layers at 70 and 350 km depth, while below, we set two non-inverted layers for the stability of the inversion, and four non-inverted layers above in the Earth's crust (from surfaces down to 50 km). Every layer of the compiled initial velocity model was assigned a constant value of the seismic velocities from the IASP91 velocity model.

We performed a thorough analysis in order to select the optimal inversion parameters. The damping parameter determines how much noise present in the data is mapped in the resolved model. Underestimation of damping would result in noise fitting while overestimation would reduce lateral velocity variations. The damping value was determined while running inversions with different values of damping and investigating trade-off between the data variance and model variance (Fig. 4). From the curve one may find that the optimal value for damping is 80. However, here we present results obtained using a damping value of 120, which is more conservative and obviates the velocity anomalies of shorter wavelengths compared to the results obtained using a damping of 80. As we aim to resolve regional-scale velocity variations in

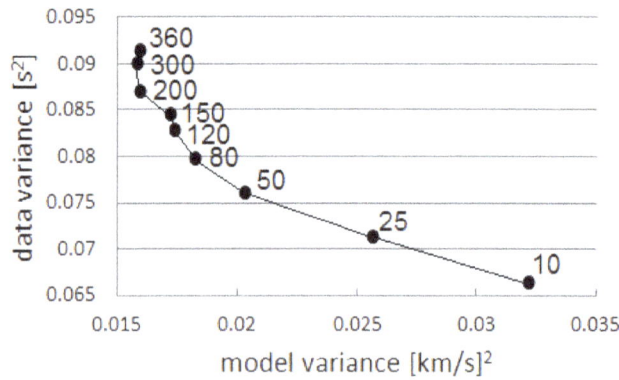

Figure 4. Trade-off between the data variance and model variance obtained with different damping values from 10 to 360. The presented results were obtained using a damping value of 120.

the study area, the larger damping value was used. We also found that three iterations are enough for inversion, because for higher number of iterations, the model and the RMS error change insignificantly.

5   Crustal travel time corrections

As discussed previously (see Sect. 1.2), the structure of the crust in the study area varies significantly, as well as the thickness of the sedimentary cover, which is up to about 20 km in the Polish Basin. In order to obtain the upper mantle structure, it is important to remove the effects, which are created by the Earth's crust, from the inversion results. The crustal TT corrections for individual seismic stations were compiled as follows:

$$TT_{model} - TT_{iasp} = TT_{diff},   \tag{5}$$

where TT_{model} is TT through the crustal velocity model, TT_{iasp} is TT through the IASP91 velocity model, and TT_{diff} is TT difference. We used two sets of the crustal TT corrections: (1) the first set was compiled using the EuCRUST-07 (Tesauro et al., 2008) 3-D crustal model for Europe with model grid of $1° \times 1°$; (2) the second set was compiled using the precise 3-D crustal model for Poland (Majdański, 2012) with model grid of $0.3°$ of latitude and $0.5°$ of longitude, and results of some DSS profiles. The crustal model by Majdanski (2012) was compiled using all available information from the DSS profiles carried out around Poland. However, outside the crustal model there is not much data to be used, thus, the territories not covered by the model by Majdański (2012) were assigned with constant values that were estimated using the interpreted results (full velocity profiles) below shot point SP9 in the EUROBRIDGE'95 profile and shot point SP2 in the CELEBRATION09 profile. The value obtained from the EUROBRIDGE'95 profile was used for the stations deployed in Lithuania, and value obtained from the CELE-BRATION09 profile was used for the stations deployed in

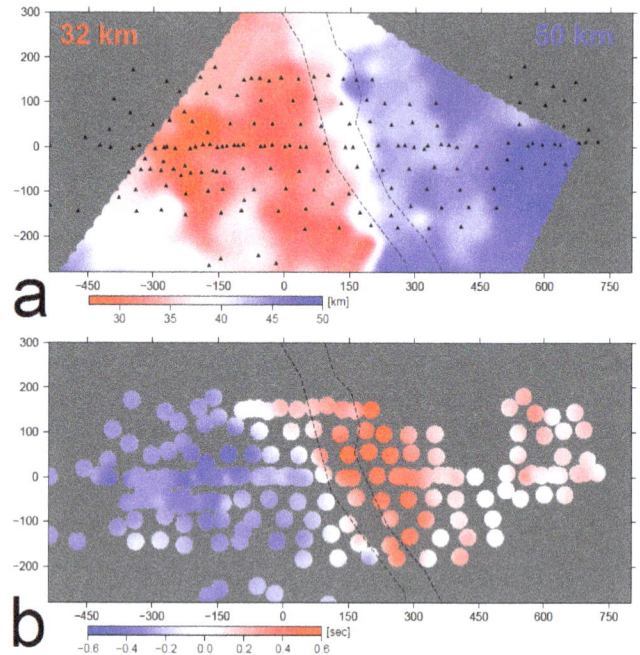

Figure 5. (a) Moho map of the precise 3-D crustal model by Majdański (2012), which was used to estimate the crustal TT corrections. The Moho depth for the areas outside the model was defined using results of some DSS projects: the area to the east was assigned 50 km and, to the west 32 km. (b) Estimated crustal TT corrections in the individual seismic stations. The values are expressed in seconds with respect to the IASP91 velocity model.

Germany and the Czech Republic (the constant depths of the Moho boundary of 50 km and 32 km, respectively, were assigned as well) (Fig. 5a). The crustal TT corrections were calculated assuming the vertical ray propagation in the crust. Regarding the incidence angles in our data set, the assumed vertical propagation in the crust causes $< 2\%$ shortening of the raypaths, thus, the effect in the results on velocity amplitudes is negligible.

In order to estimate the effect of the crustal TT corrections on the velocity amplitudes, we performed inversion with the real data set without (Fig. 6a) and with the crustal corrections applied (Fig. 6b, c). In the inversion results with the EuCRUST-07 model (Fig. 6b), we observe a "high–low–high" distribution of velocity variations in the study area, and artificially high signal amplitudes of up to $\pm 12\%$, especially around the TESZ, where the thickness of the sediments is significantly larger compared to the surroundings. This result is not consistent with our knowledge about the possible geological conditions in the study area (see Sect. 1.2), and obviously it is not what we may expect from a decent set of crustal TT corrections. Thus, we concluded that this set of the crustal corrections is too robust and is not applicable in our study.

The inversion results (Fig. 6c) obtained with the second set of crustal TT corrections (Fig. 5b) based on the crustal

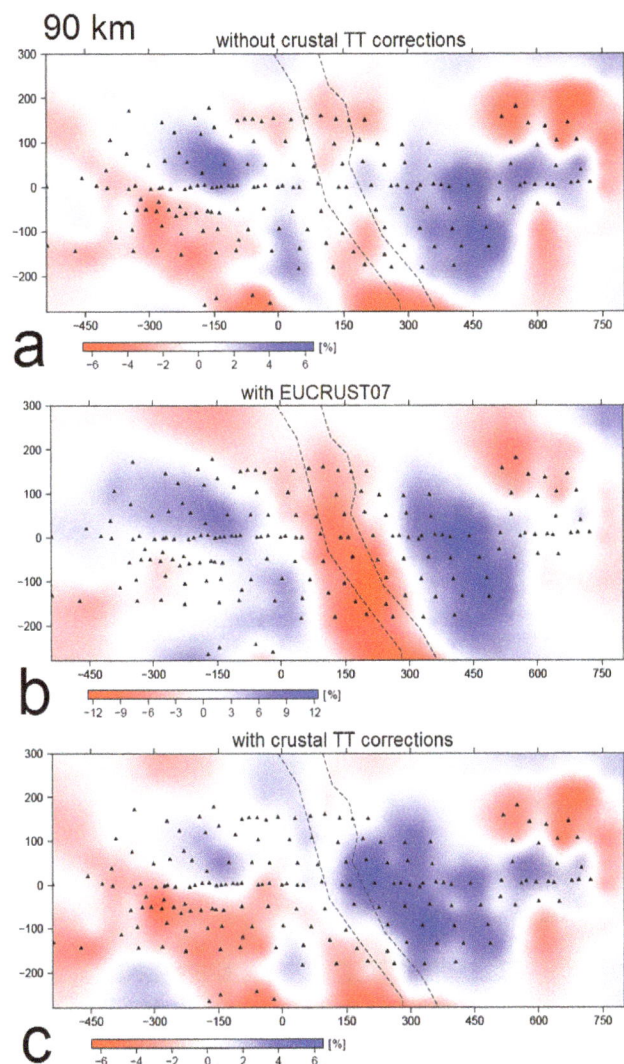

Figure 6. Horizontal slices at a depth of 90 km of the inversion results (**a**) without crustal TT corrections, (**b**) with the EuCRUST-07 model, and (**c**) with crustal TT corrections compiled using the model by Majdański (2012) and the result of some DSS projects. Triangles indicate seismic stations. Dashed lines indicate the TESZ.

plitudes turn positive, which indicates significant correction for the thick sedimentary cover. Moreover, one may indicate reduced positive anomalies in the western part of the study area. We observe no obvious artifacts in the results, which are quite consistent with what we expect from the previous studies (see Sect. 1.2). Thus, we concluded that this set of the crustal TT corrections is reasonable, and it was used in our study. However, the introduced crustal TT corrections bring in some additional effects to the results. We observe this effect down to about 180 km, while in the deeper parts, it is negligible. The effects from the crustal TT corrections were also reported in other studies; e.g., Sandoval et al. (2003) observe the effect down to about 200 km.

6 Resolution and synthetic tests

To estimate the resolution of the inversion results, we use the hit matrix and the checkerboard test. The two methods combined enable us to define the resolution fairly well. The hit matrix is based on calculation of the number of rays that transverse a particular cell. The inversion with the synthetic checkerboard model shows which parts of the target area can be and cannot be resolved with the same configuration as the observed data set. In our study, we compiled the synthetic velocity model of the checkerboard structure with blocks of 200 km in horizontal directions and four layers thick with a ±4 % velocity difference with respect to the IASP91 velocity model (Fig. 7a). The synthetic data set was compiled by adding Gaussian-distributed perturbations (up to ±0.4 s) to all observed TT. The inversion results obtained with the synthetic data set show a reasonably well-resolved checkerboard-type structure (Fig. 7b). However, in the vertical slices in Fig. 7b, we observe the vertical smearing dipping to the east, which is most likely due to the majority of rays coming from the regions located to the east of the study area (Fig. 3). Moreover, the synthetic structure in the western part is better resolved than in the eastern part (Fig. 7b), due to the larger number of top-quality picks in the data of the stations deployed to the west of the TESZ. The further estimate of the resolution is derived from the diagonal elements of the resolution matrix (Fig. 8), which provides a relative measure of the resolution: the low values show areas of low resolution and the high values show areas of high resolution. The inversion was performed using the larger damping value (i.e., 120), but we still obtain quite large velocity perturbations (up to 6.5 %) that are related to the small values of the diagonal elements of the resolution matrix observed in Fig. 8, which suggests quite sparse data coverage and considerable vertical smearing in some parts of the study area, which is consistent with the results of the checkerboard test (Fig. 7b). We will discuss the resolution of the areas that are directly beneath the seismic array because, outside the array, we have no ray coverage and, thus, zero resolution. Fig. 8 indicates the highest resolution (dark color) in the southwestern part

model by Majdański (2012) and some DSS studies do not reproduce the shapes of the thick sediments in the TESZ, as is obvious in Fig. 6b, but show two distinct structures on both sides of the TESZ: the higher velocities to the east and lower velocities to the west. Compared to the results obtained without (Fig. 6a) and with (Fig. 6c) the crustal TT corrections, one may find quite similar patterns of velocity distribution; however, there are some significant differences. As the color scale is the same in both Fig. 6a and c, one may notice somehow reduced amplitudes of the velocity perturbations in Fig. 6c. As expected, the negative amplitudes are reduced in the northeastern part of the study area (western Lithuania), where the sedimentary basin up to 2 km thick is present, and in the northern and central part of the TESZ, the negative am-

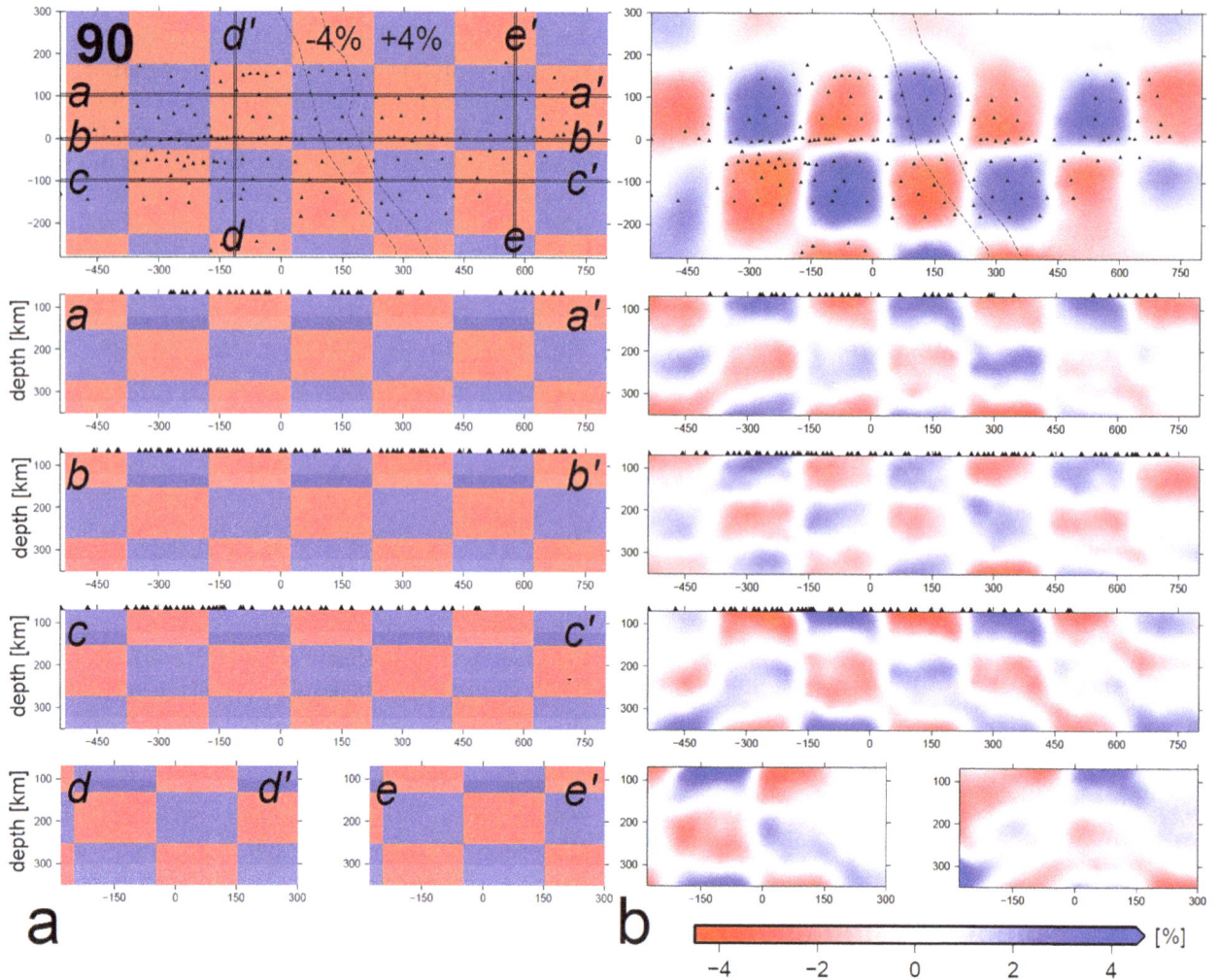

Figure 7. Checkerboard test. Horizontal slice at a depth of 90 km and vertical slices along the depicted transects. (**a**) Initial velocity model. (**b**) Inversion results with the synthetic data set. Triangles mark the seismic stations. Dashed lines indicate the position of the TESZ.

of the study area, from the top of the inverted layers down to about 250 km, which coincides with areas of the densest station coverage (i.e., the larger number of picks in the data set) and good crossing of the seismic ray paths. In the rest of the areas below the seismic array, we obtain a fair resolution (lighter grey). On the vertical slice along the main transect (Fig. 8), we observe a "dark color" below 300 km, which is an artifact from the inversion and does not indicate good resolution, because at these depths the rays do not cross.

We also performed a synthetic test with a robust "geologically possible" velocity model in order to find out whether our data set with current station configuration is capable to resolve the introduced large scale structures. We also aim to test whether the obtained inversion result would be similar to the one obtained with the real data set, because this could invoke some reasonable speculations for interpretation. Based on the previous geophysical and petrophysical studies (Wilde-Piórko et al., 2010; Griffin et al., 2003), we com-

piled a synthetic 3-D velocity model with geologically possible structure. The main features of the synthetic "geological" velocity model (Fig. 9a) are: (1) the lower and the higher seismic velocities to the west and to the east from the TESZ, respectively, (2) the shape of the LAB of a ramp type dipping to the northeasterly direction, and (3) the deep cratonic roots for the EEC (in the northeastern part of the study area). Small TT perturbations were added to the synthetic TT, as for the checkerboard test. The inversion result obtained with the synthetic data set (Fig. 9b) shows the lower and the higher velocity areas to the west and to the east from the TESZ, respectively. In the results we also observe the clear ramp shape of the LAB and the higher velocity anomaly at the bottom of the velocity model in the northeastern part of the study area (Fig. 9b). In the results, one may also notice that we do not resolve the same average velocities with respect to the input velocity model, however, the total ratio of positive and negative amplitudes of perturbations is similar in both the

90 km depth

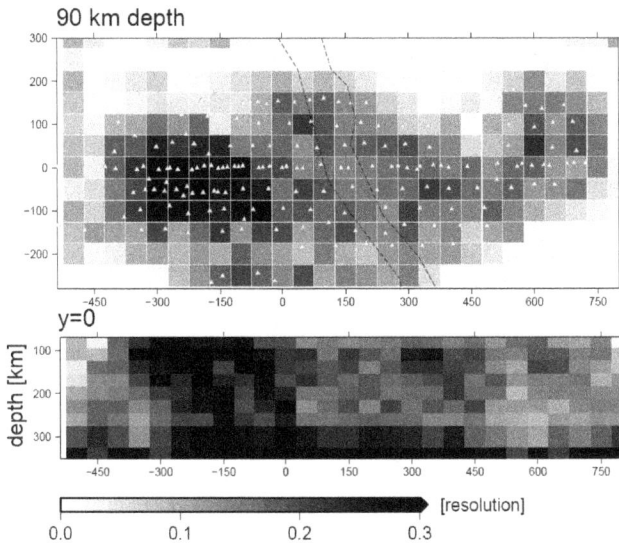

Figure 8. Diagonal elements of the resolution matrix in horizontal slice at 90 km depth and vertical slice along the main transect of the study area. The low and high values of resolution indicate poorly resolved and well-resolved areas, respectively. Triangles mark seismic stations. Dashed lines indicate the TESZ.

reference model and the inversion results. This implies that the resolved negative amplitudes could be considered as neutral or slightly positive instead.

7 Results and discussion

As shown by numerous seismic studies (e.g., Knapmeyer-Endrun et al., 2013b), the LAB in Precambrian cratonic areas is not easily detected by seismic methods and can be misinterpreted with the so-called Mid-Lithospheric Boundary (MLB). The nature of the latter is still not completely understood. However, the seismic LAB can be detected beneath the younger areas and traced across boundaries of the cratons in the passive seismic experiments that sample both the cratonic and non-cratonic lithospheres. In our study, we used the data of such a passive seismic experiment and performed inversions with the compiled data set of top-quality P wave arrivals. We resolved the structure of the upper mantle from 70 km down to 350 km in the study area. The obtained model of P wave velocity variations can be used to estimate the seismic LAB and the lithosphere thickness around the TESZ. In our study, we embrace the definition of the seismic LAB as a transition between the higher and lower seismic velocities, which was discussed previously. The obtained results, of course, depend on the reference velocity model; thus, to be on the conservative side, we used the well-known IASP91 velocity model and obtained results with respect to this reference model.

In our results (Fig. 9c) we observe amplitudes of velocity variations up to ±6.5 % with respect to the IASP91 velocity

model, which is definitely too high to be explained by the geological-tectonic conditions only. In teleseismic tomography many factors contribute to the observed signal (velocity) amplitudes, such as damping value, implementation of the crustal TT corrections (about 1 % of the observed velocity contrast), temperature variations (about 1 %) and anisotropy in the study area, and distortions on the full raypaths outside the velocity model (which varies from region to region). Moreover, the used TELINV code implements the "flat-earth" model, which affects the apparent seismic velocities. Regarding the size of our velocity model and the incidence angles in our data set, the discrepancy due to the used "flat-earth" model is about 1.5 % of the observed amplitudes of velocity variations. Thus, taking into account all the above mentioned causes we should consider the amplitudes of velocity variations not ±6.5 %, but close to about ±3 %.

The inversion results with our real data set show the higher P wave velocity values with respect to the IASP91 velocity model beneath the EEC and lower ones beneath western Europe, while the TESZ appears as a transitional complex tectonic structure with significant velocity perturbations in longitudinal and transversal directions (Fig. 9c). This general finding coincides with the results by Koulakov et al. (2009) who reported the sharp transition along the TESZ from the negative amplitudes, characterizing the young tectonic features of central–western Europe, to positive ones beneath the old EEC. Moreover, a sharp transition from low to high shear-wave velocities between the Phanerozoic Europe and EEC, respectively, was observed from waveform inversion of both body and surface waves by Zielhuis and Nolet (1994). We also indicate that the LAB is more distinct beneath the Phanerozoic part of Europe than beneath the Precambrian part, which coincides with the results by Plomerova and Babuska (2010) and Knapmeyer-Endrun et al. (2013b).

To the east of the TESZ the pronounced high-velocity structure in the upper mantle is observed beneath Poland (Fig. 9c). The observed velocity perturbations down to about 120 km beneath Poland are about 2 to 3 % higher with respect to the IASP91 velocity model, while going deeper the variations are slightly smaller, which most likely indicates some effects due to the applied crustal TT corrections. The higher velocity values in this area are observed down to about 200 km, which coincides well with the studies by Wilde-Piórko et al. (2010), Majorowicz et al. (2003) and Koulakov et al. (2009). Legendre et al. (2012) found the highest velocity values in the mantle of the EEC at about 150 km depth. Further to the northeast of the TESZ, the high-velocity area goes deeper, and beneath the territory of Lithuania, we find the thickest lithosphere of about 300 km or more (Fig. 9c). Due to vertical smearing (Fig. 7b), which is intrinsic to all tomography inversions, the observed higher velocity area associated with the deep cratonic roots could be extended to the layers deeper than it really is, however, our result is in a good agreement with other observations – the obtained value of thickness of the lithosphere beneath the EEC is about 50 km

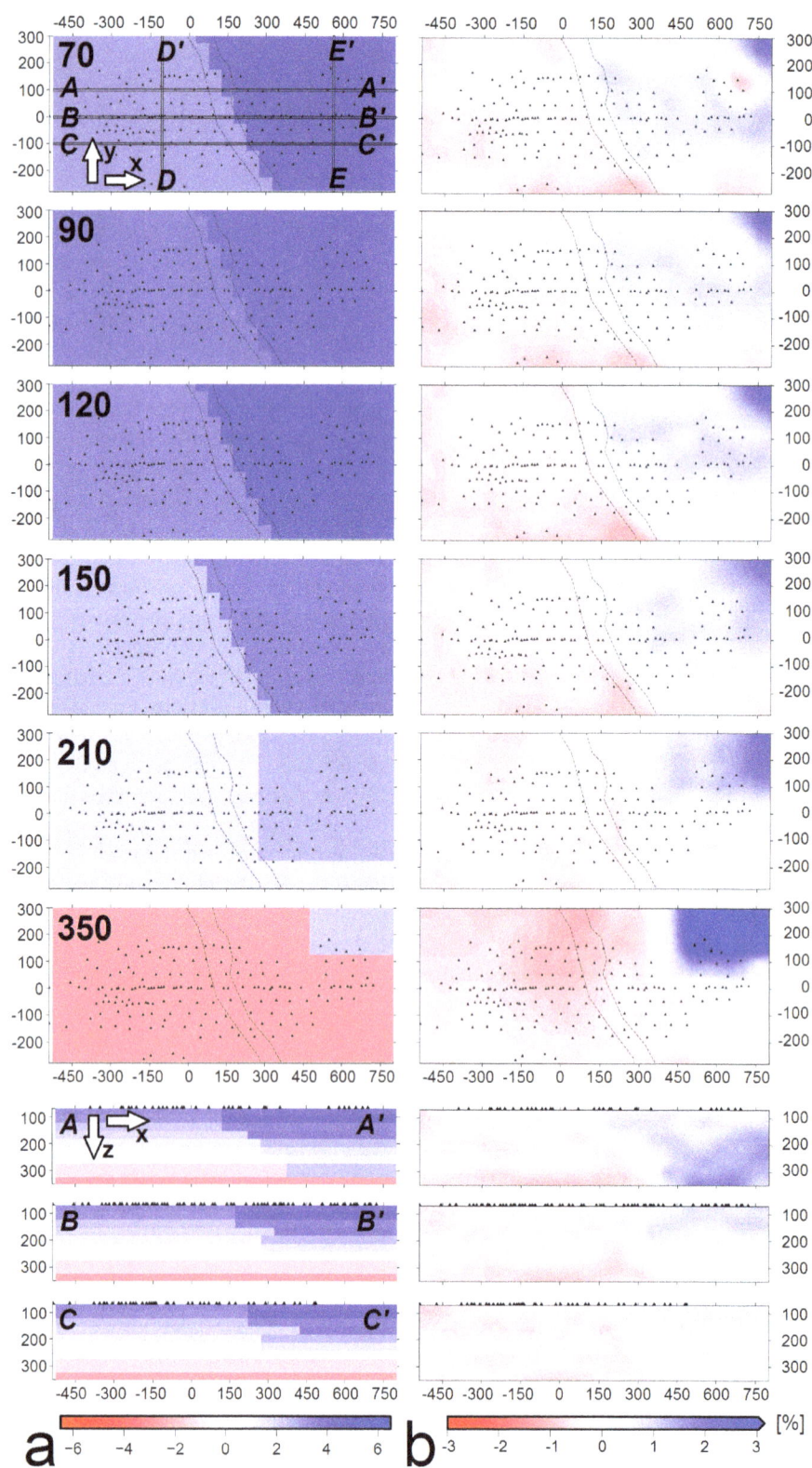

Figure 9.

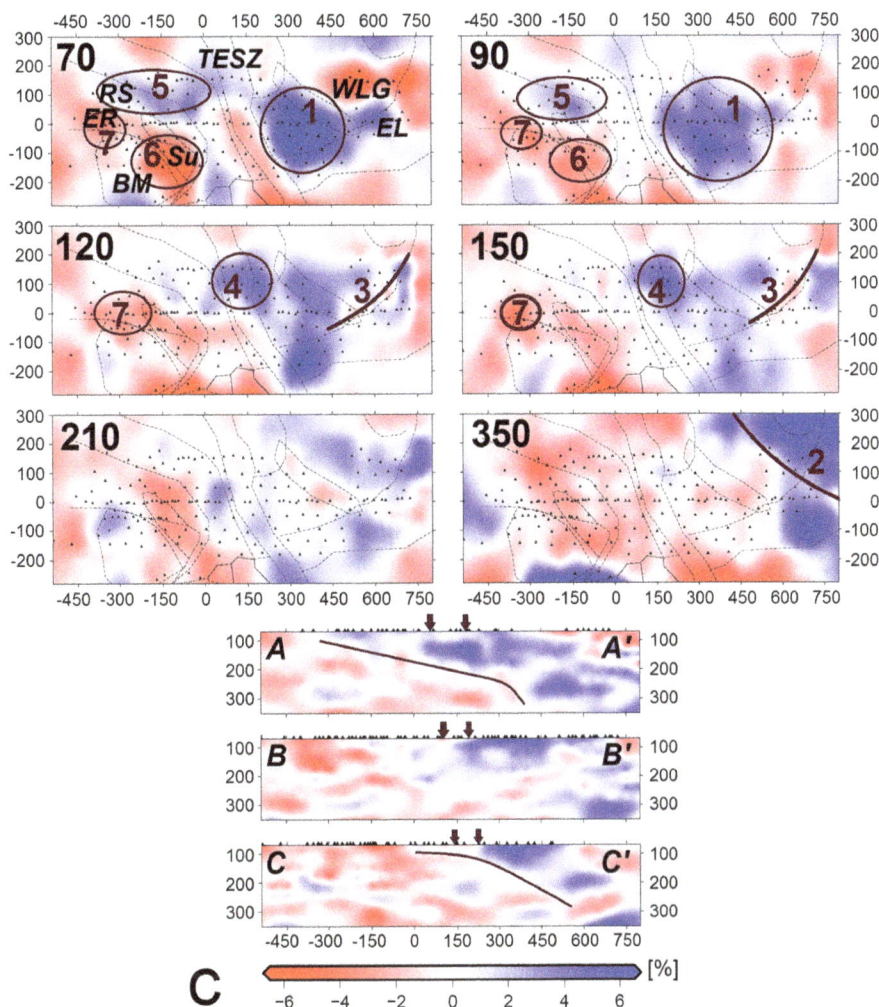

Figure 9. *P* wave velocity perturbations in horizontal slices at indicated depths (km) and vertical slices parallel to the main transect of the study area. Results obtained with the synthetic "geological" model: (**a**) input velocity model, and (**b**) inversion results. (**c**) Results obtained with the field data set. Triangles mark seismic stations. *x*, *y* and *z* indicate longitude, latitude and depth (km), respectively, in a local Cartesian coordinate system. Dashed lines on horizontal slices indicate boundaries of tectonic units (see Fig. 1b). Numbered areas mark the discussed interpreted structures: (1) high-velocity area beneath Poland (craton); (2) deep cratonic roots extending to at least 300 km or more beneath Lithuania; (3) paleosubduction boundary between the WLG and the EL; (4) high-velocity area beneath northern Poland; (5) higher velocity area along the Rheic Suture; (6) lower velocity area beneath the Sudetes Mountains and the Bohemian Massif; (7) low-velocity area beneath the Eger Rift. Solid lines on vertical slices show the interpreted seismic LAB; and brown arrows indicate the TESZ.

larger compared to the global tomography results obtained by Artemieva et al. (2006), but coincides well with results obtained from P- and S-wave tomography by Koulakov et al. (2009) who find the P wave velocities up to 2 % higher extending to at least 300 km beneath Lithuania. Thick lithosphere extending to at least 250 km depth is also found beneath the central part of the Fennoscandian Shield (Sandoval et al., 2004), but there are found no indications of the seismic LAB anywhere within 300 km beneath the EEC (Bruneton et al., 2004; Geissler et al., 2010; Legendre et al., 2012). Our study does not show the seismic LAB beneath the EEC either. The study of S receiver functions by Knapmeyer-Endrun et al. (2013b) indicates a negative conversion that could be re-

lated to a velocity decrease at 190 km to 230 km depth, which is in agreement with the depth estimates for the cratonic LAB; however, the conversion was not observed in all analyzed seismic stations in the EEC. Thus, the authors suggest that the stations might imply spatial variations in the sharpness of the corresponding velocity change.

In the northeastern part of the study area beneath Lithuania, at depths of 120–150 km, we find the lower velocities compared to the surroundings following the MLSZ (Fig. 9c) – the predicted paleosubduction zone between the WLG and EL (Motuza, 2004; Motuza, 2005; Motuza and Staškus, 2009). Our results (Fig. 10b) also indicate a slope of higher velocities dipping to the north, which agrees with the model

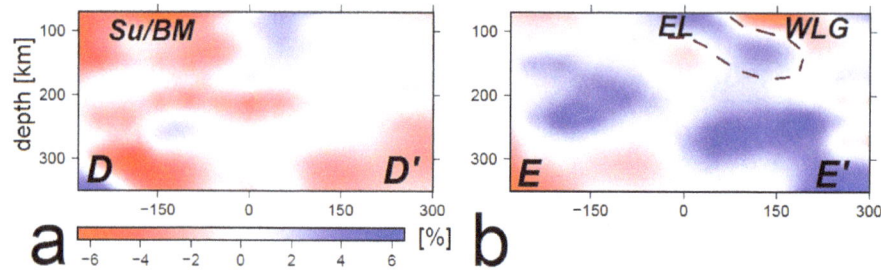

Figure 10. *P* wave velocity perturbations in vertical slices DD′ and EE′ transverse to the main transect (see Fig. 9a). **(a)** Low velocities are observed in the western part of the Bohemian Massif (BM) and the Sudetes Mountains (Su) from 70 km. **(b)** Dashed line indicates a possibly resolved paleosubduction zone under Lithuania between the WLG and the EL.

proposed by Motuza and Staškus (2009) that the EL subducted under the WLG. The anomaly is relatively small, thus its existence is questionable; however, the ray coverage in this part of the study area is reasonable (Fig. 8). We infer that this feature may indicate a slab of "frozen" paleosubduction, while the lower velocities observed below the slab along the predicted paleosubduction edge could be related to an increase in temperature.

We find an area of the higher velocities in the lithospheric mantle down to about 180 km in the northern part of the TESZ (northern Poland) (Fig. 9c). Knapmeyer-Endrun et al. (2013a) observe an increase in TT of Ps conversions across the mantle transition zone that could be caused either by a temperature reduction or an increase in water content in this mantle region. As we observe the higher velocities in this part, we propose that this anomaly could be related to thermal regime and temperature reduction. In general, the upper mantle of the northern TESZ is more of a cratonic type, while going to the south, the seismic velocities are lower.

Our results indicate the dominating negative velocity amplitudes to the west of the TESZ almost everywhere down to 350 km, except in the territory of northern Poland and Germany along the Rheic Suture, where we find the higher velocity anomaly down to about 90–100 km, while closer to the TESZ, the LAB is observed at a depth of about 120 km (Fig. 9c). The result is consistent with results obtained by Knapmeyer-Endrun et al. (2013b) and Wilde-Piorko et al. (2010), who indicate the average seismic lithosphere thickness of about 90 km, and associate the uplift of the LAB beneath western Europe and the TESZ with partial melting of the upper mantle due to thermal conditions (Wilde-Piorko et al., 2010). Moreover, the studies of Shomali et al. (2006) and Gregersen et al. (2010) carried out using data of the TOR 1996–1997 passive seismic project indicate a lithosphere thickness of about 100 km in northern Germany, which coincides well with our results for this territory. The depth of the LAB of about 100 km is a characteristic common to the Phanerozoic regions (Plomerova et al., 2002).

The observation of the lower velocity values (Fig. 9c) with respect to the IASP91 velocity model to the west of the TESZ

coincides with results by Koulakov et al. (2009) who report the negative anomalies up to 4 % for this area. In our results the large lower velocity area of about −2 to −3 % with respect to the IASP91 velocity model is observed beneath the Bohemian Massif and the rift systems in central Europe (Fig. 9c). The lithosphere thinning of 80–90 km beneath the Armorican terrains of Saxothuringian, Tepla-Barrandian and Moldanubian is reported in studies by Babuška and Plomerova (2001). Karousova et al. (2013) find an extensive low-velocity heterogeneity in the upper mantle beneath the Bohemian Massif, while Koulakov et al. (2009) report the broad negative zone (−1 to −3 %) beneath the Central Rift System and the Bohemian Massif at depths from 100 to 200 km. In our results we find the largest negative signal amplitudes under the northeastern part of the Bohemian Massif and the Sudetes Mountains from a depth of at least 70 km (Figs. 9c, 10a). Moreover, our results indicate the lower velocity anomaly under the Eger Rift (Fig. 9c). Although the Eger Rift is a relatively small structure, our data set is sufficient to resolve it, thus, we indicate the lower velocities from 70 km down to at least 180 km beneath it. This result is in a good agreement with results by Karousova et al. (2013) who indicate the most distinct low-velocity perturbations along the Eger Rift down to about 200 km, and Koulakov et al. (2009) who observe the low-velocity zone (−2 %) in this area between about 80 and 250 km. Plomerova et al. (2007) interpreted the broad low-velocity anomaly beneath the Eger Rift as an uplift of the LAB.

The asthenosphere on the western edge and on the eastern edge of the TESZ is at depths of about 150 km and 180 km, respectively. Moreover, the structure of the TESZ varies significantly, going from north to south (Fig. 9c). In the studies of Legendre et al. (2012), it is found that the mantle lithosphere beneath the TESZ shows moderately high velocities, and is of an intermediate character between that of the cratonic lithosphere and the thin lithosphere of central Europe. The studies carried out around the TESZ indicated a sharp discontinuity along the TESZ, but provided no strong evidence of the shape of the LAB beneath it due to a lack of resolution (discussed by Knapmeyer-Endrun et al., 2013a). As we used a dense network of the seismic stations (Fig. 2) (with an average spacing of 60 km and a spacing of 20 km along the main PASSEQ transect), we are able to resolve the shape of the LAB with higher precision. In the results (Fig. 9c), we indicate that, in the northern part of the study area, the higher velocities (which are associated with the seismic LAB) are observed deeper going in the northeasterly direction, which shows the ramp shape of the LAB. The angle of the deepening of the LAB is about 30 degrees. In the northern part of the TESZ, we do not recognize any separate structures or clear contact, which could be related to the different tectonic settings of Phanerozoic and Proterozoic Europe but, further to the south, we may refer to a sharp and steep contact on the eastern edge of the TESZ (Fig. 9c). In our "geological" synthetic model, we introduced and reasonably resolved the LAB ramp type dipping in the northeasterly direction as well (Fig. 9a, b), which is somehow similar to the results obtained with the real data set (Fig. 9c). Gregersen et al. (2010) compared results of different studies performed using the data of the TOR project and concluded that the transition between the two tectonic settings on both sides of the STZ is sharp and steep, with a weak tendency to the northeasterly slope. We indicate from our results (Fig. 9c) that, further to the south, the LAB is shallower, and its shape changes most probably due to younger tectonic settings (i.e., the Carpathian Mountains) in the region.

8 Conclusions

- The observed higher P wave velocity values to the east of the TESZ correspond to the older EEC and the lower ones to the west of the TESZ correspond to younger western Europe. The TESZ is resolved as a complex structure with intermediate characteristics between those of the EEC and western Europe.

- We indicate that the seismic LAB is more distinct beneath the Phanerozoic part of Europe than beneath the Precambrian part. The lower velocity anomalies from 70 km are observed under the Bohemian Massif, the Sudetes Mountains and the Eger Rift, while further north, beneath the Variscides, the depths of the LAB are about 100–120 km. Our study does not show the seismic LAB beneath the EEC, but beneath Lithuania we find the thickest lithosphere of about 300 km or more. In the TESZ, the asthenosphere is at depths of 150–180 km, which is an intermediate value between that of the EEC and western Europe.

- In the northern part of the TESZ, the upper mantle is more of a cratonic type. We infer that the LAB in the northern part of the study area is of a ramp type dipping to the northeasterly direction at an angle of about 30 degrees. Under the northern part of the TESZ, we do not recognize any contact between the Phanerozoic and Proterozoic parts of Europe, but, further to the south, we may refer to a sharp and steep contact on the eastern edge of the TESZ. Going to the south, the shape of a LAB beneath TESZ is changing, and its depth is shallower, most likely due to younger tectonic processes.

- Beneath Lithuania at depths of 120–150 km, we observe the low-velocity area that follows the boundary of the proposed paleosubduction zone between the EL and the WLG tectonic units.

Appendix A

Table A1. List of 101 teleseismic EQs used in this study.

Year	Month	Day	Time UTC	Lat.	Long.	Depth	M
2006	6	18	18:28:00	32.9995	−39.7009	8.6	6.0
2006	6	22	10:53:11	45.3023	149.4132	104.3	6.0
2006	6	27	18:07:21	6.4781	92.7356	25.8	6.3
2006	6	27	2:39:33	52.1552	176.1572	28.3	6.2
2006	6	28	21:02:09	26.8361	55.806	15.1	5.8
2006	7	6	3:57:52	39.0233	71.7719	23.7	5.8
2006	7	8	20:39:57	51.1889	−179.264	3.2	6.6
2006	7	10	7:21:36	−11.5727	−13.4176	10	5.5
2006	7	12	14:44:44	−8.5692	67.8158	10	5.7
2006	7	27	11:16:40	1.7244	97.1295	30	6.3
2006	7	29	19:53:41	23.5288	−63.876	8.5	5.8
2006	8	6	14:26:17	37.4091	74.7119	4.9	5.6
2006	8	6	18:16:39	26.2558	143.9864	23	5.9
2006	8	11	14:30:39	18.4706	−101.135	58.4	6.1
2006	8	16	18:38:58	−28.8283	61.7726	10	5.9
2006	8	24	21:50:36	51.0679	157.5354	53.5	6.5
2006	9	1	12:04:21	53.9609	−166.361	75.6	5.9
2006	9	10	14:56:06	26.39	−86.5804	10	5.9
2006	9	24	22:56:21	−17.6967	41.8104	17.2	5.7
2006	9	29	13:08:24	10.8486	−61.7653	53.4	6.1
2006	9	30	17:50:22	46.189	153.1761	19.4	6.6
2006	10	1	9:06:00	46.3193	153.3046	19.5	6.5
2006	10	9	10:01:47	20.7054	120.0645	17.3	6.3
2006	10	10	23:58:06	37.1616	142.8023	32.2	6.0
2006	10	21	18:23:20	13.3641	121.4278	18	5.9
2006	10	23	21:17:22	29.411	140.3506	29.9	6.4
2006	11	17	18:03:11	28.5876	129.8655	23.1	6.2
2006	11	29	15:38:43	53.8157	−35.435	10	5.6
2006	12	1	3:58:20	3.4573	99.103	204.2	6.3
2006	12	25	20:00:59	42.0738	76.0856	15.2	5.8
2006	12	26	12:26:20	21.8354	120.533	6.3	7.1
2006	12	30	8:30:47	13.205	51.3376	10	6.6
2007	1	9	15:49:32	59.4467	−137.138	10	5.7
2007	1	17	23:18:48	10.0815	58.7013	10	6.2
2007	2	4	20:56:57	19.3369	−78.3947	10	6.2
2007	2	19	2:33:42	1.6404	30.6974	27.3	5.6
2007	3	1	23:11:50	26.6058	−44.647	10	6.0
2007	3	6	3:49:38	−0.506	100.4824	21.2	6.4
2007	3	9	7:27:29	−11.4284	66.2758	10	5.7
2007	3	9	3:22:42	43.2206	133.5123	439.5	6.0
2007	3	13	2:59:00	26.1733	−110.697	10	6.0
2007	3	18	2:11:03	4.6505	−78.5033	1.1	6.2
2007	3	22	6:10:43	−3.342	86.7202	26.9	5.9
2007	3	25	0:41:56	37.3209	136.5686	4	6.7
2007	3	28	21:17:10	−6.2242	29.619	13.4	5.8
2007	4	3	3:35:06	36.4738	70.6405	215.5	6.2
2007	4	4	19:58:02	−17.1836	66.875	10	5.9
2007	4	5	3:56:49	37.3659	−24.6358	16.2	6.3
2007	4	10	13:56:50	13.0113	92.5102	15.3	5.5
2007	4	13	5:42:21	17.2469	−100.241	33.4	6.0
2007	4	20	1:45:55	25.6879	125.0772	9.2	6.3
2007	5	4	12:06:51	−1.3273	−15.0009	10	6.2
2007	5	5	8:51:38	34.3079	81.9875	13.4	6.1
2007	5	7	11:59:46	31.3215	97.6605	12	5.5
2007	5	16	8:56:13	20.5565	100.7342	10	6.3
2007	5	23	19:09:13	21.9055	−96.3184	1.7	5.6
2007	5	30	20:22:11	52.0987	157.2889	120.4	6.4

Table A1. Continued.

Year	Month	Day	Time UTC	Lat	Long	Depth	M
2007	6	2	21:34:58	23.0785	101.0073	11	6.1
2007	6	13	19:29:44	13.7024	−90.6465	64	6.7
2007	6	15	18:49:51	1.7332	30.7452	20.1	5.9
2007	6	18	14:29:48	34.4568	50.8578	11.4	5.5
2007	7	3	8:25:59	0.7697	−30.1971	10	6.3
2007	7	6	1:09:18	16.5781	−93.6161	120	6.1
2007	7	13	21:54:43	51.8785	−176.246	44.1	6.0
2007	7	15	13:08:00	52.4899	−168.032	12.5	6.1
2007	7	16	14:17:36	36.866	134.7943	347.1	6.8
2007	7	17	14:10:41	−2.826	36.267	14.8	5.9
2007	7	20	10:06:52	42.9111	82.2962	19.1	5.6
2007	7	29	4:54:35	53.6067	169.7092	28	5.9
2007	7	30	22:42:05	19.3104	95.541	15.9	5.6
2007	7	31	22:55:28	−0.1482	−17.7189	2.7	6.2
2007	8	2	13:37:27	12.447	47.4593	10	5.7
2007	8	2	2:37:42	46.9248	141.8324	19.9	6.2
2007	8	2	5:22:16	46.7681	141.7716	6.9	5.8
2007	8	2	3:21:44	51.3075	−179.975	37.8	6.7
2007	8	7	0:02:21	27.3494	126.7991	4.4	6.0
2007	8	13	22:23:03	−30.9737	−13.4479	10	5.5
2007	8	15	20:22:11	50.2629	−177.554	17.8	6.5
2007	8	16	14:18:25	−3.4566	−12.1013	20.9	5.5
2007	8	20	22:42:28	8.1332	−39.2186	10	6.5
2007	9	1	19:14:22	25.0103	−109.64	11.9	6.1
2007	9	3	16:14:52	45.7243	150.1509	98.6	6.2
2007	9	6	17:51:26	24.3526	122.237	56.2	6.2
2007	9	10	1:49:12	3.0475	−77.9501	27.6	6.8
2007	9	13	3:35:27	−2.156	99.5994	18.8	7.0
2007	9	13	2:30:01	−1.6595	99.61	24	6.5
2007	9	20	8:31:13	−2.0015	100.064	29.1	6.7
2007	9	26	18:39:33	−7.0062	−11.6291	10	5.6
2007	10	2	18:00:07	54.5033	−161.735	42.9	6.3
2007	10	4	12:40:29	2.5719	92.9055	34.7	6.2
2007	10	18	16:13:13	30.1823	−42.6211	12.3	5.7
2007	10	24	21:02:50	−3.9271	101.0147	28.2	6.8
2007	10	31	3:04:54	37.372	−121.798	10	5.6
2007	11	7	7:10:20	22.1583	92.3702	29.7	5.5
2007	11	27	4:26:59	16.2324	119.824	45.3	5.9
2007	12	6	17:12:03	22.7483	−45.1418	15.9	5.8
2007	12	8	19:55:18	−7.5221	37.6041	10	5.6
2007	12	12	23:39:58	52.1242	−131.437	10	5.8
2007	12	19	9:30:26	51.3295	−179.509	34.2	7.2
2007	12	25	14:04:33	38.4955	142.0641	48.1	6.1
2007	12	26	22:04:55	52.5351	−168.221	34.1	6.4

Acknowledgements. Our study is a part of the PASSEQ 2006–2008 project (Wilde-Piórko et al., 2008). The study was partly funded by the NordQuake project. The one-event files were created in the Institute of Geophysics, University of Warsaw, Poland. The figures were produced using Generic Mapping Tools (GMT) (Wessel and Smith, 1991). Data review, picking *P* wave arrivals and calculations of the theoretical *P* wave arrivals were performed using the SHM program package (http://www.seismic-handler.org/). Special thanks to Gediminas Motuza and Hanna Silvennoinen for useful discussions. Majdanski work was supported within statutory activity no. 3841/E-41/S/2014 of the Ministry of Science and Higher Education of Poland. We thank Ulrich Achauer, Andreas Fichtner, Jaroslava Plomerova and one anonymous reviewer for useful comments, which helped to improve the manuscript.

[7] PASSEQ Working Group: Monika WildePiorko[I], Wolfram H. Geissler[II], Jaroslava Plomerova[III], Marek Grad[I], Vladislav Babuška[III], Ewald Bruckl[IV], Jolanta Cyziene[V], Wojciech Czuba[VI], Richard England[VII], Edward Gaczyński[VI], Renata Gazdova[XIII], Soren Gregersen[VIII], Aleksander Guterch[VI], Winfried Hanka[VII], Endre Hegedűs[X], Barbara Heuer[IX], Petr Jedlička[III], Jurga Lazauskiene[V,XVII], G. Randy Keller[XI], Rainer Kind[IX], Klaus Klinge[XII], Petr Kolinsky[XIII], Kari Komminaho[XIV], Elena Kozlovskaya[XV], Frank Kruger[XVI], Tine Larsen[VIII], Mariusz Majdański[VI], Jiři Malek[XIII], Gediminas Motuza[XVII], Oldřich Novotny[XIII], Robert Pietrasiak[VI], Thomas Plenefisch[XII], Bohuslav Růžek[III], Saulius Sliaupa[V], Piotr Środa[VI], Marzena Świeczak[VI], Timo Tiira[XIV], Peter Voss[VIII], PawełWiejacz[VI]. [I] University of Warsaw, Warsaw, Poland; [II] Alfred Wegener Institute for Polar and Marine Research, Bremerhaven, Germany; [III] Institute of Geophysics Czech Academy of Sciences, Prague, Czech Republic; [IV] Vienna University of Technology, Vienna, Austria; [V] Geological Survey of Lithuania, Vilnius, Lithuania; [VI] Institute of Geophysics Polish Academy of Sciences, Warsaw, Poland; [VII] University of Leicester, Leicester, Great Britain; [VIII] Geological Survey of Denmark and Greenland, Copenhagen, Denmark; [IX] GeoForschungsZentrum Potsdam, Potsdam, Germany; [X] Eötvös Loránd Geophysical Institute, Budapest, Hungary; [XI] University of Oklahoma, Norman, USA; [XII] Seismological Central Observatory, Erlangen, Germany; [XIII] Institute of Rock Structure and Mechanics Czech Academy of Sciences, Prague, Czech Republic; [XIV] University of Helsinki, Helsinki, Finland; [XV] University of Oulu, Oulu, Finland; [XVI] University of Potsdam, Potsdam, Germany; [XVII] University of Vilnius, Vilnius, Lithuania.

Special Issue: The Lithosphere-Asthenosphere Boundary (LAB) Dilemma
Edited by: U. Achauer, J. Plomerova, and R. Kind

References

Aki, K., A. Christoffersson, and Husebye, E. S.: Determination of the three-dimensional seismic structure of the lithosphere, J. Geophys. Res., 82, 277–296, 1977.

Arlitt, R.: Teleseismic body wave tomography across the Trans-European Suture Zone between Sweden and Denmark, PhD theses, ETH, Swiss Federal Institute of Technology Zürich, Swiss, 1999.

Artemieva, I. M.: Global $1° \times 1°$ thermal model TC1 for the continental lithosphere: implications for lithosphere secular evolution, Tectonophysics, 416, 245–277, 2006.

Artemieva, I. M.: Dynamic topography of the East European craton: Shedding light upon lithospheric structure, composition and mantle dynamics, Global and Planetary Change, 58, 411–434, 2007.

Artemieva, I. M., Thybo, H., and Kaban, M. K.: Deep Europe today: Geophysical synthesis of the upper mantle structure and lithospheric processes over 3.5 Ga, in: European Lithosphere Dynamics, edited by: Gee, D. and Stephenson, R., Geological Society London, Special Publication, 32, 11–41, 2006.

Babuška V. and Plomerova J.: Subcrustal Lithosphere Around the Saxothuringian-Moldanubian Suture Zone – a model derived from anisotropy of seismic wave velocities, Tectonophysics, 332, 185–199, 2001.

Babuška, V., Plomerova, J., and Fischer, T.: Intraplate Seismicity in the Western Bohemian Massif (Central Europe): A possible correlation with a paleoplate junction, J. Geophysics, 44, 149–159, 2007.

Banka, D., Pharao, T. C., Williamson, J. P., and the TESZ Project Potential Field Core Group: Potential field imaging of Paleozoic orogenic structure in northern and central Europe, Tectonophysics, 360, 23–45, 2002.

Belka, Z., Ahrendt, H., Franke, and W., and Wemmer, K.: The Baltica-Gondwana suture in central Europe: evidence from K–Ar ages of detrital muscovites and biogeographical data, in: Orogenic Processes: Quantification and Modelling of the Variscan Belt, edited by: Franke, W., Haak, V., Oncken, O., and Taner, D., Geological Society, London, 87–102, 2000.

Beller, S., Kozlovskaya, E., Achauer, U., and Tiberi, Ch.: Joint inversion of teleseismic and gravity data beneath the Fennoscandian Shield, EGU General Assembly 2013, Geophys. Res. Abstracts, 15, EGU 2013-4771-2, 2013.

Blundel, D., Freeman, R., and Mueller, S. (Eds.): A Continent Revealed – The European Geotraverse. European Science Foundation, Cambridge University Press, Cambridge, 287 pp., 1992.

Bogdanova, S. V., Gorbatschev, R., Stephenson, R. A., and Guterch A. (Eds.): EUROBRIDGE: Palaeoproterozoic accretion of Fennoscandia and Sarmatia, Tectonophysics, 339, 1–2, 2001.

Bogdanova, S., Gorbatschev, R., Grad, M., Janik, T., Guterch, A., Kozlovskaya, E., Motuza, G., Skridlaite, G., Starostenko, V., Taran, L., and the EUROBRIDGE and POLONAISE Working Groups: EUROBRIDGE: new insight into the geodynamic evolution of the East European Craton, in: European Lithospheric Dynamics. Memoirs Number 32, edited by: Gee, D. G. and Stephenson, R. A., Geological Society, London, 599–625, 2006.

Bogdanova, S. V., Bingen, B., Gorbatschev, R., Kheraskova, T. N, Kozlov, V. I., Puchkov, V. N., and Volozh, Yu. A.: The East European Craton (Baltica) before and during the assembly of Rodinia, Precam. Res., 160, 23–45, 2008.

Bruneton, M., Pedersen, H. A., Farra, V., Arndt, N. T., Vacher, P., Achauer, U., Alinaghi, A., Ansorge, J., Bock, G., Friederich, W., Grad, M., Guterch, A., Heikkinen, P., Hjelt, S.-E., Hyvönen, T. L., Ikonen, J.-P., Kissling, E., Komminaho, K., Korja, A., Kozlovskaya, E., Nevsky, M. V., Paulssen, H., Pavlenkova, N. I., Plomerová, J., Raita, T., Riznichenko, O. Y., Roberts, R. G., Sandoval, S., Sanina, I. A., Sharov, N. V., Shomali, Z. H., Tiikkainen, J., Wielandt, E., Wilegalla, K., Yliniemi, J., and Yurov, Y. G.:

Complex lithospheric structure under the central Baltic Shield from surface wave tomography, J. Geophys. Res., 109, B10303, doi:10.1029/2003JB002947, 2004.

Cotte, N., Pedersen, H. A., and TOR Working Group: Sharp contrast in lithospheric structure across the Sorgenfrei–Tornquist zone as inferred by Rayleigh wave analysis of TOR1 project data, Tectonophysics, 360, 75–88, 2002.

Czuba, W., Grad, M., Luosto, U., Motuza, G., Nasedkin, V., and POLONAISE P5 Working Group: Crustal structure of the East European craton along the POLONAISE'97 P5 profile, Acta Geoph. Pol., 49, 145–168, 2001.

Dadlez, R., Grad, M., and Guterch, A.: Crustal structure below the Polish Basin: Is it composed of proximal terranes derived from Baltica?, Tectonophysics, 411, 111–128, 2005.

Dörr W., Belka Z., Marheine D., Schastok J., Valverde Vaquero P., and Wiszniewska J.: U–Pb and Ar–Ar geochronology of anorogenic granite magmatism of the Mazury complex NE Poland, in: Precambrian Research, edited by: Rämo, T., Special issue, 119, 101–102, 2002.

Eaton, D. W., Darbyshire, F., Evans, R. L., Grütter, H., Jones, A. G., and Yuan, X.: The elusive lithosphere–asthenosphere boundary (LAB) beneath cratons, Lithos, 109, 1–22, 2009.

EUROBRIDGE Seismic Working Group: Seismic velocity structure across the Fennoscandia-Sarmatia suture of the East European Craton beneath the EUROBRIDGE profile through Lithuania and Belarus, Tectonophysics, 314, 193–217, 1999.

EUROBRIDGE Working Group, and EUROBRIDGE'95: Deep seismic profiling within the East European Craton, Tectonophysics, 339, 153–175, 2001.

Franke, W. and Zelazniewicz, A.: The Eastern Termination of the Variscides: Terrane Correlation and Kinematic Evolution in Orogenic processes: quantification and modelling in the Variscan Belt, Special Publications, edited by: Franke W., Haak W., Oncken O., and Tanner, D., Geol. Soc., London, 179, 63–85, 2000.

Garetskii, R. G., Boborykin, A. M., Bogino, V. A., German, V. A., Veres, S. A., Klushin, S. V., and Shafaruk, V. G.: Deep seismic sounding on the territory of Belorussia, Geophys. J. Internat., 8, 439–448, 1990.

Gee, D. G. and Stephenson, R. A.: The European lithosphere: an introduction, in: European Lithophere Dynamics, edited by: Gee, D. G. and Stephenson, R. A., Geol. Soc. Lond. Mem., 32, 1–9, 2006.

Geissler, W. H., Sodoudi, F., and Kind, R.: Thickness of the central and eastern European lithosphere as seen by S receiver functions, Geophys. J. Int., 181, 604–634, 2010.

Grad, M. and Tripolsky, A.: Crustal structure from P and S seismic waves and petrological model of the Ukrainian shield, Tectonophysics, 250, 89–112, 1995.

Grad, M., Keller, G. R., Thybo, H., Guterch, A., and POLONAISE Working Group: Lower lithospheric structure beneath the Trans-European Suture Zone from POLONAISE'97 seismic profiles, Tectonophysics, 360, 153–168, 2002.

Grad, M., Janik, T., Guterch, A., Sroda, P., Czuba, W., and EUROBRIDGE'94–97, POLONAISE'97 and CELEBRATION 2000 Seismic Working Groups: Lithospheric structure of the western part of the East European Craton investigated by deep seismic profiles, Geol. Quart., 50, 9–22, 2006.

Gregersen, S., Pedersen, L. B., Roberts, R. G., Shomali, H., Berthelsen, A., Thybo, H., Mosegaard, K., Pedersen, T., Voss, P.,

Kind, R., Bock, G., Gossler, J., Wylegala, K., Rabbel, W., Woelbern, I., Budweg, M., Busche, H., Korn, M., Hock, S., Guterch, A., Grad, M., Wilde-Piorko, M., Zuchniak, M., Plomerova, J., Ansorge, J., Kissling, E., Arlitt, R., Waldhauser, F., Ziegler, P., Achauer, U., Pedersen, H., Cotte, N., Paulssen, H., and Engdahl, E. R.: Important findings expected from Europe's largest seismic array. Eos, Trans. Am. Geophys. Union, 80, 1–6, 1999.

Gregersen, S., Voss, P., Nielsen, L. V., Achauer, U., Busche, H., Rabbel, W., and Shomali, Z. H.: Uniqueness of modeling results from teleseismic P wave tomography in Project TOR, Tectonophysics, 481, 99–107, 2010.

Griffin, W. L., O'Reilly, S. Y., Abe, N., Aulbach, S., Davies, R. M., Pearson N. J., Doyle, B. J., and Kivi, K.: The origin and evolution of Archean lithospheric mantle, Precam. Res., 127, 19–41, 2003.

Guterch, A. and Grad, M.: Seismic structure of the Earth's crust between Precambrian and Variscan Europe in Poland, Publs. Inst. Geophys. Pot. Acad. Sc., M-18, 67–73, 1996.

Guterch, A. and Grad, M.: Lithospheric structure of the TESZ in Poland based on modern seismic experiments, Geol. Quart., 50, 23–32, 2006.

Guterch, A., Grad, M., Thybo, H., Keller, G. R., and the POLONAISE Working Group: POLONAISE'97 – an international seismic experiment between Precambrian and Variscan Europe in Poland, Tectonophysics 314, 101–121, 1999.

Guterch, A., Grad, M., Keller, G. R., and POLONAISE'97, CELEBRATION 2000, ALP 2002, SUDETES 2003 Working Groups: Huge contrasts of the lithospheric structure revealed by new generation seismic experiments in Central Europe, Przegld Geologiczny, 52, 2004.

Hansen, T. M. and Balling, N.: Upper-mantle reflectors: modeling of seismic wavefield characteristics and tectonic implications, Geophys. J. Int, 157, 664–682, 2004.

Johnston, A. C., Coppersmith, K. J., Kanter, L. R., and Cornell, C. A.: The earthquakes of stable continental regions. Elektric Power Institute, Report in Vol. 1: Assessment of Large Earthquake Potential, no. TR-102261-V1, 1994.

Jones, A. G., Plomerova, J., Korja, T., Sodoudi, F., and Spakman, W.: Europe from the bottom up: A statistical examination of the central and northern European lithosphere-asthenosphere boundary from comparing seismological and electromagnetic observations, Lithos, 120, 14–29, 2010.

Karousova, H., Plomerova, J., and Babuska, V.: Upper-mantle structure beneath the southern Bohemian Massif and its surroundings imaged by high-resolution tomography, Geophys. J. Int., 194, 1203–1215, doi:10.1093/gji/ggt159, 2013.

Knapmeyer-Endrun, B., Krüger, F., Legendre, C. P., Geissler, W. H., and PASSEQ Working Group: Tracing the influence of the Trans-European Suture Zone into the mantle transition zone, Earth Planet. Sci. Lett., 363, 73–87, 2013a.

Knapmeyer-Endrun, B., Krüger, and PASSEQ Working Group: Imaging the lithosphere-asthenosphere boundary across the transition from Phanerozoic Europe to the East-European Craton with S-receiver functions, Geophys. Res. Abstr., 15, EGU2013-6972, 2013b.

Korja, T.: How is the European Lithosphere Imaged by Magnetotellurics?, Surv. Geophys., 28, 239–272, .2007.

Kostyuchenko, S. L., Egorkin, A. V., and Solodilov, L. N.: Structure and genetic mechanisms of the Precambrian rifts of the East European Platform in Russia by integrated study of seismic, gravity, and magnetic data, Tectonophysics, 313, 9–28, 1999.

Koulakov, I., Kaban, M. K., Tesauro M., and Cloetingh S.: P- and S-velocity anomalies in the upper mantle beneath Europe from tomographic inversion of ISC data, Geophys. J. Int., 179, 345–366, 2009.

Legendre, C. P., Meier, T., Lebedev, S., Friederich, W., and Viereck-Götte, L.: A shear wave velocity model of the European upper mantle from automated inversion of seismic shear and surface waveforms, Geophys. J. Int., 191, 282–304, 2012.

Majdański, M.: The structure of the crust in TESZ area by kriging interpolation, Acta Geophys., 60, 59–75, 2012.

Majorowicz, J. A., Čermak, V., Šafanda, J., Krzywiec, P., Wróblewska, M., Guterch, A., and Grad, M.: Heat flow models across the Trans-European Suture Zone in the area of the POLONAISE'97 seismic experiment, Phys. Chem. Earth, 28, 375–391, 2003.

Malinowski, M., Grad, M., Guterch, A., and CELEBRATION 2000 Working Group: Three-dimensional seismic modelling of the crustal structure between East European Craton and the Carpathians in SE Poland based on CELEBRATION 2000 data, Geophys. J. Int., 173, 546–565, 2008.

Matzel, E. and Grand, S. P.: The anisotropic structure of the East European platform, J. Geophys. Res., 109, B01302, doi:10.1029/2001JB000623, 2004.

McKenzie, D. P.: Some remarks on the heat flow and gravity anomalies, J. Geophys. Res., 72, 6261–6273, 1967.

Menke, W.: Geophysical data analysis: Discrete inverse theory, Academic Press, Inc., Orlando, Fl., 260 pp., 1984.

Motuza, G.: Žemės plutos bei kristalinio pamato sandaros ir sudėties raida, in: Lietuvos žemės gelmių raida ir ištekliai, edited by: Baltrūnas, V., UAB Petro ofsetas, Vilnius, 11–40, 2004.

Motuza, G.: Structure and formation of the crystalline crust in Lithuania, Mineralogical Society of Poland, Special Papers, 26, 67–79, 2005.

Motuza, G. and Staškus, V.: Seniausios Lietuvos uolienos, Geologijos akiračiai, ISSN 1392-0006, 3/4, 41–47, 2009.

Nolet, G. and Zielhuis, A.: Low S velocities under the Tornquist–Teisseyre zone: evidence from water injection into the transition zone by subduction, J. Geophys. Res., 99, 15813–15820, 1994.

Pharaoh, T. C.: Paleozoic terranes and their lithospheric boundaries within the Trans-European Suture Zone (TESZ): a review, Tectonophysics, 314, 17–41, 1999.

Pharaoh, T. C. and TESZ Project Core Group: EUROPROBE Trans-European Suture Zone project. British Geological Survey, EUROPROBE News 12, June 2000, 2000.

Pharaoh, T. C., England, R. W., Verniers, J., and Zelainiewicz, A.: Introduction: geological and geophysical studies in the Tram-European Suture Zone, Geol. Mug., 134, 585–590, 1997.

Plomerova, J. and Babuska, V.: Seismic anisotropy of the lithosphere around the Trans-European suture zone (TESZ) based on teleseismic body-wave data of the TOR experiment, Tectonophysics, 360, 89–114, 2002.

Plomerova, J. and Babuska, V.: Long memory of mantle lithosphere fabric – European LAB constrained from seismic anisotropy, Lithos, 120, 131–143, 2010.

Plomerova J., Babuška V., Sileny J., and Horalek, J.: Seismic anisotropy and velocity variations in the mantle beneath the Saxothuringicum-Moldanubicum con- tact in central Europe, in: Geodynamics of Lithosphère and Earh's Mantle: Seismic Anisotropy as a Recird of the Past and Present Dynamic Processes, edited by: Plomerova, J., Liebermann, R. C., and Babuška, V., Pure and Appi. Geoph., Special issue, 151 pp., 1998.

Plomerova, J., Kouba, D., and Babuska, V.: Mapping the lithosphere-asthenosphere boundary (LAB) through changes in surface-wave anisotropy, Tectonophysics, 358, 175–185, 2002.

Plomerova, J., Achauer, U., Babuska, V., Vecsey, L., and BOHEMA Working Group: Upper mantle beneath the Eger Rift (Central Europe): plume or asthenosphere upwelling?, Geophys. J. Int., 169, 675–682, 2007.

Plomerova, J., Babuska, V., Kozlovskaya, E., Vecsey, L., and Hyvonen, L. T.: Seismic anisotropy – a key to resolve fabrics of mantle lithosphere of Fennoscandia, Tectonophysics, 462, 125–136, 2008.

Praus, O., Pěčová, J., Petr, V., Babuska, V., and Plomerova, J.: Magnetotelluric and seismological determination of the lithosphere–asthenosphere transition in Central Europe, Phys. Earth Planet Int., 60, 212–228, 1990.

Rämö, O. T., Huhma, H., and Kirs, J.: Radiogenic Isotopes of the Estonian and Latvian Rapakivi Granite Suite: New Data from the Concealed Precambrian of the East European Craton, Precam. Res., 79, 209–226, 1996.

Sandoval, S., Kissling, E., Ansorge, J., and the SVEKALAPKO STWG: High- Resolution body wave tomography beneath the SVEKALAPKO array: I. A-priori 3-D crustal model and associated traveltime effects on teleseismic wavefronts, Geoph. J. Int., 153, 75–87, 2003.

Sandoval Castano, S.: The Lithosphere-Asthenosphere System beneath Fennoscandia (Baltic Shield) by Body-wave Tomography, A dissertation submitted to the Swiss Federal Institute of Technology Zurich, 2002.

Shomali, Z. H., Roberts, R. G., Pedersen, L. B., and the TOR Working Group: Lithospheric structure of the Tornquist Zone resolved by nonlinear P and S teleseismic tomography along the TOR array, Tectonophysics, 416, 133–149, 2006.

Skridlaite, G. and Motuza, G.: Precambrian domains in Lithuania: evidence of terrane tectonics, Tectonophysics, 339, 113–133, 2001.

Sroda, P. and the POLCRUST and PASSEQ Working Groups: Seismic anisotropy and deformations of the TESZ lithosphere near the East European Craton margin in SE Poland at various scales and depths, EGU General Assembly 2014, Geophys. Res. Abstracts, Vol. 16, EGU2014-6463-1, 2014.

Steck, L. K. and Prothero, W. A. J.: A 3-D raytracer for teleseismic body-wave arrival times, Bull. Seism. Soc. Am., 81, 1332–1339, 1991.

Tesauro, M., Kaban, M. K., and Cloetingh, S. A. P. L.: EuCRUST-07: A new reference model for the European crust, Geophys. Res. Lett., 35, L05313, doi:10.1029/2007GL32244, 2008.

Thybo, H.: Crustal structure and tectonic evolution of the Tornquist Fan region as revealed by geophysical methods, Bull. Geol. Soc. Den., 46, 145–160, 2000.

Thybo, H., Janik, T., Omelchenko, V. D., Grad, M., Garetsky, R. G., Belinsky, A. A., Karatayev, G. I., Zlotski, G., Knudsen, M. E., Sand, R., Yliniemi, J., Tiira, T., Luosto, U., Komminaho, K., Giese, R., Guterch, A., Lund, C.-E., Kharitonov, O. M., Ilchenko, T., Lysynchuk, D. V., Skobelev, V. M., and Doody, J. J.: Upper lithospheric seismic velocity structure across the Pripyat Trough and the Ukrainian Shield along the EUROBRIDGE'97 profile, Tectonophysics, 371, 41–79, 2003.

Vecsey, L., Plomerova, J., Babuska, V., and PASSEQ Working Group: Structure of the mantle lithosphere around the TESZ – from the East European Craton to the Variscan Belt. EGU General Assembly 2013, Geophys. Res. Abstracts, Vol. 15, EGU2013-3133, 2013.

Vinnik, L. P. and Ryaboy, V. Z.: Deep structure of the East European platform according to seismic data, Phys. Earth Planet. Inter., 25, 27–37, 1981.

Wagner, G. A., Gögen, K., Jonckheere, R., Wagner, I., and Woda, C.: Dating the Quarternary volcanoes Komorní Hurka (Kammerbühl) and Železná Hurka (Eisenbühl), Czech Republic, by TL, ESR, alpha-recoil and fission track chronometry, Z. Geol. Wiss., 30, 191–200, 2002.

Weiland, C. M., Steck, L. K., Dawson, P. B., and Korneev, V. A.: Nonlinear teleseismic tomography at Long Valley caldera, using three-dimensional minimum travel time ray tracing, J. Geophys. Res., 100, 20379–20390, 1995.

Wessel, P. and Smith, W.: Free software helps map and display data, EOS, Trans. Am. Union 72, 441 pp., 1991.

Wilde-Piorko, M., Grad, M., and TOR Working Group: Crustal structure variation from the Precambrian to Palaeozoic platforms in Europe imaged by the inversion of teleseismic receiver functions – project TOR, Geophys. J. Internat., 150, 261–270, 2002.

Wilde-Piórko, M., Geissler, W.H., Plomerová, J., Grad, M., Babuška, V., Brückl, E., Čyžienė, J., Czuba, W., Eengland, R., Gaczyński, E., Gazdova, R., Gregersen, S., Guterch, A., Hanka, W., Hegedűs, E., Heuer, B., Jedlička, P., Lazauskienė, J., Randy Keller, G., Kind, R., Klinge, K., Kolinsky, P., Komminaho, K., Kozlovskaya, E., Krűger, F., Larsen, T., Majdański, M., Málek, J., Motuza, G., Novotný, O., Pietrasiak, R., Plenefish, Th., Růžek, B., Šliaupa, S., Środa, P., Świeczak, M., Tiira, T., Voss, P., and Wiejacz, P.: PASSEQ 2006–2008: PASsive Seismic Experiment in Trans-European Suture Zone, Stud. Geophys. Geod., 52, 439–448, 2008.

Wilde-Piórko, M., Świeczak, M., Grad, M., and Majdański, M.: Integrated seismic model of the crust and upper mantle of the Trans-European Suture zone between the Precambrian craton and Phanerozoic terranes in Central Europe, Tectonophysics, 481, 108–115, 2010.

Winchester, J. A. and the PACE TMR Network Team: Paleozoic amalgamation of Central Europe: new results from recent geological and geophysical investigations, Tectonophysics, 360, 5–21, 2002.

Wüstefeld, A., Bokelmann, G., and Barruol, G.: Evidence for ancient lithospheric deformation in the East European Craton based on mantle seismic anisotropy and crustal magnetics, Tectonophysics, 481, 16–28, 2010.

Yliniemi, J., Kozlovskaya, E., Hjelt, S. E., Komminaho, K., and Ushakov, A.: Structure of the crust and uppermost mantle beneath southern Finland revealed by analysis of local events registered by the SVEKALAPKO seismic array, Tectonophysics, 394, 41–67, 2004.

Zhu, H., Bozdağ, E., Peter, D., and Tromp, J.: Structure of the European upper mantle revealed by adjoint tomography, Nat. Geosci., 5, 493–498, 2012.

Zielhuis, A. and Nolet, G.: Deep seismic expression of the ancient plate boundary in Europe, Science, New Series, 265, 79–81, 1994.

2

Fault evolution in the Potiguar rift termination, equatorial margin of Brazil

D. L. de Castro and F. H. R. Bezerra

Programa de Pós-Graduação em Geodinâmica e Geofísica, Universidade Federal do Rio Grande do Norte, Natal, Brazil

Correspondence to: D. L. de Castro, (david@geologia.ufrn.br)

Abstract. The transform shearing between South American and African plates in the Cretaceous generated a series of sedimentary basins on both plate margins. In this study, we use gravity, aeromagnetic, and resistivity surveys to identify architecture of fault systems and to analyze the evolution of the eastern equatorial margin of Brazil. Our study area is the southern onshore termination of the Potiguar rift, which is an aborted NE-trending rift arm developed during the breakup of Pangea. The basin is located along the NNE margin of South America that faces the main transform zone that separates the North and the South Atlantic. The Potiguar rift is a Neocomian structure located at the intersection of the equatorial and western South Atlantic and is composed of a series of NE-trending horsts and grabens. This study reveals new grabens in the Potiguar rift and indicates that stretching in the southern rift termination created a WNW-trending, 10 km wide, and ∼ 40 km long right-lateral strike-slip fault zone. This zone encompasses at least eight depocenters, which are bounded by a left-stepping, en echelon system of NW–SE- to NS-striking normal faults. These depocenters form grabens up to 1200 m deep with a rhomb-shaped geometry, which are filled with rift sedimentary units and capped by postrift sedimentary sequences. The evolution of the rift termination is consistent with the right-lateral shearing of the equatorial margin in the Cretaceous and occurs not only at the rift termination but also as isolated structures away from the main rift. This study indicates that the strike-slip shearing between two plates propagated to the interior of one of these plates, where faults with similar orientation, kinematics, geometry, and timing of the major transform are observed. These faults also influence rift geometry.

1 Introduction

One-third of the present-day passive margins of the world were formed as transform margins. These margins are characterized by abrupt ocean–continent transition, steep continental slopes, and high bathymetric gradients inherited from the nearly vertical transform plate boundary (Basile et al., 2013). Classic models of rifting and lithospheric thinning (McKenzie, 1978) cannot explain the evolution of these transform margins, which are characterized by oblique rifting and often postrift tectonic inversions (e.g., Vagnes et al., 1998).

This type of transform margin was responsible for the generation of the South Atlantic, which was the result of a diachronic breakup between the African and South American plates. While the eastern margin of South America was subjected to orthogonal rifting, the equatorial margin was subjected to a transform motion between these plates (Matos, 2000). The equatorial margin of South America and West African is characterized by Mesozoic-Cenozoic basins formed along more than 2000 km along the South Atlantic (Matos, 2000). Rifting in the equatorial Atlantic margin occurred during the early Barremian to Aptian, and the first oceanic crust probably accreted during the late Aptian (Basile et al., 2005). This margin and the associated South Atlantic in the South American plate also encompass the Chain, Romanche, and Saint Paul transform faults, which are several hundreds of kilometers long (Fig. 1a).

In this context, the Neocomian Potiguar Basin, which lies at the intersection of the eastern and equatorial Atlantic margins of Brazil (Fig. 1b), is a key point for both piercing points and continental breakup evolution (Ponte et al., 1977). The Potiguar Basin, however, developed along a triple junction, which encompasses faults both parallel and oblique to the

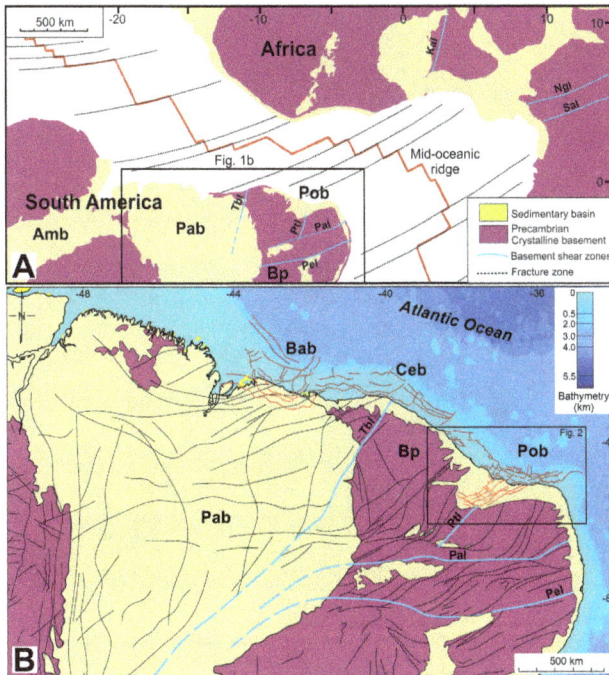

Figure 1. (a) Schematic reconstruction of northeastern Brazil and western Africa at chron C34 (84 Ma) showing the main prerift piercing point and sedimentary basins (Amb – Amazon; Pab – Parnaíba; Pob – Potiguar) in both margins (adapted from Moulin, 2010). Bp – Borborema Province. Precambrian lineaments: Tbl – Transbrasiliano; Ptl – Portalegre; Pal – Patos; Pel – Pernambuco; Kal – Kandi; Ngl – Ngaoundere; Sal – Sanaga. **(b)** Detailed view of the equatorial margin showing the tectonic setting of the Potiguar Basin. The red faults along the margin were formed during the breakup of Pangea; the black traces in the Parnaíba Basin area represent major Precambrian basement terrain boundaries, basement foliations, and shear zones. Basins: Bab – Barreirinhas; Ceb – Ceará; Pob – Potiguar; Pab – Parnaíba.

margin (Bertani et al., 1990). In addition, this basin has links with the Benue Basin in Nigeria in the counterpart African plate (De Castro et al., 2012). Conventional rifting processes cannot explain this basin and others along the margin.

It follows that, despite the general knowledge of the transform margin evolution, several scientific gaps remain, which have important implications for the predrift misfit of the plates (Conceição et al., 1988; Unternehr et al., 1988; De Castro et al., 2012). First, most of the Precambrian fabric is NE-oriented at the margin (e.g., De Castro et al., 2012, 2014), but the equatorial margin trends mainly EW. Second, several rifts exhibit fault systems that are not explained by an orthogonal stretching perpendicular to the rift trend (Bonini et al., 1997). Third, the role of the transform plate motion in the interior of the plate is still uncertain and lacks precise documentation.

We focus on recently published regional magnetic and gravity maps of the Potiguar Basin (De Castro et al., 2012), which show areas at the SW rift boundary, whose geophys-

ical signatures suggest the presence of unidentified buried grabens. The geophysical and geological knowledge of this rift's internal geometry and boundaries were established by Bertani et al. (1990), Matos (1992), and Borges (1993), and few changes have been added to the rift architecture proposed more than 20 years ago. Here, we examine how faults evolve at rift terminations and whether their geometry is inherited from basement fabric. We used a multidisciplinary geophysical survey, which included acquisition, processing, and inversion of magnetic, gravity, and geoelectrical data. In the present study, we investigated the architecture of these structures observed in the study by De Castro et al. (2012) at the southern onshore termination of the Potiguar rift (Fig. 2).

This work may provide new insights that can contribute to a better understanding of the process of continental rifts and transform margin evolution. This study indicates that several faults that form graben boundaries in the Potiguar rift, oblique to the main transform margin, have the same orientation, kinematics, and geometry as the main transform. It also indicates that deformation related to this margin propagates and influences rift development inside the plate. This may have similarities in other basins of the equatorial margin and in other transform margins elsewhere.

2   Tectonic setting

The deformation during the breakup of South America–Africa shifted from extension in the eastern margin to right-lateral shearing in the equatorial margin, where several NE-trending intracratonic basins were formed (Matos, 1992, 2000). The onset of rifting in the equatorial Atlantic occurred at ∼ 140 Ma in the Neocomian. The early stages of rifting have a dominantly half-graben geometry, controlled by NE-striking bounding faults, which reactivated the NE-trending Precambrian fabric (Matos, 1992; Souto Filho et al., 2000). A series of NW-trending depocenters were also formed in the equatorial margin during this period (Matos, 2000). Two dominant directions of stretching occurred: NW–SE and EW (Matos, 1992). Rifting was aborted in the early to the late Barremian (125 Ma), which is coeval with the oldest sediments of the African margin in the Benue Basin (Matos, 1992; Nóbrega et al., 2005). After that period, the equatorial and Southern Atlantic oceans united in the late Albian (105 Ma) (Koutsoukos, 1992) and a subsequent thermal subsidence occurred, allowing the deposition of a transitional unit that was capped by siliciclastic and carbonate postrift sedimentary units (Bertani et al., 1990).

The onshore Potiguar rift comprises an area ∼ 150 km long and ∼ 50 km wide, with an internal geometry of half grabens, which are bounded by NE-trending normal faults and NW-trending transfer faults, dipping to the NW and N, respectively. The former reactivated, whereas the latter cut across Precambrian shear zones. The Potiguar rift is limited in the east by the Carnaubais fault, in the west by the Areia

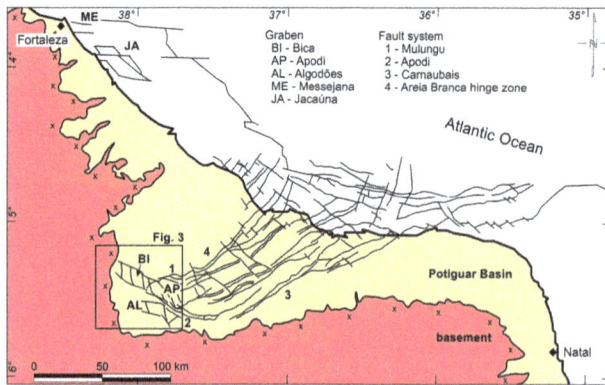

Figure 2. Simplified geologic map of the Potiguar Basin in NE Brazil (adapted from Angelim et al., 2006). The rift structures in the maps of Figs. 2 and 4 are inferred from interpretation of seismic sections and well logs, conducted by Matos (1992) and Borges (1993). The grabens located at the SW rift termination are derived from the present geophysical survey.

Branca hinge zone, and in the south by the Apodi fault. The main axis of the onshore Potiguar rift is NE–SW (Fig. 2) (Bertani et al., 1990). The NE–SW-oriented flat to lystric normal faults control the rift internal geometry, whereas NW–SE-trending faults acted as accommodation zones and transfer faults in response to the extensional deformation (Matos, 1992).

The main depocenters reach maximum depths of 6000 m, and their basin infill was deposited in a typical continental environment (Araripe and Feijó, 1994). Furthermore, a few grabens occur away from the main depocenters. The best examples are the Jacaúna and Messejana grabens at the western part of the Potiguar Basin (Fig. 2). They are transtensional structures bounded by E–W-trending transfer faults and NW-trending normal faults (Matos, 1992).

In the Potiguar Basin, the rift sequence of Neocomian age is covered by a transitional Aptian marine unit, and later by the Aptian–Campanian fluvial and marine transgressive sequence, followed by the regional progradation of Paleogene clastic and carbonate deposits. The limit between *syn*-rift and postrift units is well marked by an angular unconformity that separates the *syn*-rift units from the postrift units (Souto Filho et al., 2000; Pessoa Neto et al., 2007). In the SW border of the Potiguar rift, the siliciclastic (lower) and carbonate (upper) sequences overlap the rift zone, represented here by the Apodi and Algodões grabens (Fig. 3). Faulting also affected the postrift units from the Late Cretaceous to the Quaternary (Bezerra and Vita-Finzi, 2000; Kirkpatrick et al., 2013). These faults reactivate the Precambrian shear zones or basin-bounding faults as well as cut across preexisting structures (Bezerra et al., 2011).

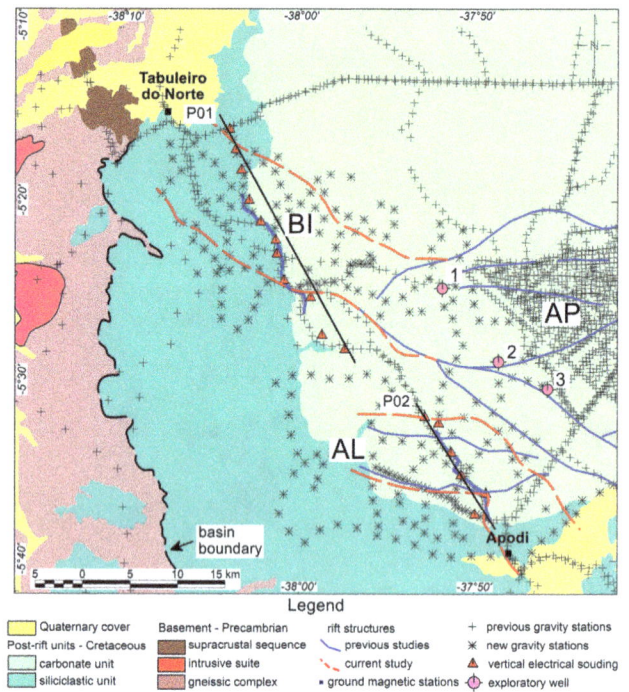

Figure 3. Geologic map of the SW border of the Potiguar rift with the location of the geophysical data sets. (Grabens: BI – Bica; AP – Apodi; and AL – Algodões. Profiles: P01 and P02 (black lines). Exploratory wells: 1, 2, and 3).

3 Geophysical data set

3.1 Magnetics

The aeromagnetic survey in the Potiguar Basin Project was flown between 1986 and 1987 by the Brazilian Petroleum Company (Petrobras) at nominal flight height of 500 m along flight lines oriented 340–160° and spaced 2.0 km apart (MME/CPRM, 1995). We leveled and interpolated the aeromagnetic data into a 500 m grid, using the bidirectional method for the purposes of digital analysis. We further applied filtering and source detection techniques to the magnetic data such as regional–residual separation, reduction to magnetic pole, 3-D analytic signal, and Euler deconvolution.

In addition, we carried out a magnetic ground survey along two profiles (P01 and P02, Fig. 3) to obtain an enhanced magnetic response of the buried structures. We measured 593 stations, spaced each 40 m, using an ENVI PRO MAG (proton precession) magnetometer in the base stations and a rover with a Geometrics G-858 (cesium vapor) magnetometer.

The reduced-to-pole residual magnetic map is marked by a rugged relief, with positive and negative anomalies of short to medium wavelengths and amplitudes that reach values of between −125 and 215 nT (Fig. 4a). The dominant magnetic trends are NE–SW-oriented but show E–W inflections in the W and central parts of the study area, revealing the NE–SW and E–W directions of the crystalline basement fabric.

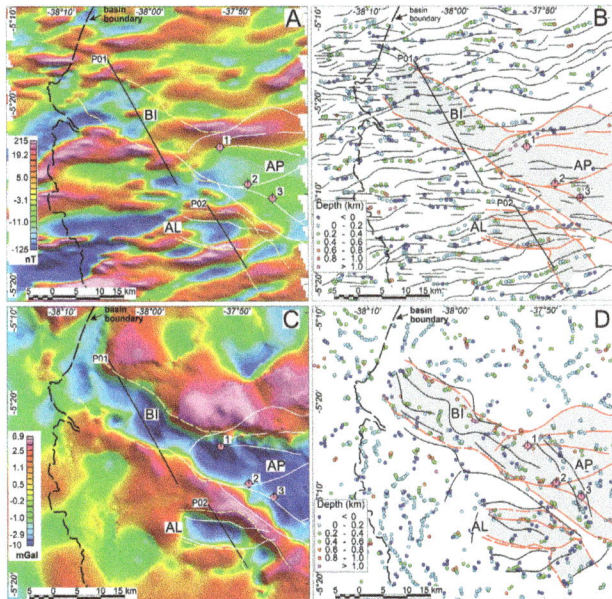

Figure 4. (a) Residual component of the magnetic field reduced to the pole and **(b)** major magnetic lineaments and Euler solutions; **(c)** residual gravity anomaly map and **(d)** major gravity lineaments and Euler solutions. (Grabens (grey zones): BI – Bica; AP – Apodi; and AL – Algodões. Profiles: P01 and P02 (black lines). Exploratory wells: 1, 2, and 3). Solid and dashed white and red traces: rift structures from previous and current studies, respectively.

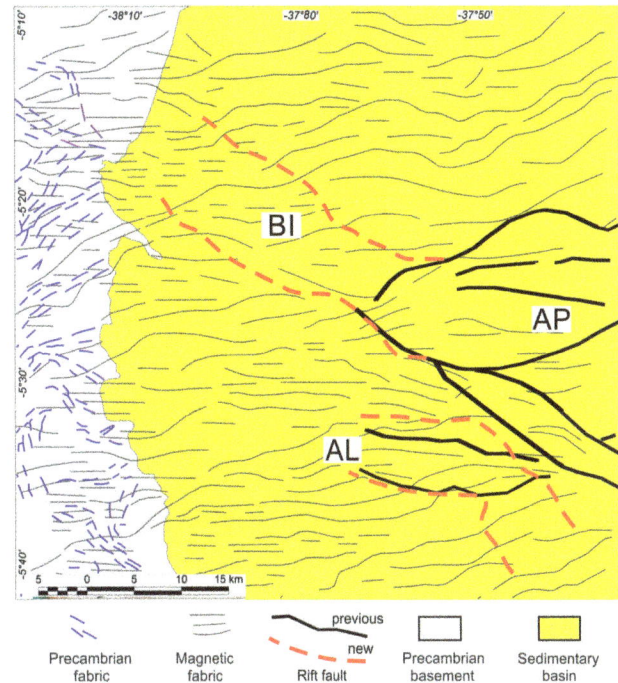

Figure 5. Comparison between Precambrian structural fabric derived from remote sensing and NE–SW- to E–W-trending magnetic lineaments. Grabens: BI – Bica; AP – Apodi; and AL – Algodões.

The magnetic lineaments cut across the Precambrian fabric (metamorphic foliations and shear zones) (Fig. 5). Inside the rift structures (BI, AP, and AL in Fig. 4a), the magnetic surface is smooth and the anomalies are almost negative, denoting the low magnetic content of the Cretaceous sedimentary infill. A slight NW–SE-oriented lineament coincides with the Apodi fault.

Figure 4a exhibits the magnetic lineaments extracted from the phase of the 3-D analytical signal and the solutions of magnetic sources location and depth analysis using the 3-D Euler deconvolution method (Reid et al., 1990). The optimal parameters to apply the Euler deconvolution for the study area were a structural index of 0 to calculate solutions for source body with contact geometry, search window size of 5.0 km, and maximum tolerance of 15 % for depth uncertainty of the calculated solution. The NE–SW main magnetic trend is followed by the Euler solutions, whose sources are concentrated at depths lower than 1.0 km (Fig. 4b). It is worth mentioning that only a few solutions are coincident with the rift faults. This suggests that the lateral contacts between basin structures and the basement units provide incipient contrasts of the magnetic susceptibility.

3.2 Gravity

This study integrated 1743 gravity data points (Fig. 3), which included 234 new gravity stations and 1509 data points provided by the Brazilian Petroleum Agency (ANP). This data set was interpolated with a grid cell size of 500 m using minimum curvature technique (Briggs, 1974). Afterwards, we removed the regional component from the gravity field by applying a Gaussian regional–residual filter with a 0.8 cycles m^{-1} standard deviation. Figure 4c exhibits the resulting residual gravity map, where the NW–SE-trending strips of negative anomalies mark a series of grabens. The most northwesterly gravity minimum, here named Bica graben (BI in Fig. 4), represents an extension of the Apodi graben (AP in Fig. 4). Alternatively, a less dense, intrabasement gravity source could be the causative bodies for this anomaly. However, the gravity response of the Apodi graben, with NW–SE elongated minima surrounded by positive anomalies, is accurately reproduced in the Bica region. It is unlikely that basement units generated such an anomaly, especially inserted in a structural framework with a main NE–SW direction (Fig. 4b). Furthermore, magnetic data also corroborate the presence of a thickened basin infill in this area, since the magnetic anomalies and Euler solutions show no intrabasement source.

In the SE portion of the study area, the Algodões graben comprises two gravity minima, separated by a slight positive anomaly (AL in Fig. 4c). The 20 km long gravity low is oriented in the NW–SE direction parallel to the main trend of the Bica and Apodi grabens. The gravity anomalies suggest that the eastern segment of the rift is extended southeastwards in comparison with the limits drawn by Borges (1993) based on reflection seismic lines. Other short-wavelength gravity minima occur in the NW and NE parts of the study area (Fig. 4c). Nevertheless, the presence of a graben is not expected in those cases. Despite the lack of appropriate station coverage in those areas, different gravity trends and partially outcropped granitic and supracrustal units lead us to such an interpretation.

Figure 4d exhibits the gravity lineaments extracted from the residual anomaly map and the solutions of gravity source detection using the 3-D Euler deconvolution method. The Euler deconvolution parameters applied to gravity data are the same as those applied to the magnetic data. Differently from the magnetic case, the gravity lineaments preferentially trend in the NW–SE direction, following the main rift faults. In turn, the Euler solutions reveal narrow (less than 1500 m depth) gravity sources oriented in the NW–SE direction in the rift zone (shaded area in Fig. 4d). The faulted borders of the grabens are delimited by the Euler solutions. On the other hand, Euler solutions are oriented N–S and E–W in the SW and northern parts of the study area, respectively. Some of these solutions are related to the intrabasement gravity sources and structures, but most of them are biased by the scarce and irregular distribution of gravity stations, concentrated along roads (Fig. 3).

3.3   Geoelectrical sounding

Seventeen geoelectrical surveys were carried out along two profiles crossing the rift structures (P01 and P02 in Figs. 3 and 4). The vertical electrical soundings (VESs) were measured to define different geoelectrical layers and the internal geometry of the grabens. The soundings were spaced 2.0 to 3.0 km, and all measurements were taken using a Schlumberger electrode array with current electrode half spacing (AB/2) ranging between 1.5 and 1200 m. The resistivity equipment comprises a DC–DC converter 12/1000, with maximum power of 500 W, and a digital potential receiving unit, which were able to provide the apparent resistivity with high accuracy.

We constructed two geoelectrical pseudosections using the resistivity measurements and the half spacing between the current electrodes (Fig. 6). The study indicates four geoelectrical units in both sections. The deepest unit represents the crystalline basement with a resistivity up to 50 Ω m. Directly overlying the bedrock occurs a low-resistivity layer (<35 Ω m), which is interpreted as the siliciclastic rift unit. In profile 01, the lateral increase of resistivity between VESs 8 and 9 indicates the faulted border of the Bica graben and, consequently, the SE limit of this geoelectrical layer (Fig. 6a). The geoelectrical layers show a generalized increase in resistivity from this area as far as the SE end of profile 01 and in all of profile 02. This pattern could be explained as a decrease in the moisture content caused by the presence of a low-permeability carbonate layer on the top of the sedimentary infill. Along profile 02, the rift sequence reaches its high-

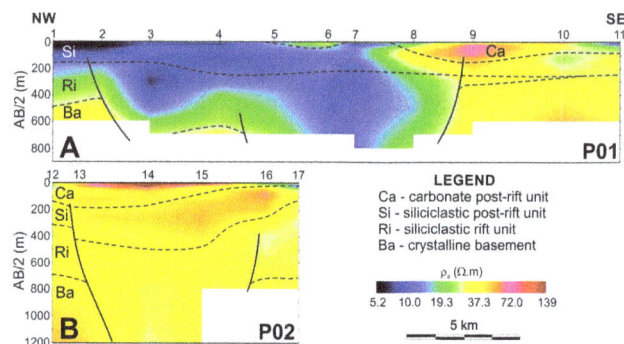

Figure 6. Interpreted apparent resistivity cross sections of profiles P01 (top) and P02 (bottom). VES locations: 1 to 17.

est thickness in the Algodões depocenter between VESs 13 and 16 (Fig. 6b).

The intermediary geoelectrical layer (Si in Fig. 6) is characterized by very low resistivities (<18 Ω m), where the siliciclastic unit of the postrift sequence outcrops (Figs. 3 and 6). In the SE part of profile 01 (VESs 8 to 11), this layer resistivity reaches 55 Ω m, where it is overlapped by a more resistive layer (>140 Ω m), the carbonate unit. Its thickness varies from 150 to 230 m along profile 01, whereas this layer is 350 m thicker over the main depocenter in profile 02 (Fig. 6), suggesting local reactivation of rifting faults during fluvial and marine transgression in the postrift phase. The uppermost carbonate unit also exhibits a thickening in the Algodões rift zone along profile 02.

4   Gravity–geoelectric joint inversion

We applied an algorithm developed by Santos et al. (2006) in two transects, crossing the Bica (P01) and Algodões (P02) grabens (Figs. 3 and 4) to identify the resistivity interfaces and subsurface electrical resistivity distribution within the rifting areas. This algorithm is based on the simulated annealing technique to jointly invert gravity and resistivity (VES) data for mapping the internal architecture of the basin and its layered infill. Using seismic and well log data to constrain this joint-inversion procedure, De Castro et al. (2011) obtained good results for the rift internal architecture applying the Santos algorithm in a regional transect across the Potiguar Basin.

Gravity lows suggest semi-grabens with depocenters located between 4 and 21 km and 3 and 9 km in P01 and P02, respectively (Fig. 7a and d). The footwalls are represented by magnetic maxima, and the depocenters by negative magnetic anomalies (Fig. 7b and e). We also calculated a 2-D Euler deconvolution along the profiles (Fig. 7c and f) to guide the gravity–geoelectrical joint inversion, providing the expected rift geometries and locations of intrabasement heterogeneities. The structural indexes of 0.5 to gravity and 2.0 to magnetic data are the best ones to describe the

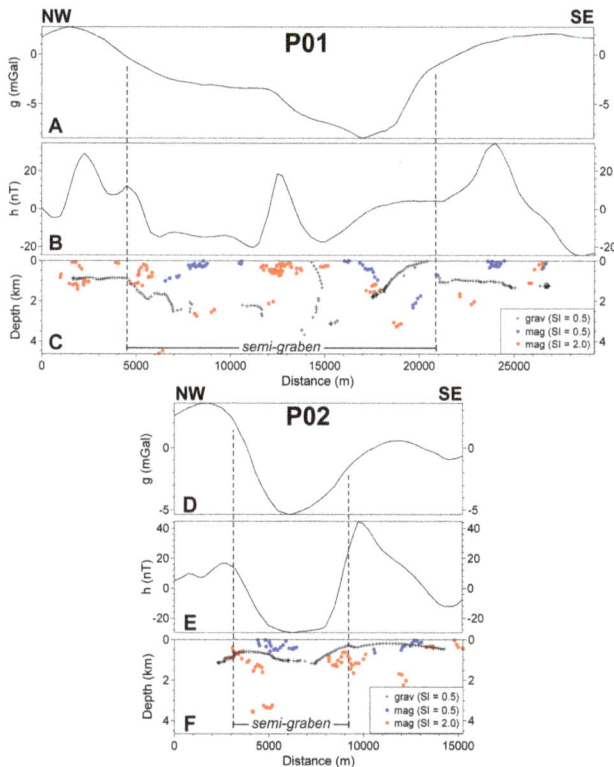

Figure 7. Gravity (**a, d**) and magnetic (**b, e**) anomalies and Euler solutions (**c, f**) of profiles P01 (top) and P02 (bottom).

expected behavior of the faulted borders of the grabens in depth. The structural index of 0.5 applied to magnetic data is more suitable to indicate basement heterogeneities.

In transect P01, the alignment pattern of gravity Euler solutions marks both the NW and SE edges of the Bica graben (crosses in Fig. 7c). This set of gravity Euler solutions suggests a semi-graben in agreement with the geoelectrical section (Fig. 6). Unlike the gravity data, clouds of magnetic Euler solutions indicate shallow causative sources within the basin (circles in Fig. 7c), although few solutions are coincident with gravity Euler solutions at the boundaries of the graben. A similar result was obtained in the Algodões graben (P02 in Fig. 7f). However, the gravity Euler solutions are flatter than expected for the fault that limits the NW rift edge, which suggests that the border faults of the Algodões rift exhibit a low dip angle. Additionally, magnetic Euler solutions mark intrabasement sources at the graben shoulders (red circles in Fig. 7f).

In order to apply the joint inversion, we adopted a four-layer model for transect P01, representing the basement, rift and postrift units, and a thin soil layer. In profile 02, the uppermost postrift sequence could be divided into two layers, since the siliciclastic and carbonate units are well defined along all VESs (Fig. 6b). Each layer was discretized in 31 (profile 01) or 17 (profile 02) cells with widths of 1.0 km. At both ends of the profile, the cells are extended 10 km to avoid

edge effects in the calculated gravity anomalies. The density values of the layers in kilograms per cubic meter ($kg\,m^{-3}$) were, from base to top, 2750 for the bedrock (basement), 2500 (rift sequence), 2300 (siliciclastic unit), 2450 (carbonate unit), and 2000 (superficial dry soil). In profile 01, a density of $2350\,kg\,m^{-3}$ was assumed for the postrift unit, encompassing the siliciclastic and carbonate units. We performed 25 density measurements on selected samples that represented sedimentary and basement rocks. Densities obtained by De Castro (2011) in the well logs located at the eastern border of the Potiguar rift were also considered in the models. The density measurements increase with depth and represent sediment compaction.

Initially, a 1-D inversion method was applied in each VES to obtain estimates of the resistivity of the 2-D model layers, as well to establish search limits of resistivity and depth for each model cell. Estimates of the resistivity and thickness values were calculated from the original data by using the IPI2Win software developed by Bobachev (2003). The inversion method uses a variant of the Newton algorithm of the least number of layers or the regularized fitting minimizing algorithm using Tikhonov's approach (Tikhonov and Arsenin, 1977) to solve problems. Iterations using this code were carried out automatically and interactively (semi-automated) until the calculated model satisfied a minimum difference between measured and calculated data.

Figures 8 and 9 present the internal geometry and density–resistivity distribution of the final models obtained by the joint inversion for each profile. In general, both gravity and geoelectric data have good degrees of fit in comparison with the calculated gravity anomaly and DC curves, respectively. The grabens identified in the geoelectrical sections (Fig. 6) and by gravity Euler solutions (Fig. 7) were reconstituted by gravity–geoelectric modeling. In profile 01, the SE border of the Bica half graben, revealed by the calculated model, is controlled by a normal fault with 40° dip to the NW and vertical offset of almost 1200 m (Fig. 8). A basement high bounds the 8 km wide main depocenter to the NW. Outside the rift, the basin infill sharply decreases to less than 200 m thick. The postrift unit exhibits a slight thickening northwestward (Ca/Si in Fig. 8). Likewise, the Algodões graben shows half-graben geometry, reaching a maximum depth of 1150 m (Fig. 9). However, the postrift siliciclastic unit is thickened on the central portion of the rift (Si in Fig. 9), unlike the flattened postrift deposition in the Bica graben, which suggests a tectonic reactivation in the Algodões graben during the deposition of the postrift unit.

5 3-D gravity modeling

The study employed a 3-D model of the gravity anomaly that used the approach proposed by De Castro et al. (2007). The algorithm simulates gravity anomalies of vertical rectangular prisms in the observed field using a quadratic function

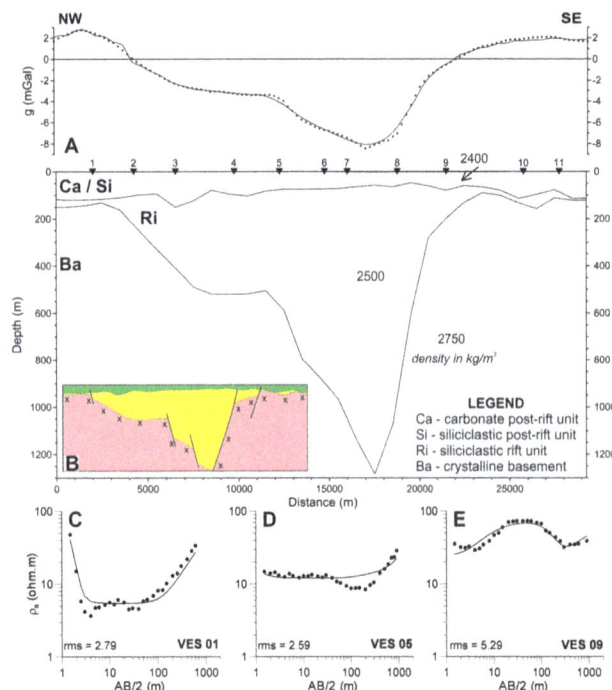

Figure 8. Observed (dots) and calculated (solid line) gravity anomaly across profile 01 (**a**) and the final model response obtained from joint inversion method (**b**). Comparison of three VES data (dots) and model responses of the gravity–geoelectric joint inversion (**c–e**). rms: VES misfit (per cent).

Figure 9. Observed (dots) and calculated (solid line) gravity anomaly across profile 02 (**a**) and the final model response obtained from joint inversion method (**b**). Comparison of three VES data (dots) and model responses of the gravity–geoelectric joint inversion (**c–e**). rms: VES misfit (per cent).

to account for the increase in density with depth within the basin (Rao and Babu, 1991). The new approach used here takes into account the possibility that basement rocks that underlie sedimentary basins have variable density. This approach separates basin and basement gravity during the modeling process, which provides the shape of the low-density basin, without the gravity effects of the heterogeneous basement (Jachens and Moring, 1990; Blakely, 1996).

The calculated thickness of basin-filling deposits depends on the density–depth function used in the modeling (Blakely et al., 1999). In the study area, the coefficients of the density function within the basin were fitted by the least-squares method and were extracted from the joint-inverted final density models. Nevertheless, the linear coefficient of the quadratic function represents the density contrast in surface and guides the modeling process. The chosen superficial contrast was $-270\,\mathrm{kg\,m^{-3}}$, which provides good agreement with joint-inverted models (Figs. 8 to 10 and Table 1). However, the calculated depths of this 3-D model do not match with the basin infill thickness at exploratory wells in the Apodi graben (location in Figs. 3, 4, and 10). Using a lower-density contrast ($-200\,\mathrm{kg\,m^{-3}}$), the resulting gravity model provided depths for the basement top that are consistent with the depth found in the exploratory well 3 (Table 1). The high misfit for well 1 points out that the density contrast increases westward to the Apodi graben boundary, getting closer to the density

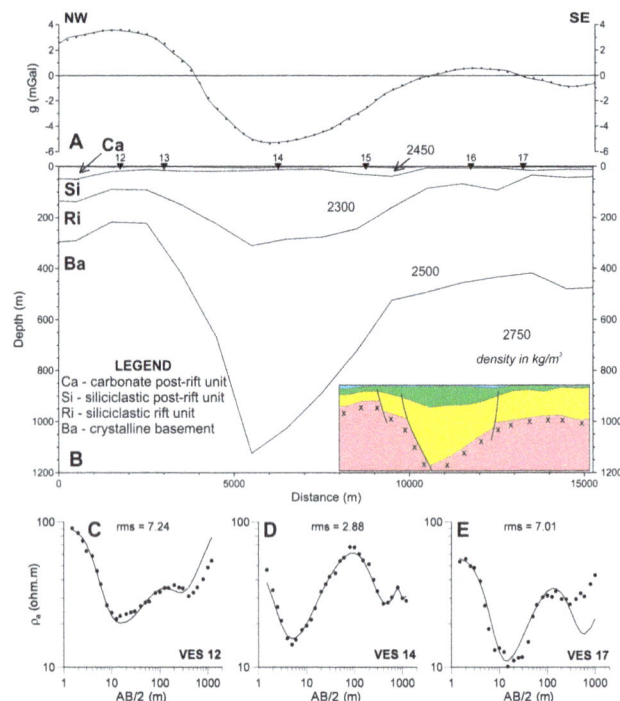

data set for the Bica graben. In summary, gravity modeling reveals that the densities are higher where the basin infill is thicker, probably due to more intense sediment compaction in these areas. Since the major interest of this research is focused on the Bica and Algodões grabens, Figure 10 shows the 3-D gravity model obtained using the density contrast of $-270\,\mathrm{kg\,m^{-3}}$, which was more consistent with the results of the joint inversion. Assuming a basement density of $2750\,\mathrm{kg\,m^{-3}}$, the modeling yielded average densities of these sedimentary units of about $2480\,\mathrm{kg\,m^{-3}}$.

6 Architecture and kinematics of the Potiguar rift termination

The present study indicates that the southern termination of the main rift is more complex than the previous investigations have indicated. The analysis of the magnetic lineaments and the basement foliation and shear zones indicate that the basement fabric did not exert control in fault geometry at the rift termination, as already observed in the NE-trending lystric faults by De Castro et al. (2012) along the main rift. The new rift termination is characterized by a WNW-trending, 10 km wide, and ~ 40 km long fault zone. Inside this fault zone, stretching created a series of NW–SE- to N–S-trending depocenters. Based on gravity maps, we

Table 1. Comparison between the depths to basement obtained from joint-inverted approach (A) and in exploratory wells (B) and those depths obtained by the 3-D gravity modeling using density contrast of $-270\,\mathrm{kg\,m^{-3}}$ (C) and $-200\,\mathrm{kg\,m^{-3}}$ (D) and the respective misfits.

Graben	Location	Depth (m)					
		A	B	C	Misfit (%)	D	Misfit (%)
Bica	Profile 01	1110	–	1130	1.8	5902	431.7
Algodões	Profile 02	1030	–	1050	1.9	5446	428.7
Apodi	Well 1	–	1898	720	62.1	4622	143.5
	Well 2	–	>3703	709	–	4300	–
	Well 3	–	4424	881	80.1	4865	9.9

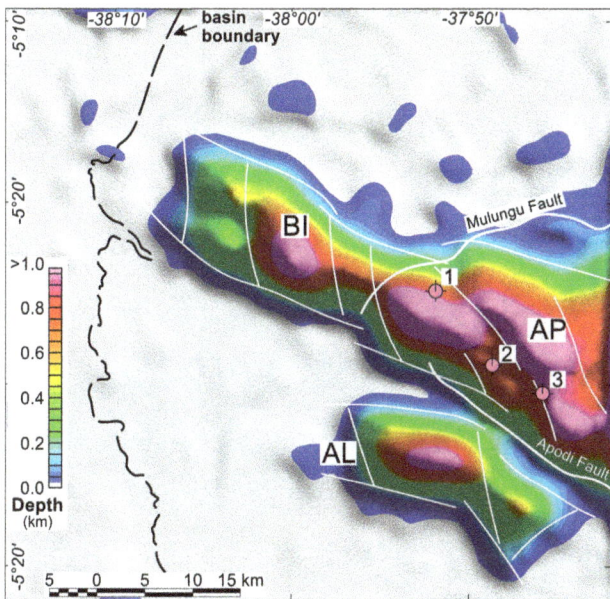

Figure 10. Basement contour map of the SW border of the Potiguar rift derived from 3-D-gravity modeling with major fault segments (thin white traces). Thick white traces: rift limits from previous studies. Grabens: BI – Bica; AP – Apodi; and AL – Algodões.

interpreted the depocenters to exhibit an en echelon geometry and fault segments 35 km long.

The depocenters form two main grabens, the Algodões and the Bica grabens. The former was described by Matos (1992), whereas the latter is a new structure presented for the first time in the present study. Both grabens are separated from the main rift by horsts, and their main axes are at a high angle to the NE-trending Potiguar rift. Both grabens are composed of *syn*-rift and postrift sedimentary units. The *syn*-rift units are bounded by rift faults, whereas the postrift units cap the whole basin. In addition, both grabens do not exhibit present-day topographic expression, and most of the faults that cut across the rift units die out in the postrift layers.

The 3-D gravity model reveals a NW-trending rift geometry for the Bica graben (BI in Fig. 10), beyond the previously mapped limits of the Potiguar rift. This graben is ~30 km

long and ~15 km wide, and it is limited by segmented NW-trending oblique-slip faults. NS-oriented, en echelon faults split the graben into four depocenters, whose greatest thickness reaches 1130 m. One of these rift borders is the Apodi fault (2 in Fig. 2), previously described in the study of Bertani et al. (1990) as a normal fault. The Mulungu fault was also identified by Bertani et al. (1990) and Matos (1992) (1 in Fig. 2). This rift geometry is roughly similar to the internal architecture of the Apodi graben (AP in Fig. 10). The Algodões graben comprises an E–W-trending structure 25 km long and 8 km wide, which bends in the NW–SE direction in its eastern part (AL in Fig. 10). Unlike in others grabens, the Apodi fault system exerts structural control on the northern rift border of the Algodões graben (Fig. 10). Furthermore, an incipient basement high separates the Algodões graben into two depocenters. The occurrence of this structure is well recorded in the magnetic, gravity, and geoelectrical data (Figs. 4 and 6). The local basin infill is up to 1050 m deep.

The deformation was partitioned between the WNW-striking rift strike-slip faults and the internal N–S- to NW–SE-striking, en echelon normal faults. The lack of surface expression of the faults in the study area indicates that they were mainly active during rifting. The study also indicates that the WNW-trending faults that border the Bica and Algodões grabens and their relationship with the NS-trending faults are consistent with an oblique-slip dextral component of displacement of the former. The NW–SE- to NS-trending en echelon faults occur in both grabens and are consistent with this oblique-slip dextral component movement of the WNW-striking faults. The maximum vertical throws of the NW-trending border faults are ~1100 m, and they decrease eastward in the Algodões graben and westward in the Bica graben (Fig. 10). Relative geometry of the two fault sets indicate that the structures were formed by transtensional shearing.

7 Discussion

7.1 The evolution of the Potiguar rift in the context of the transform margin

The reactivation of the Precambrian tectonic fabrics originated the main NE-trending basin-bounding fault (De Castro et al., 2012). In this context, the preexisting structures in the upper lithosphere exert the main control of fault reactivation during continental rifting (De Castro et al., 2012). However, this study indicates that the southern rift termination cut across the existing Precambrian fabric. The en echelon depocenters in the southern rift termination are consistent with the *syn*-transtensional phase of the equatorial margin (Matos, 2000), which also cut across the preexisting basement fabric along the margin.

The Potiguar rift experienced two phases of extension: the first was a NW-trending extension in the Neocomian and the second was an E–W-trending rift extension in the Barremian (Matos, 1992). During the first rift stage, the Apodi fault marked the rift termination as a normal fault (Matos, 1992). The stretching observed at the Potiguar rift in the present study is consistent with the second phase of rift extension, where the Apodi fault moved as a right-lateral shear zone. It suggests that this rift termination developed after the main rift trend was aborted (Matos, 1992). This rift termination also coincides with the development of the Jacaúna and Messejana grabens (Fig. 2), which were formed by EW-trending extension (Matos, 1992), and with the onset of rifting in the equatorial margin (Matos, 2000).

Crustal extension in the first rift phase was distributed across the NE-trending rift faults of the Potiguar rift. Afterwards, during the evolution of the Potiguar rift termination, fault movement was partitioned between the master faults and the internal graben faults. This pattern of rift termination is different from the one observed at the other small basins to the south of the Potiguar rift, where the rift border and intrarift faults are roughly orthogonal to the rift stretching (De Castro et al., 2007, 2008).

The dextral shear of the border faults of the small grabens adjacent to and to the west of the Potiguar main rift roughly coincides with the major transform movement of Africa and South America along the equatorial margin (Fig. 1). The transtension of the equatorial margin is consistent with the NW-trending depocenters and right-lateral shear of the southern termination of the Potiguar rift.

It follows that the main Potiguar graben trends NE–SW and developed along preexisting basement structures (De Castro et al., 2012). However, the right-lateral shearing of the Apodi fault was dissipated by transtensional opening and formation of en echelon grabens at the Potiguar rift southern termination (Fig. 11a, b). The slip of the Apodi fault is synthetic to the right-lateral shear of the equatorial margin. The right-lateral movement of the Apodi fault continued during deposition of the postrift units (Fig. 11c).

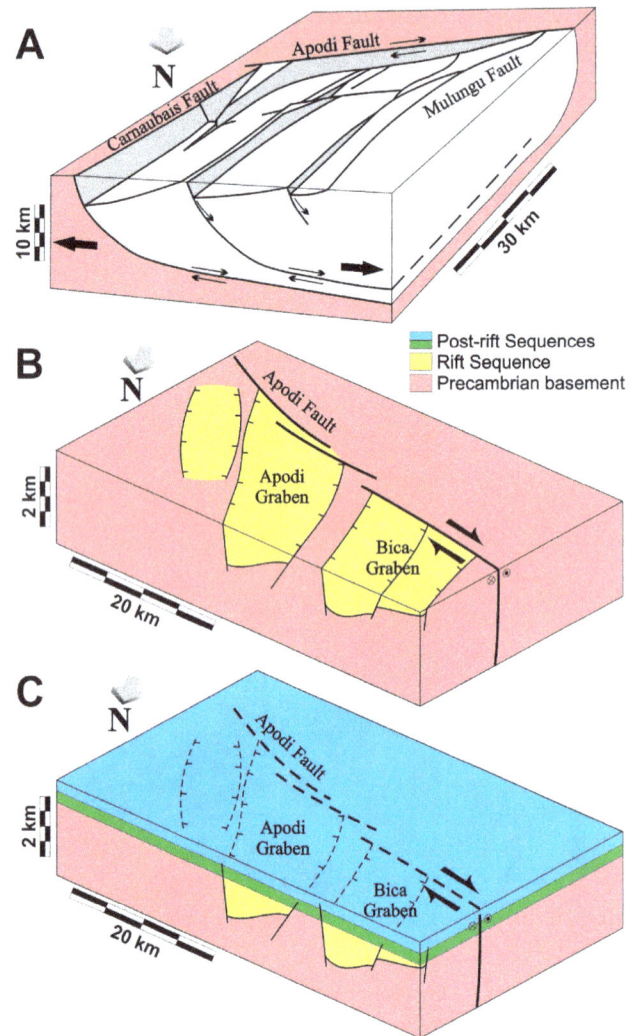

Figure 11. Illustration of (**a**) the main framework of the Potiguar Basin proposed by Matos (1992) (modified from Rodrigues et al., 2014); (**b**) the geometry and kinematics of the Apodi faults and the N–S-trending graben during the rift phase and (**c**) the postrift phase.

Our study also indicates that some rift faults affect upper layers of the postrift phase, such as the Late Cretaceous Açu and Jandaíra formations. However, the precision of the methods we used in the present study does not allow a more precise description and analysis of the upper continuation of these faults. Therefore, additional deformation in the postrift phase with tectonic inversion and oceanward tilting, as already observed along the equatorial margin of South America (Basile et al., 2013), must be addressed by further studies.

7.2 Implications for other transform margins

A few modern examples indicate the influence of transform boundaries inside intraplate settings, generating fault-controlled depressions and volcanism. Faults linked to the transform fracture zones produced topographic depression, which allowed the deposition of sediments during the Cenozoic in East Antarctica (Cianfarra et al., 2009). Likewise, the E–W-trending faults at the southern termination of the Potiguar rift also produced depressions, which resulted in the generation of grabens. In both cases, these structures form fault-controlled, elongated troughs.

In SE Australia, the N–S-trending Tasman fracture zone induced volcanism inside the Australian plate. The volcanic site occupies an area at least 200–450 km away from the main trace of the transform fracture zone. The faults inside the plate have also the same geometry, trend, kinematics, and timing of the main Tasman fracture zone. The magmatism is the most recent volcanic activity in the Australian continent, spanning from the Pliocene to the Holocene. The propagation of the transform fracture zone also cut across preexisting basement structures such as folds and foliations when they present unfavorable orientations for reactivation under the present stress field (Lesti et al., 2008). This pattern clearly indicates the propagation of the deformation of the main transform boundary inside the intraplate setting, where the deformation progressively fades away. This example could indicate that faulting related to the equatorial margin of South America could occur even farther south of the Potiguar rift termination and that this kind of influence could be expected along the entire equatorial margin of Brazil.

Likewise Storti et al. (2007) concluded that plate boundaries related to fracture zones in the Ross Sea region of Antarctica have been slowly deforming during the Cenozoic, and they proposed the potential intraplate influence of oceanic transform fracture zone shear inside the plate. This type of transfer of transform shear from the plate boundary to the plate interior is possible if the oceanic spreading velocity is larger than the velocity of the plate (Salvini et al., 1997). These studies, however, have argued for a collinear continuation of transform boundaries inside the plate. Our study indicates that this kind of influence could also occur in zones parallel to and at least 200 km away from the transform boundary. This could also be the case of the entire equatorial margin of South America, but further investigations are needed.

8 Conclusions

Previous studies indicate that the Potiguar rift lies at the intersection of the equatorial margin and the eastern margin of South America and is overburden by the postrift sedimentary units. It encompasses a series of NE-trending horsts and grabens. This study extends the investigation of previous works by focusing on the fault evolution at the rift termination using gravity, magnetic, and resistivity data. This study indicates that stretching of the southern end of the Potiguar rift was accommodated by both a ∼ 40 km long strike-slip and a system of minor left-stepping, en echelon normal faults. We documented two small rhomb-shaped grabens at

the rift termination. They are NW–SE- to NS-trending full grabens developed during rifting. The grabens were developed along NW-trending oblique-slip faults. Depocenters in the grabens were split by en echelon NS-trending normal faults. Faults of these grabens die out in the postrift sedimentary units. The rifting coincided with the development of the equatorial margin, which was subjected to right-lateral transform movement during this period. The southern part of the Potiguar rift termination is consistent with the transfer of transform shear from the plate boundary to the plate interior, influencing rift development inside the plate. This kind of deformation may have occurred along the entire equatorial margin of South America and may occur elsewhere along other transform plate boundaries.

Acknowledgements. We thank two anonymous reviewers and the topical editor, Federico Rossetti, for their positive criticism, which greatly improved our manuscript. This project was funded by the Brazilian Research Council (CNPq), grant 470891/2010-6 awarded to D. L. de Castro. We also thank the Brazilian Agency of Petroleum and Gas (Agência Nacional do Petróleo, Gás Natural e Biocombustíveis – ANP) for providing well logs and seismic lines. Both authors hold grants from CNPq (PQ grant).

Edited by: F. Rossetti

References

Angelim, L. A. A., Medeiros, V. C., and Nesi, J. R.: Programa Geologia do Brasil – PGB, Projeto Mapa Geológico e de Recursos Minerais do Estado do Rio Grande do Norte, Mapa Geológico do Estado do Rio Grande do Norte, Escala 1 : 500 000, Recife: CPRM/FAPERN, 2006.

Araripe, P. T. and Feijó, F. J.: Bacia Potiguar. Boletim de Geociências da Petrobras, 8, 127–141, 1994.

Basile, C., Mascle, J., and Guiraud, R.: Phanerozoic geological evolution of the Equatorial Atlantic, J. Af. Earth Sci., 43, 275–282, 2005.

Basile, C., Maillard, A., Patriat, M., Gaullier, V., Loncke, L., Roest, W., Mercier de Lépinay, M., and Pattier, F.: Structure and evolution of the Demerara Plateau, offshore French Guiana: Rifting, tectonic inversion and post-rift tilting at transform–divergent margins intersection, Tectonophysics, 591, 16–29, 2013.

Bertani, R. T., Costa, I. G., and Matos, R. M. D.: Evolução tectono-sedimentar, estiloestrutural e habitat do petróleo na Bacia Potiguar., In: Origem e evolução de Bacias Sedimentares, Petrobras, Rio de Janeiro, edited by: Gabaglia, G. P. R., Milani, E. J., 291–310, 1990.

Bezerra, F. H. R. and Vita-Finzi, C.: How active is a passive margin? Paleoseismicity in northeastern Brazil, Geology, 28, 591–594, 2000.

Bezerra, F. H. R., Do Nascimento, A. F., Ferreira, J. M., Nogueira, F. C. C., Fuck, R. A., Brito Neves, B. B., and Sousa, M. O. L.: Review of active faults in the Borborema Province, Intraplate South America Integration of seismological and paleoseismological data, Tectonophysics, 510, 269–290, 2011.

Blakely, R. J.: Potential Theory in Gravity and Magnetic Applications, 2nd Edn., Cambridge University Press, London, p. 441, 1996.

Blakely, R. J., Jachens, R. C., Calzia, J. P., and Langenheim, V. E.: Cenozoic basins of the Death Valley extended terrane as reflected in regional-scale gravity anomalies, Geol. S. Am. S., 333, p. 16, 1999.

Bobachev, A.: IPI2Win-1D automatic and manual interpretation software for VES data, available at: http://geophys.geol.msu.ru/ipi2win.htm (last access: 19 August 2007), 2003.

Bonini, M., Souriot, T., Boccaletti, M., and Brun, J. P.: Successive orthogonal and oblique extension episodes in a rift zone: Laboratory experiments with application to the Ethiopian Rift, Tectonics, 16, 347–362, 1997.

Borges, W. R. E.: Caracterização Estrutural daPorção SW do Rifte Potiguar. MSc thesis, Universidade Federal de Ouro Preto, Ouro Preto, Brazil, p. 146, 1993.

Briggs, I. C.: Machine contouring using minimum curvature, Geophysics, 39, 39–48, 1974.

Cianfarra, P., Forieri, A., Salvini, F. Tabacco, I. E., and Zirizotti, A.: Geological setting of the Concordia Trench-Lake system in East Antarctica, Geophys. J. Int., 177, 1305–1314, 2009.

Conceição, J. C. J., Zalán, P. V., and Wolff, S.: Mecanismo, Evolução e Cronologia do Rift Sul-Atlântico, Boletim de Geociências da PETROBRAS, 2, 255–265, 1988.

De Castro, D. L.: Gravity and magnetic joint modeling of the Potiguar rift basin (NE Brazil): basement control during Neocomian extension and deformation, J. S. Am. Earth Sci., 31, 186–198, 2011.

De Castro, D. L., Oliveira, D. C., and Castelo Branco, R. M. G.: On the Tectonics of the Neocomian Rio do Peixe rift basin, NE Brazil: Lessons from gravity, magnetics and radiometric data, J. S. Am. Earth Sci., 24, 186–202, 2007.

De Castro, D. L., Bezerra, F. H. R., and Castelo Branco, R. M. G.: Geophysical evidence of crustal-heterogeneity control of fault growth in the Neocomian Iguatu basin, NE Brazil, J. S. Am. Earth Sci., 26, 271–285, 2008.

De Castro, D. L., Pedrosa, N. C., and Santos, F. A. M.: Gravity-geoelectric joint inversion over the Potiguar rift basin, NE Brazil, J. Appl. Geophys., 75, 431–443, 2011.

De Castro, D. L., Bezerra, F. H. R., Sousa, M. O. L., and Fuck, R. A.: Influence of Neoproterozoic tectonic fabric on the origin of the Potiguar Basin, northeastern Brazil and its links with West Africa based on Gravity and Magnetic Data, J. Geodyn., 54, 29–42, 2012.

De Castro, D. L., Fuck, R. A., Phillips, J. D., Vidotti, R. M., Bezerra, F. H. R., and Dantas, E. L.: Crustal structure beneath the Paleozoic Parnaíba Basin revealed by airborne gravity and magnetic data, Brazil, Tectonophysics, 614, 128–145, 2014.

Jachens, R. C., and Moring, B. C.: Maps of thickness of Cenozoic deposits and the isostatic residual gravity over basement for Nevada, US Geological Survey Open-File Report 90-404, scale 1 : 1 000 000, 1990.

Kirkpatrick, J. D., Bezerra, F. H. R., Shipton, Z. K., Do Nascimento, A. F., Pytharouli, S. I., Lunn, R. J., and Soden, A. M.: Scale-dependent influence of pre-existing basement shear zones on rift faulting: a case study from NE Brazil, J. Geol. Soc. London, 170, 237–247, 2013.

Koutsoukos, E. A. M.: Late Aptian to Maastrichtian foraminiferal biogeography and palaeoceanography of the Sergipe basin, Brazil, Palaeogeogr. Palaeocl., 92, 295–324, 1992.

Lesti, C., Giordano, G., Salvini, F., and Cas, R.: Volcano tectonic setting of the intraplate, Pliocene-Holocene, Newer Volcanic Province (southeast Australia): Role of crustal fracture zones, J. Geophys. Res., 113, B07407, doi:10.1029/2007JB005110, 2008.

Matos, R. M. D.: The northeast Brazilian rift system, Tectonics, 11, 766–791, 1992.

Matos, R. M. D.: Tectonic Evolution of the Equatorial South Atlantic, Geophysical Monography AGU, 115, 331–354, 2000.

McKenzie, D. P.: Some remarks on the development of sedimentary basins, Earth. Planet. Sc. Lett., 40, 25–32, 1978.

MME/CPRM: Catálogo General de Produtos e Serviços, Geologia, Levantamentos Aerogeofísicos – Database AERO. Rio de Janeiro, p. 359, 1995.

Moulin, M., Aslanian, D., and Unternehr, P.: A new starting point for the South and Equatorial Atlantic Ocean, Earth-Sci. Rev., 98, 1–37, 2010.

Nóbrega, M. A., Sa, J. M., Bezerra, F. H. R., Hadler Neto, J. C., Iunes, P. J., Oliveira, S. G., Saenz, C. A. T., and Lima Filho, F. P.: The use of apatite fission track thermochronology to constrain fault movements and sedimentary basin evolution in northeastern Brazil, Radiat. Meas., 39, 627–633, 2005.

Pessoa Neto, O. C., Soares, U. M., Silva, J. G. F., Roesner, E. H., Florência, C. P., and Souza, C. A. V.: Bacia Potiguar, Boletim de Geociências da Petrobras, 15, 357–369, 2007.

Ponte, F. C., Fonseca, J. R., and Morales, R. E.: Petroleum geology of the eastern Brazilian Continental Margin, AAPG Bull., 61, 1470–1482, 1977.

Rao, D. B., and Babu, N. R.: A Fortran-77 computer program for three-dimensional analysis of gravity anomalies with variable density contrast, Comput. Geosci., 17, 655–667, 1991.

Reid, A. B., Allsop, J. M., Granser, H., Millet, A. J., and Somerton, I. W.: Magnetic interpretation in three dimensions using Euler deconvolution, Geophysics, 55, 80–91, 1990.

Rodrigues, R. S., De Castro, D. L., and Reis Jr., J. A.: Characterization of the Potiguar Rift Structure based on Euler Deconvolution, Brazil. J. Geophys., 32, 109–121, 2014.

Salvini, F., Brancolini, G. Busetti, M. Storti, F. Mazzarini, F., and Coren, F.: Cenozoic geodynamics of the Ross Sea region, Antarctica: Crustal extension, intraplate strike-slip faulting and tectonic inheritance, J. Geophys. Res., 102, 669–696, 1997.

Santos, F. A. M., Sultan, S. A., Represas, P., and El Sorady, A. L.: Joint inversion of gravity and geoelectrical data for groundwater and structural investigation: application to the northwestern part of Sinai, Egypt. Geophys. J. Int., 165, 705–718, 2006.

Souto Filho, J. D., Correa, A. C. F., Santos Neto, E. V., and Trindade, L. A. F.: Alagamar-Açu petroleum system, onshore Potiguar Basin, Brazil: A numerical approach for secondary migration. In: Petroleum systems of South Atlantic margins, AAPG Memoir., edited by: Mello, M. R. and Katz, B. J., 73, 151–158, 2000.

Storti, F., Salvini, F., Rossetti, F., and Phipps Morgan, J.: Intraplate termination of transform faulting within the Antarctic continen, Earth Planet. Sc. Lett., 260, 115–126, 2007.

Tikhonov, N. and Arsenin, V. Y.: Solution of Ill-Posed Problems, V. H. Winston and Sons, Baltimore, MD, ISBN: 0-470-99124-0, 258 pp., 1977.

Unternehr, P., Curie, D., Olivet, J. L., Goslin, J., and Beuzart, P.: South Atlantic fits and intraplate boundaries in Africa and South America, Tectonophysics, 155, 169–179, doi:10.1016/0040-1951(88)90264-8, 1988.

Vagnes, E., Gabrielsen, R. H., and Haremo, P.: Late Cretaceous–Cenozoic intraplate contractional deformation at the Norwegian continental shelf: timing, magnitude and regional implications, Tectonophysics, 300, 29–46, 1998.

3

A 3-D shear velocity model of the southern North American and Caribbean plates from ambient noise and earthquake tomography

B. Gaite[1], A. Villaseñor[1], A. Iglesias[2], M. Herraiz[3,4], and I. Jiménez-Munt[1]

[1]Institute of Earth Sciences Jaume Almera, ICTJA-CSIC, Lluis Sole i Sabaris s/n, 08028 Barcelona, Spain
[2]Institute of Geophysics, Universidad Nacional Autónoma de México, Mexico City, Mexico
[3]Department of Geophysics and Meteorology, Universidad Complutense de Madrid, Madrid, Spain
[4]Institute of Geosciences (UCM, CSIC), Madrid, Spain

Correspondence to: B. Gaite (bgaite@ictja.csic.es)

Abstract. We use group velocities from earthquake tomography together with group and phase velocities from ambient noise tomography (ANT) of Rayleigh waves to invert for the 3-D shear-wave velocity structure (5–70 km) of the Caribbean (CAR) and southern North American (NAM) plates. The lithospheric model proposed offers a complete image of the crust and uppermost-mantle with imprints of the tectonic evolution. One of the most striking features inferred is the main role of the Ouachita–Marathon–Sonora orogeny front on the crustal seismic structure of the NAM plate. A new imaged feature is the low crustal velocities along the USA-Mexico border. The model also shows a break of the east–west mantle velocity dichotomy of the NAM and CAR plates beneath the Isthmus of the Tehuantepec and the Yucatan Block. High upper-mantle velocities along the Mesoamerican Subduction Zone coincide with inactive volcanic areas while the lowest velocities correspond to active volcanic arcs and thin lithospheric mantle regions.

1 Introduction

Crustal seismic models are important for several reasons. The first is the significant impact that crustal corrections have in mantle tomography (Bozdağ and Trampert, 2008; Lekić et al., 2010; Panning et al., 2010). Another is the strong dependency of earthquake location accuracy on the crustal velocity model.

Surface-wave earthquake-based global and regional tomography usually uses long period velocity measurements ($T \geq 20$ s), sensitive to the lower crust and mantle structure. On the contrary, surface-wave local tomography constrains the upper-crustal seismic structure in narrow regions. Therefore there is a gap in imaging the whole crust at a continental scale with surface waves generated by earthquakes or active sources. Ambient noise tomography (ANT) overcomes this problem (e.g., Sabra et al., 2005; Shapiro et al., 2005) and has been applied to obtain crustal shear velocity models in different tectonic regions (e.g., Bensen et al., 2009; Zheng et al., 2011). Also, the increasing number of broadband seismic station deployments in the last decade has facilitated a higher path density.

Recent global shear wave velocity models from surface waves image the crust and uppermost mantle with 2° or 1° resolution (e.g., Shapiro and Ritzwoller, 2002; Pasyanos et al., 2013; Schaeffer and Levedev, 2013; Auer et al., 2014). In the area of this study, there are some regional and continental mantle seismic models from earthquake tomography (e.g., Vdovin et al., 1999; Godey et al., 2003; Schaeffer and Lebedev, 2014) that cover Mexico, the Gulf of Mexico (GOM), and part of the Caribbean. There have also been several local-scale crustal structure studies (e.g., Campillo et al., 1996; Shapiro et al., 1997; Iglesias et al., 2010). Despite this, the seismic structure of the upper-crust of the whole region is not well defined from surface waves. One way to widen the period range to constrain the seismic structure from the crust to the mantle is to combine phase velocity from ANT and earthquake tomography (e.g., Yang and Ritzwoller, 2008; Yao et al., 2008; Zhou et al., 2012; Córdoba-Montiel et al., 2014). In this study we combine Rayleigh-wave group velocity from

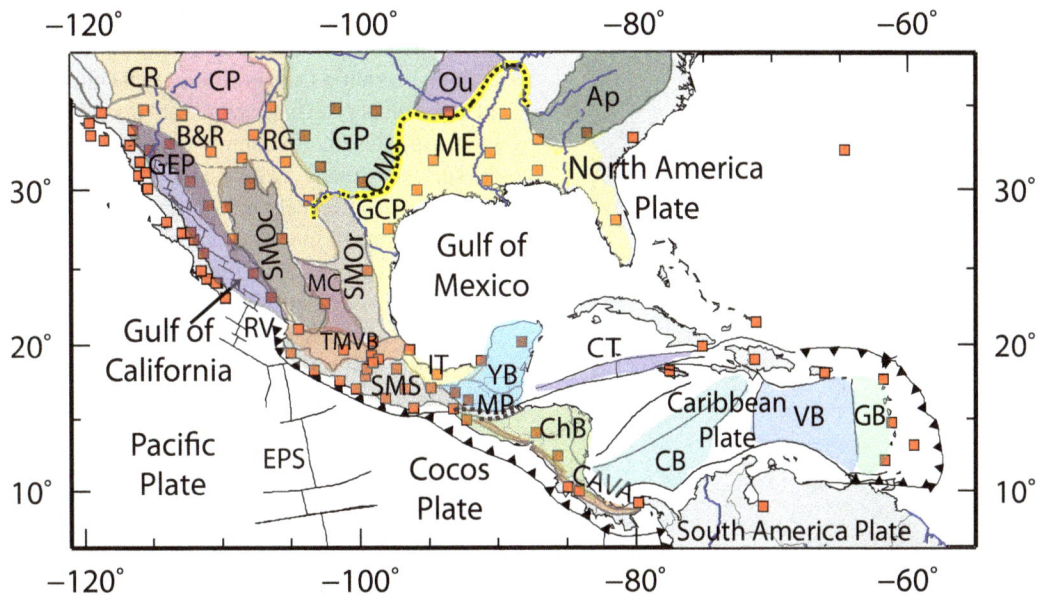

Figure 1. Simplified tectonic map of the study area: physiographic provinces shown as gray lines (Sedlock, 1993; M. Moschetti, personal communication, 2011; Marshall, 2007); stations as red squares; and plate boundaries as black lines (Bird, 2003). Ap denotes Appalachian Plateau Province; B&R Basin and Range; CAVA Central America Volcanic Arc; CB Colombian Basin; ChB Chortis Block; CP Colorado Plateau; CR Colorado River; CT Cayman Trough; EPS East Pacific Rise; GB Grenada Basin; GCP Gulf Coastal Plain; GEP Gulf Extensional Province; GP Great Plains; IT Isthmus of Tehuantepec; ME Mississippi Embayment; MC Mesa Central; MP Motagua–Polochic fault system; Ou Ouachita Province; RG Rio Grande; RV Rivera Plate; SMOc Sierra Madre Occidental; SMOr Sierra Madre Oriental; SMS Sierra Madre del Sur; TMVB Trans-Mexican Volcanic Belt; VB Venezuela Basin; and YB Yucatan Block. Blue lines indicate main rivers. Highlighted yellow dashed black line indicates the Ouachita–Marathon–Sonora orogenic belt (OMS). Its extension into Mexico is taken from Handschy et al. (1987). The GEP location is taken from Zhang et al. (2007).

earthquake tomography and ANT to obtain short periods to constrain the lower-crust seismic structure. The final objective is to obtain a crust and uppermost-mantle vertically polarized shear-wave velocity model to image the area as a whole. To achieve this goal we invert Rayleigh-wave phase velocity from ANT simultaneously with group velocity combined from ANT and earthquake tomography in Mexico, the Gulf of Mexico, and the Caribbean.

2   Data

The data set used in this study consists of continuous recordings from nearly 100 broadband seismic stations of the Mexican and US national networks, other global and regional networks, and temporary deployments. One of the most important contributions of this study comes from the increased station coverage in the region since the beginning of the 21st century. The Mexican broadband National Seismic Network (IG) has expanded its coverage towards the north and the south of the country; the regional Caltech network (CI) has increased the coverage in California; and the deployment of the U.S. Geological Survey (USGS) Caribbean Network (McNamara et al., 2006) has significantly improved the station coverage in the Caribbean. The availability of data from several high-density temporal broadband networks, such as

the NARS array in Baja California (Trampert et al., 2003) and the USArray Transportable Array in the continental US, has also increased the station density in the western and northern boundaries of the region. Figure 1 shows the distribution of the 103 broadband stations used in this study superimposed on a map showing the main tectonic features and physiographic provinces of the area. We analyze 117 earthquakes of $M \geq 5.5$, shallower than 40 km depth, and with epicenter-to-station path lengths ranging from hundreds to less than 10 000 km (Fig. 2).

3   Methods

3.1   Earthquake tomography

We determine fundamental mode Rayleigh-wave group velocity dispersion curves from the earthquake records applying FTAN (Frequency Time ANalysis) with the PGSWMFA program from Ammon (1998). We invert these group velocity measurements to obtain 2-D group velocity models by the method of Barmin et al. (2001). This inversion procedure attempts to minimize a penalty function (Eq. (15) of Barmin et al., 2001) that depends on three damping parameters. These parameters are: α the data misfit damping, σ the width of the Gaussian kernel and β the penalty parameter to low path den-

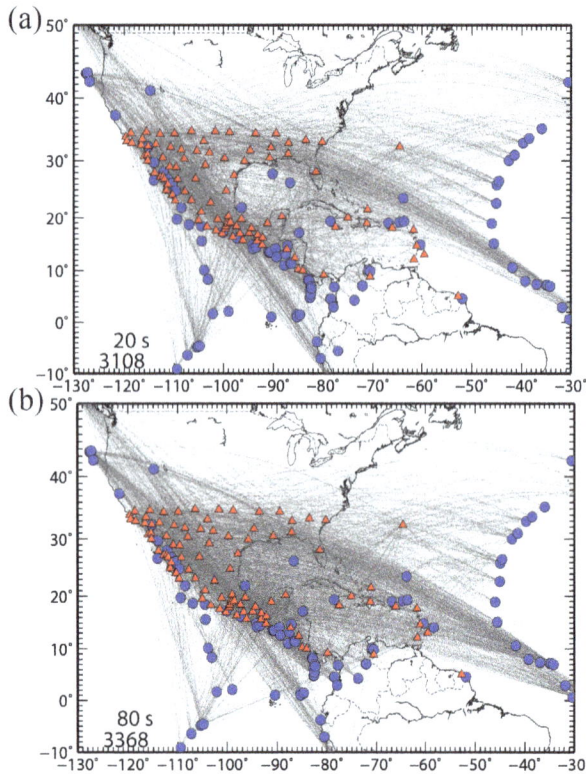

Figure 2. Path distribution of Rayleigh-wave group velocities at (**a**) 20 s and (**b**) 80 s period. Red triangles denote broadband seismic station locations and blue circles the earthquake epicenters. The number on each map indicates the number of paths.

sity regions. We perform a large number of inversions varying the value of the damping parameters. We test α values from 650 to 2000 combined with different values of σ (from 100 to 500) and β (from 1 to 100). The final values used are selected as a compromise between good data fit, stability of the features of the computed models and small model roughness. We follow a two-step tomographic inversion similar to the one described in Gaite et al. (2012). At each step we select the damping parameters. In the first step, we invert all the dispersion curves to obtain dispersion maps with damping parameters $\alpha = 2000$, $\sigma = 400$ and $\beta = 1$. In the second, we remove outliers and re-invert the remaining data, in this case with $\alpha = 1000$, $\sigma = 500$ and $\beta = 1$. We mark an observation as outlier when:

$$\delta t > 3(\text{SD}), \tag{1}$$

where δt is the travel time residual, and SD is the standard deviation. The percentage of rejected outliers lies around 0.8 % of the initial selection. Figure 2 shows the path coverage at 20 and 80 s periods. Mexico, the GOM and the western part of the Caribbean plate are well covered across all periods, whereas the eastern part of the Caribbean plate is well covered for periods longer than 20 s.

From this second step we obtain group velocity maps for periods from 20 to 100 s on a $1° \times 1°$ grid (Fig. 3). The tomographic inversion used is similar to a Gaussian beam method and considers propagation of "fat" rays along the great circle. Following this, the frequency-dependent spatial sensitivity of the surface waves is described by Gaussian lateral sensitivity kernels. These kernels help to provide an accurate estimate of spatial resolution. To compute the spatial resolution we follow the method described by Barmin et al. (2001) with modifications of Levshin et al. (2005). Firstly, we construct a resolution kernel at each node of the model grid, which is a row of the resolution matrix. Secondly, we fit this kernel with a 2-D Gaussian function. Finally we compute the scalar spatial resolution as twice the standard deviation of the Gaussian. We obtain a spatial resolution of the group velocity maps less than or equal to 200 km for periods from 20 to 100 s in the whole area of interest (Fig. 4). This value is lower than twice the distance between the model grid points (1°). This means that the minimum spatial resolution we can obtain is 2° and is limited by the distance between the nodes of the grid. Only at the edges of the inverted area do we obtain a 500 km spatial resolution.

3.2 Ambient noise tomography

We use Rayleigh waves' group and phase velocity dispersion curves from 8 to 50 s obtained from ambient noise tomography on a $1° \times 1°$ grid with a resolution of 250 km in Mexico and its surrounding area from our previous study, Gaite et al. (2012). To compute ANT we used 2 and a half years of continuous vertical component seismic records from the same stations used in this study. Firstly, we computed 1-day long ambient noise cross-correlations between each station pair and stacked them along their available time period. Secondly, we measured phase and group velocity of the fundamental mode Rayleigh wave. Finally, we inverted the dispersion curves to obtain phase and group velocity maps with the same method used for earthquake records in this study. The path coverage at periods shorter than 20 s is mostly limited to mainland North America that is well covered from 10 s.

3.3 Combination of ANT and earthquake tomography

We combine group velocity measurements from ambient noise and earthquake tomography on each node of a $1° \times 1°$ grid to obtain group velocities from 8 to 100 s period. We follow a similar method to that described by Yao et al. (2008) to combine the measurements. First, we select group velocity measurements with resolution better than 250 km from ANT and 500 km from earthquake tomography. After that, we compose the group velocity dispersion as:

$$U = \begin{cases} U_{\text{ANT}}, T < 20 \text{ s} \\ (U_{\text{ANT}} + U_{\text{eq}})/2, 20 \text{ s} \leq T < 50, \text{ if } |U_{\text{ANT}} - U_{\text{eq}}| \leq 0.2 \text{ km s}^{-1} \\ U_{\text{eq}}, 20 \text{ s} \leq T < 50 \text{ s}, \text{ if } |U_{\text{ANT}} - U_{\text{eq}}| > 0.2 \text{ km s}^{-1} \\ U_{\text{eq}}, T \geq 50 \text{ s} \end{cases}, \tag{2}$$

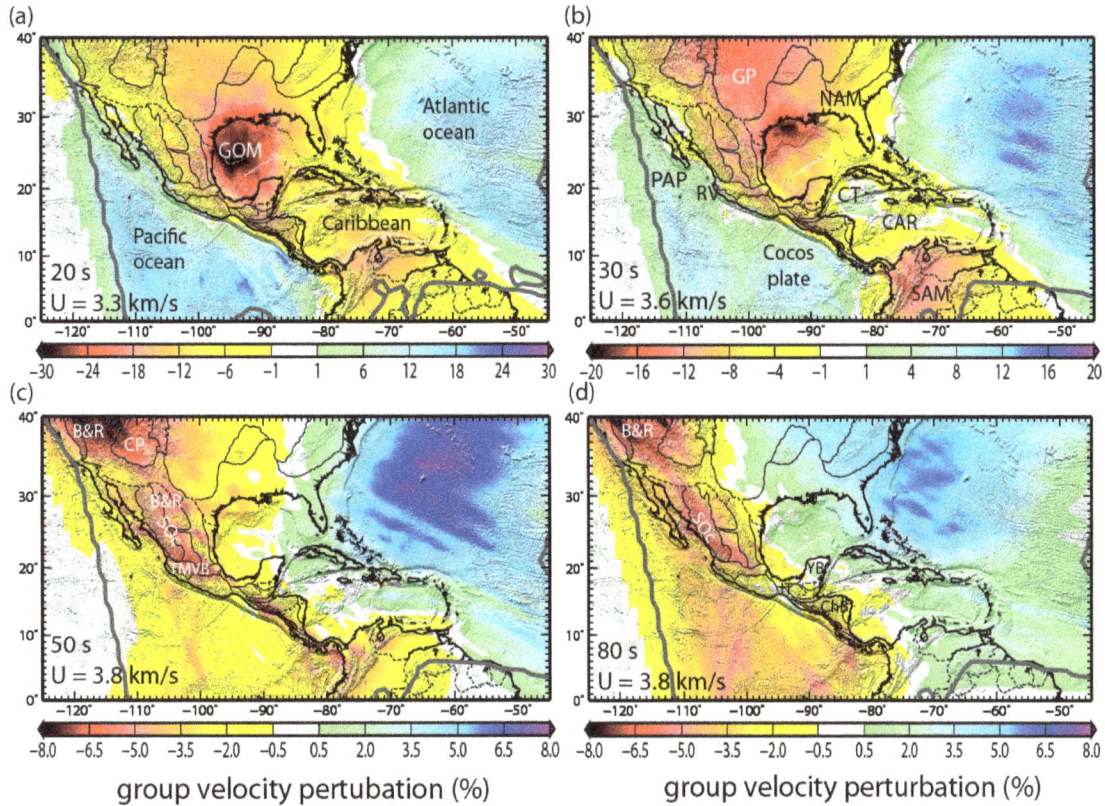

Figure 3. Rayleigh-wave group velocity perturbation maps at (**a**) 20, (**b**) 30, (**c**) 50, and (**d**) 80 s period. The velocity perturbation (%) is computed with respect to the mean average velocity of the whole inversion area at each period and is indicated in each frame. Thick gray lines indicate the 450 km resolution contour and thin gray lines the tectonic provinces. B&R denotes Basin and Range; ChB Chortis Block; CP Colorado Plateau; CT Cayman Trough; GOM Gulf of Mexico; GP Great Plains; IT Isthmus of Tehuantepec; NAM North American plate; PAP Pacific plate; RV Rivera plate; SAM South America plate; SMOc Sierra Madre Occidental; SMOr Sierra Madre Oriental; TMVB Trans-Mexican Volcanic Belt; and YB Yucatan Block.

where T is the period and U_{ANT} and U_{eq} are the group velocities obtained for ANT and earthquake tomography, respectively (Fig. 5). The averaged difference between velocities obtained from ANT and from earthquakes varies from 0.09 to 1 % in their common range of period (from 20 to 50 s) (Fig. 6). This upper limit is slightly larger than in other studies (~ 0.1–0.5 %) that compare phase velocity measurements (e.g., Lin et al., 2008; Yang and Ritzwoller, 2008; Yao et al., 2008; Ritzwoller et al., 2011; Zhou et al., 2012). Our larger difference might be due in part to the fact that we compare group instead of phase velocities. Phase velocity measurements are more stable than group velocities.

3.4 Shear wave velocity model

We simultaneously invert group and phase velocity measurements for a 1-D shear wave velocity structure at each grid point by using a simple parameterization of the medium consisting of 3 constant velocity layers over a half-space. The model parameters (4 velocities and 3 thicknesses) can vary across a wide range to obtain an optimized solution for the

whole variety of tectonic domains in the study area. We consider the media as a Poisson solid, i.e.:

$$\lambda = \mu \qquad \upsilon = 1/4, \tag{3}$$

where λ and μ are the Lamé parameters and υ is the Poisson ratio. We determine the density as per Berteussen (1977):

$$\rho = 0.32 \cdot v_{\mathrm{p}} + 0.77, \tag{4}$$

where v_{p} is the P-wave velocity.

We use a modified code from Iglesias et al. (2001) to jointly invert phase and group velocities. This code solves the forward model with the subroutine SURFACE85 (Herrmann, 1987) and inverts with the simulated annealing algorithm (Goffe et al., 1994; Goffe, 1996). Simulated annealing is a global optimization method. The algorithm scans the possible solutions space to find the optimum model by reducing the searching vector length when it is close to a minimum and allowing misfit increases to avoid local minimums. The algorithm determines as the optimum model that which minimizes the misfit during a certain number of searching iterations. To assure the inversion of high quality dispersion

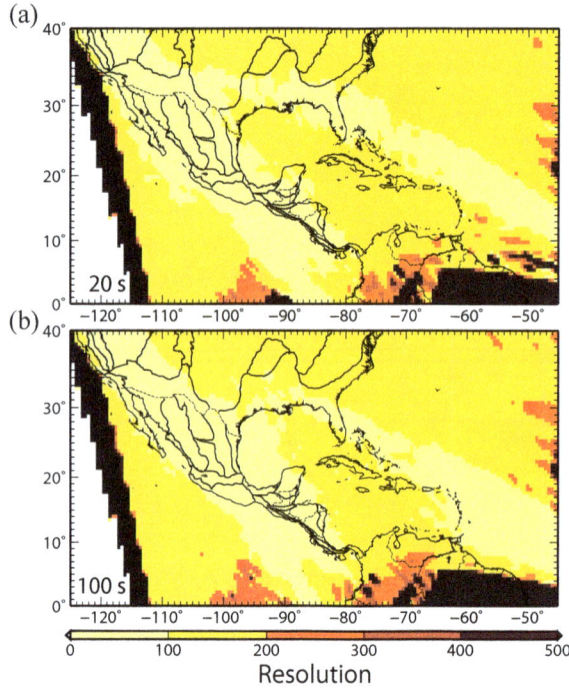

Figure 4. Estimated resolution in km for group velocity maps at **(a)** 20 s and **(b)** 100 s period.

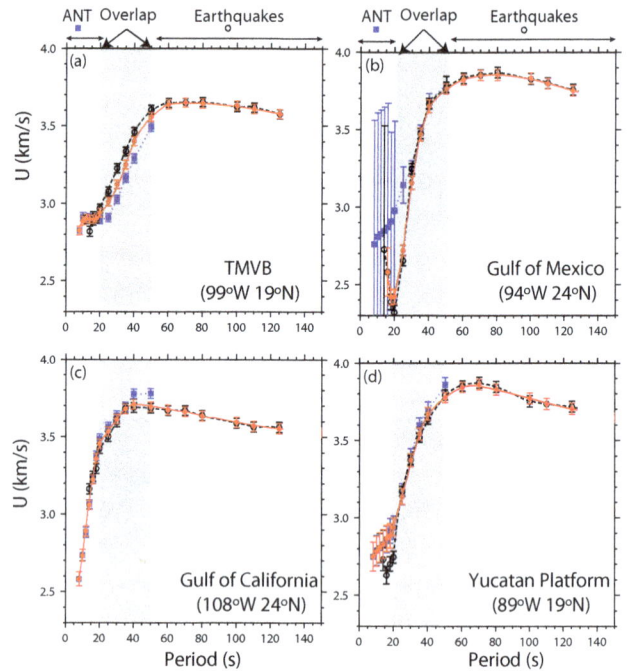

Figure 5. Examples of joining group velocity obtained from ANT (blue squares) and from earthquake tomography (empty circles) at four nodes of the inversion region representing different tectonic settings. The error bars denote resolution normalized by 2500 km at each period. The gray area limits the velocity overlapping and joining period range. Filled circles and continuous red lines indicate the combined dispersion curve.

curves, we only invert dispersion curves with velocity measurements at more than 3 discrete frequencies. By doing this we avoid inverting nodes with high resolution at narrow frequency ranges. We select as optimum models only those with velocity increasing with depth.

The misfit of the dispersion measurements is computed as:

$$\text{misfit} = \begin{cases} 0.5 \cdot e_C^M + 1.5 \cdot e_U^N \; if \; M < N \\ e_C^M + e_U^N \; if \; M = N \\ 1.5 \cdot e_C^M + 0.5 \cdot e_U^N \; if \; M > N \end{cases} \quad (5)$$

where e_C^M and e_U^N are the errors computed in a L2 sense for M phase and N group dispersion measurements, respectively. The mean misfit for all inverted nodes is $0.2 \, \text{km s}^{-1}$ (Fig. 7a). Figure 7b shows the geographical distribution of the model misfit. The highest misfit values lay offshore, in regions with low path coverage and outside the area of interest in this work. The largest misfit values in the area of interest are on the easternmost part of the GOM and the Yucatan platform.

As the final step, we combine the 1-D shear models from each node to produce a 3-D shear wave velocity model.

4 Results and discussion

The 3-D shear-wave velocity model obtained from inverting Rayleigh-wave group velocities (10 to 100 s) and phase velocities (10 to 50 s) is sensitive to velocity changes from 5 down to 70 km depth. The inversion fits periods ≤ 80 s better than the longer ones (Fig. 8). According to the procedure described above, velocities at periods around 10 s, sensitive to shallower portions of the crust, are obtained from ANT with equal or higher resolution than 250 km. The short period dispersion results are obtained for the whole of Mexico and some parts of the CAR plate and the southern US (white contour in Fig. 9a). This means that the shear velocity model constrains the shallow crust of Mexico better than the crust of the GOM and the Caribbean plate. The lateral resolution of the model is about 220–250 km and comes from the spatial resolution of the surface-wave velocity maps. This model offers a crust and uppermost mantle image of the whole area. Its agreement with the main known tectonic characteristics and the recovery of the major crustal features obtained in previous local studies provides reliability on our results and the confidence to interpret them on regions with a lack of shear-wave lithospheric information. The crustal and uppermost-mantle seismic structure features revealed by the model correlate well with traces of different tectonic evolution stages of the region. Deeper insights on the kinematics and dynamics within this region might be obtained from azimuthal anisotropy. It is out of the current scope, but a natural extension of this research.

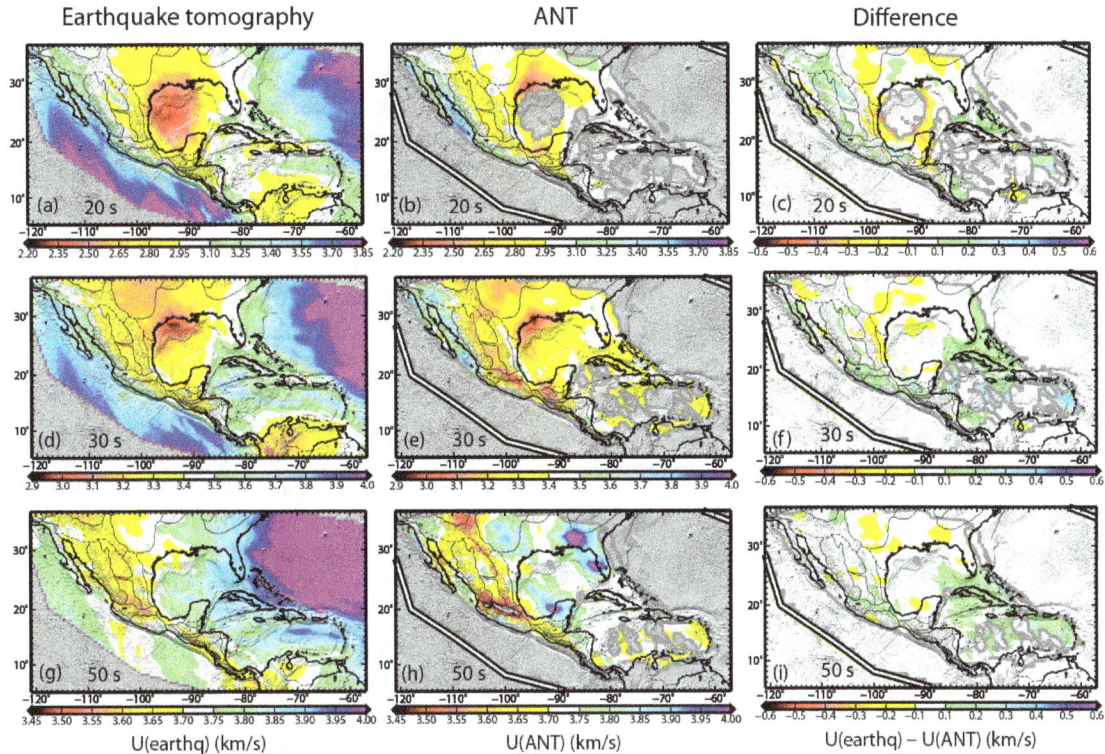

Figure 6. Rayleigh-wave group velocity maps from: earthquake tomography with resolution $\leq 500\,\mathrm{km}$ **(a, d, g)**; ANT with resolution $\leq 250\,\mathrm{km}$ **(b, e, h)**; and their difference in the common area **(c, f, i)** at 20, 30, and 50 s period. The white line outlined by black marks the ANT inversion area.

4.1 Crust

The model identifies different velocities between the Yucatan and Chortis continental blocks at 30 km depth (Fig. 9d). This seismological lower crustal difference agrees with the different origin and tectonic evolution proposed by several studies from geologic evidence and paleotectonic reconstructions (e.g., Burke, 1988; Rogers et al., 2007; Pindell and Kennan, 2009). It also reveals crustal heterogeneity on the Caribbean plate oceanic basins (Colombia, Venezuela, and Grenada) (Fig. 9c), despite the lower resolution of the model over this plate. The model also exhibits a high contrast between the upper and lower crustal velocities of the inland North American plate (Fig. 10).

4.1.1 Basins and shallow basement

Low upper-crust velocities (Fig. 9a) correspond to sedimentary basins along the Gulf Coastal Plain, the Gulf of California, the USA-Mexico border and the Motagua–Polochic fault system, while high velocities correlate with mountain ranges (e.g., the Sierra Madre Oriental, Sierra Madre Occidental, and Sierra Madre del Sur). These low velocities are observed down to approximately 5 km beneath the Gulf Coastal Plain, the Rio Grande drainage basin and the Colorado river mouth, but they reach down even further to 12 km

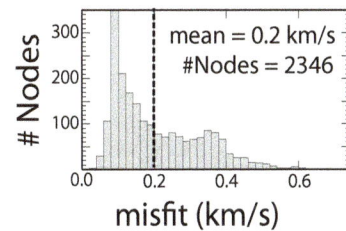

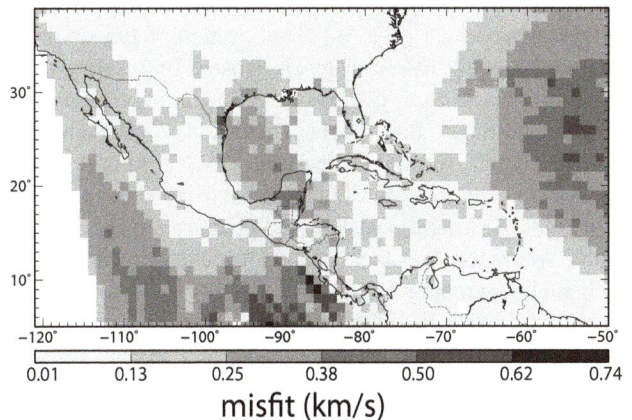

Figure 7. **(a)** Histogram and **(b)** map of the misfit of inverted dispersion curves at each node.

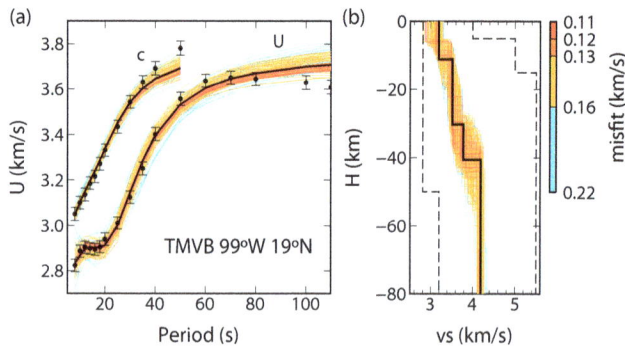

Figure 8. Example of 1-D inversion of phase and group velocity at one node of the grid situated on the TMVB. **(a)** Dispersion curves (group and phase) obtained from the combination of ANT and earthquake based tomography (circles). Their error bars are calculated as the resolution of the tomography on this node and period normalized by a factor of 2500 (in km). Accepted models are shown as colored lines, and the best fitted dispersion curves as black lines. **(b)** Dashed lines show the feasible region in the inversion; the colored lines are the models with misfits less than or equal to two times the best fitting (0.11 km s^{-1}); and the black line indicates the shear-velocity model that best fits the observed dispersion curves. The scale shows the color code of the misfit.

beneath the Mississippi embayment (Figs. 9a, b, 10a). This low velocity anomaly beneath the Mississippi embayment agrees well with the sediment thickness model of Laske et al. (2013) and the velocity model of Bensen et al. (2009). Our model also shows low velocities along the USA-Mexico border with the lowest values coincident with the Rio Grande drainage basin, the major Holocene coastal depocenter west of the Mississippi delta.

The Ouachita–Marathon–Sonora orogen is a 3000 km long belt of deformed Paleozoic rocks bordering the southern margin of the Laurentian (North American) craton (Moreno et al., 2000; Poole et al., 2005). The eastern part of this belt encloses low velocity areas beneath the Mississippi and Rio Grande embayment (Fig. 9a). The location of the southern Laurentia margin has been much debated (e.g., Moreno et al., 2000). Poole et al. (2005) localized it along Chihuahua, Sonora, and Baja California, but Dickinson (2009) considers it still a genuine frontier of geoscience. Our results at 12 km depth (Fig. 9b) show the highest inland velocities (~ 3.6–3.74 km s^{-1}) along the eastern and central margins of Laurentia, where the Appalachian and Ouachita orogens expose their rock assemblages. These velocities extend toward the west and south coinciding with the southern limit of the Great Plains and the north of Sierra Madre Oriental (SMOr), following the Ouachita–Marathon–Sonora orogen. This high crust velocity signature of the Laurentia margin is not distinguished further west in our model.

4.1.2 Present and ancient crustal extension

The extension in western North America during the late Oligocene to early Pliocene has evolved from the continental-scale Basin and Range Province, to a more limited region known as Gulf Extensional Province (GEP), and finally, the deformation has been limited to the west of the GEP forming the Gulf of California rift (Aragón-Arreola et al., 2005; and references therein). The marine incursion over the rift formed the Gulf of California (GofC). At present, the GofC hosts a zone of oblique extension that records the transition from oceanic spreading centers and transform faulting in the south (Londslade, 1989; Lizarralde et al., 2007) to the diffuse continental deformation in the north (Oskin and Stock, 2003; González-Fernández et al., 2005). We obtain a heterogeneous shear-wave velocity distribution along the GofC in accordance with its different tectonic stages and with results from several local studies (Aragón-Arreola and Martín-Barajas, 2007; Persaud et al., 2007; Wang et al., 2009; Zhang and Paulssen, 2012). Seismological data show a significant difference in crustal thickness between the Sierra Madre Occidental core and its margins. Several studies estimated the crustal thickness at the center of the Sierra Madre Occidental around 36–40 km (Gomberg et al., 1989; Couch et al., 1991). It thins towards the south and west to 25 km at the coast (Persaud et al., 2007) where the crust has been thinned by extension that led to the formation of the Gulf of California. Our model shows thinner crust beneath the GofC ($< 20 \text{ km}$) than in contiguous areas (Baja California Peninsula and SMOc). We obtain ~ 30 km crustal thickness beneath the SMOc and it thickens toward the east to ~ 35 km under SMOr (Fig. 10b). Crustal thickness differences under SMOc and SMOr between the results of this study and previous studies are within the range of our vertical resolution. Bouguer anomaly changes are the result of density variations at different depths. Negative anomalies are related to low densities, which at large scale can be due to large sediment basins, thick crust, or shallow asthenosphere. Positive Bouguer anomalies denote high density rocks and may be thin crust. Figure 11 shows the Bouguer gravity anomaly map for the study area. It has been computed applying a complete Bouguer correction to free-air satellite data (Sandwell and Smith, 1997) using the code FA2BOUG (Fullea et al., 2008) with a reduction density of 2670 kg m^{-3}. The observed changes in crustal thickness between the SMOc core and its margins correlate well with the large negative Bouguer anomaly values at the center and less negative at its western part (Fig. 11).

One of the novelties of this velocity model is that it clearly draws the limits of the GEP province as high lower-crust velocities in contrast with low velocities in the surrounding areas. For example, at 25 km depth the contour between high ($> 4.0 \text{ km s}^{-1}$) and low velocities ($< 3.5 \text{ km s}^{-1}$) is narrow and sharp, indicating a limit between extended and unextended crust (Fig. 9c). Defining the GEP province like this, it

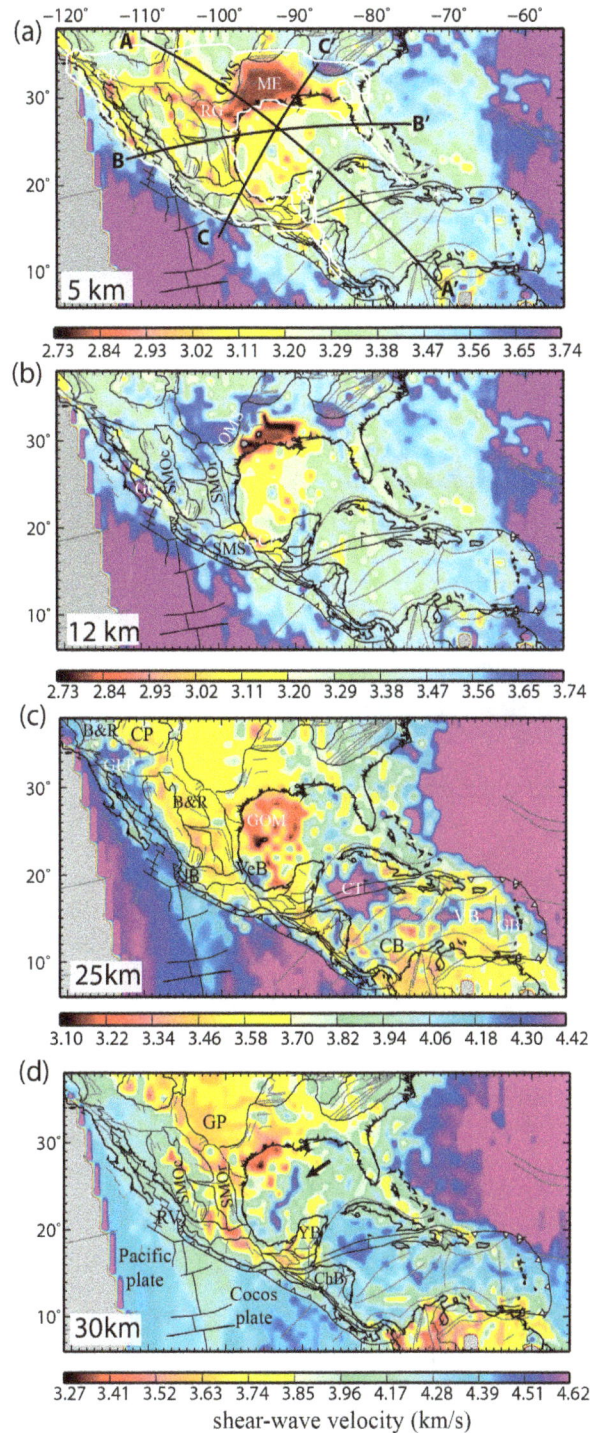

Figure 9. Shear wave velocity maps at different depths ((**a**) 5, (**b**) 12, (**c**) 25, and (**d**) 30 km). Faults, ridges, fracture zones, and basin limits are denoted as gray lines (CGMW/UNESCO, 2000). (**a**) Thick black lines indicate the cross-sections shown in Fig. 10 and the white line contours the area with ANT resolution equal to or lower than 250 km at 10 s period. B&R denotes Basin and Range; CB Colombian Basin; ChB Chortis Block; CP Colorado Plateau; CR Colorado River; CT Cayman Trough; GB Grenada Basin; GCP Gulf Coastal Plain; GEP Gulf Extensional Province; GOM Gulf of Mexico; GP Great Plains; IT Isthmus of Tehuantepec; JB Jalisco Block; ME Mississippi Embayment; OMS Ouachita–Marathon–Sonora orogenic belt; RG Rio Grande; RV Rivera Plate; SMOc Sierra Madre Occidental; SMOr Sierra Madre Oriental; SMS Sierra Madre del Sur; TMVB Trans-Mexican Volcanic Belt; VB Venezuela Basin; VeB Veracruz Basin; and YB Yucatan Block.

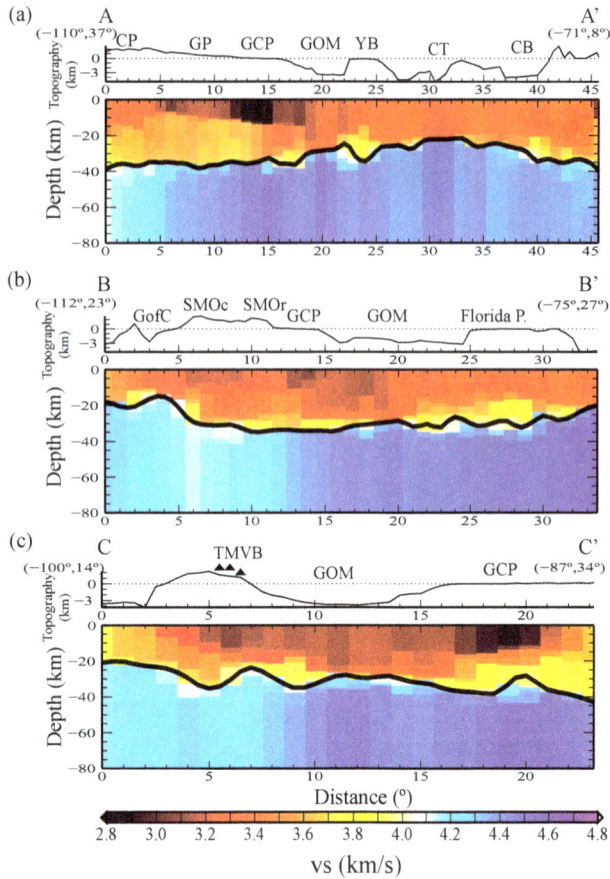

Figure 10. Shear wave velocity along the cross-sections delineated in Fig. 9; **(a)** A-A', **(b)** B-B', and **(c)** C-C'. The figure shows moho depth (thick black line), topography (thin black line above the velocity profile), and sea level (dashed line). CB denotes Colombian Basin; CP Colorado Plateau; CT Cayman Trough; Florida P. Florida Peninsula; GCP Gulf Coastal Plain; GofC Gulf of California; GOM Gulf of Mexico; GP Great Plains; SMOc Sierra Madre Occidental; SMOr Sierra Madre Oriental; TMVB Trans-Mexican Volcanic Belt; and YB Yucatan Block.

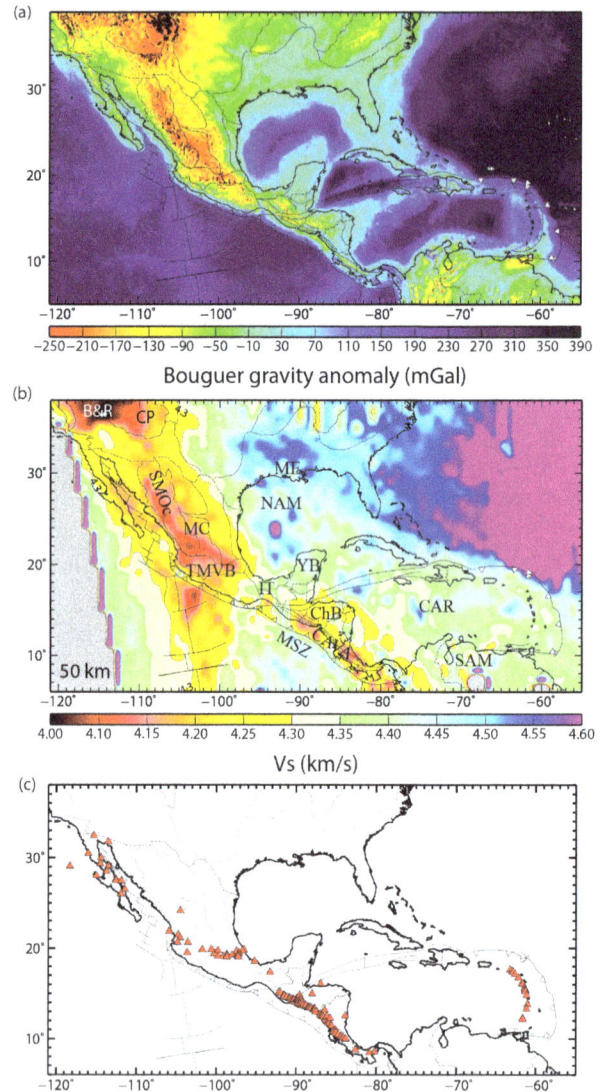

Figure 11. (a) Bouguer gravity anomaly map. **(b)** Shear wave velocity map at 50 km depth. **(c)** Map with the location of volcanoes (red triangles) exhibiting current unrest or eruptions during the Holocene (Siebert and Simkin, 2002). B&R denotes Basin and Range; CAR Caribbean plate; CAVA Central America Volcanic Arc; ChB Chortis Block; CP Colorado Plateau; IT Isthmus of Tehuantepec; ME Mississippi Embayment; MSZ Mesoamerican Subduction Zone; NAM North American plate; SAM South American plate; SMOc Sierra Madre Occidental; TMVB Trans-Mexican Volcanic Belt; and YB Yucatan Block.

comprises the US B&R and the western part of SMOc, where Ferrari et al. (2007) indicated a signature of the active extension related with the subduction of the Farallon plate under the NAM plate. We obtain a similar high velocity structure beneath the western part of the TMVB that coincides with the area enclosed by the triple graben (Luhr et al., 1985) on the Jalisco Block where the Rivera plate subducts. The thin crust observed in this area is evidence of an extension process, coherent with the proposed Jalisco Block rifting from the North American plate (Luhr et al., 1985; Allan et al., 1991). Another noteworthy feature is that our results highlight a different crustal seismic structure between the US and Mexican Basin and Range provinces.

Widely accepted Gulf of Mexico reconstruction models fit its opening from 158 to 130 Ma (e.g., Pindell and Kennan, 2009). During the extension of the GOM, fragments de-

tached from NAM, migrating to the south, and forming the Yucatan Block and the northern portion of SAM plate. The GOM tectonic evolution comprises seafloor spreading, and Yucatan Block rifting and rotation (30–40° clockwise) from its origin location, attached to south-central US, to its present location. The GOM sediment seismic structure has been extensively explored for hydrocarbons and is well known; however, the underlying crust and mantle velocity distribution are still poorly understood (Swayer et al., 1991). Whole images of the GOM crustal seismic structure come from compilations of local-experiments (e.g., Swayer et al., 1991; Bird et al., 2005). Besides, the large basin's sediment thickness made deep-penetration observations difficult (Swayer et al., 1991). In fact, the short period ambient noise cross-correlations from paths crossing the GOM had a very low signal-to-noise ratio (Gaite et al., 2012). Therefore, we define the GOM seismic structure from tomographic results of 20 s period and longer which means that its shallow crust shear wave velocity structure is not as well defined as in mainland North America. In spite of this limitation, our results show a sharp difference between crustal velocities west and east of $-90°$ longitude (Fig. 9). Previous tomographic studies (e.g., Vdovin et al., 1999) associate low Rayleigh and Love wave group velocity at 20 s period on the western part of the GOM with a large accumulation of sediments. Our results confirm this correspondence: we find very low shear-wave velocities ($\sim 3.2\,\mathrm{km\,s^{-1}}$) down to 20 km depth that coincide with the sediment thickness on the Gulf of Mexico reported by Divins (2003) from isopach maps, ocean drilling results, and seismic reflection profiles. We obtain an average crustal thickness beneath the GOM of 25–30 km that coincides with the results of Bird et al. (2005) from gravimetric data and a compilation of seismic reflection experiments in particular areas of the GOM. At 30 km depth our results show a narrow NNE high velocity area (Fig. 9d) indicating a thinner crust than at the rest of the GOM. This high velocity feature should be interpreted with caution because the lack of path coverage at periods shorter than 20 s to constrain the uppermost crust and the large misfits in the western GOM (Fig. 7). This feature might be related with the gulf opening during the Jurassic, since it matches with the youngest crust in the gulf (Müller et al., 2008) and roughly with the recent gravity results of Sandwell et al. (2014). However, its orientation does not coincide with the ENE direction of the extinct ridge proposed by Pindell and Kennan (2009), the results by Swayer et al. (1991), and with the GOM largest gravity anomalies by Bird et al. (2005).

Some local seismic experiments of receiver functions infer thin crust beneath the Veracruz Basin (e.g., Melgar and Pérez-Campos, 2011; Zamora-Camacho et al., 2010). Our results confirm these observations, revealing high velocities ($\sim 4.2\,\mathrm{km\,s^{-1}}$) at 25 km depth offshore of the Veracruz Basin (Fig. 9c).

4.2 Upper mantle

Several tomographic continental-scale studies (e.g., Alsina et al., 1996; Van der Lee and Nolet, 1997; Vdovin et al., 1999; Godey et al., 2003; Bedle and van der Lee, 2009) image the dichotomy between the low mantle seismic velocities of the western North American and Caribbean plates and the high velocities of their eastern parts. Our model shows this velocity contrast from 50 km depth (Fig. 11) with great

detail due to the large number of stations used in Mexico and the Caribbean. We find low shear-wave velocities in the western US, along Mexico and below the Chortis Block, and high velocities in the central-east US, the Gulf of Mexico, the Isthmus of Tehuantepec, the Yucatan Block, the central and eastern parts of the Caribbean plate, and on the northern South American plate. At 50 km depth, the $4.30\,\mathrm{km\,s^{-1}}$ velocity contour roughly follows the western boundary of the Great Plains, the northeast of the Sierra Madre Oriental, and the western part of the Gulf of Mexico toward the Isthmus of Tehuantepec. This contour resembles the $4.55\,\mathrm{km\,s^{-1}}$ velocity contour at 80 km depth obtained by Bensen et al. (2009), which lies close to the Rocky Mountain Front in the southern US. The west–east mantle dichotomy symmetry breaks beneath the eastern part of the Sierra Madre del Sur, the Isthmus of Tehuantepec, and the Yucatan Block, whose high velocities contrast with the lower ones of the surrounding areas. This symmetry break supports the aforementioned different origin of the Yucatan Block in comparison with the other Mexican terrains and the Chortis Block.

Along the Mesoamerican Subduction Zone high velocities at 50 km depth coincide with a lack of active volcanism in certain areas (e.g., south of Sierra Madre del Sur, part of the Isthmus of Tehuantepec), while low velocities correspond to active volcanic arcs (e.g., TMVB and CAVA). Regional and global seismic tomographic studies (Grand, 1994; Alsina et al., 1996; Van der Lee and Nolet, 1997; Bijwaard and Spakman, 2000; Ritzwoller et al., 2002; Ritsema et al., 2004) suggest that the lithospheric mantle has been mostly removed and replaced by asthenospheric mantle in the region between the Gulf of California and the Mesa Central, and from the US Basin and Range Province to latitude 20° N. This is in agreement with the low velocities estimated at 50 km depth (Fig. 11). We also obtain low velocities along the Gulf of California oceanic ridge. Negative Bouguer gravity anomalies coincide with low shear-wave velocities at 50 km depth on the north of the Basin and Range, west of the Colorado Plateau and Mesa Central (Fig. 11). This coherence may be the effect of a thin lithosphere (e.g., B&R, Colorado Plateau) or may support the presence of magmas from a mantle wedge below the Mesa Central crust inferred by Nieto-Samaniego et al. (2005). However, we did not find such a straightforward relation between negative Bouguer gravity anomalies and low mantle velocities in every region (for example, at the westernmost part of SMOc and TMVB). This different pattern on the gravity field may be due to the combination of the contrary effects of thin crust and thin lithospheric mantle.

5   Conclusions

We invert group and phase velocities of fundamental mode Rayleigh waves to obtain a vertically polarized 3-D shear-wave velocity model (3DVSAM) of the crust and uppermost mantle of Mexico, the Gulf of Mexico and the Caribbean plate. We combine surface wave velocities from ANT and earthquake tomography. The model offers a picture of the seismic structure from 5 to 70 km depth of the region as a whole. Our model agrees with present and past tectonic processes in the region, coincides with crustal features showed in local studies, images with high detail the uppermost mantle, and exhibits some new seismological features. This model may be useful to constrain tectonic evolution models, localize regional earthquakes, simulate ground motions, and correct crustal effects in mantle tomography studies, among other possible applications.

The 3-D crustal and uppermost mantle shear-wave velocity model 3DVSAM is available to download at: https://sites.google.com/site/earthsciencesbgaite.

Author contributions. B. Gaite designed and carried out the data processing. A. Villaseñor designed the research and collected the earthquake records. A. Iglesias developed the inversion code. I. Jiménez-Munt computed the gravity anomaly. All the authors interpreted the results. B. Gaite prepared the manuscript with contributions from the co-authors.

Acknowledgements. Seismic data come from the Mexican National Seismological Service (NSSM), and from the networks CI, CU, G, GE, II, IU, LI, NR, OV, TA and US through the IRIS Consortium. We acknowledge C. Valdés, A. Cárdenas, C. Cárdenas, I. Rodríguez, J. Pérez, J. Estrada, and S. I. Franco for providing NSSM data.

The maps and graphs were drawn with the Generic Mapping Tools (Wessel and Smith, 1998).

Funds provided by the REPSOL CO-DOS project supported B. Gaite. This is a contribution of the Team Consolider-Ingenio 2010 TOPO-IBERIA (CSD2006-00041). We thank D. García and A. Ugalde for their useful comments for improving this manuscript. We also are grateful to Taka'aki Taira and two anonymous reviewers for their constructive and helpful suggestions for enhancing the paper.

Edited by: T. Taira

References

Allan, J. F., Nelson, S. A., Luhr, J. F., Carmichael, I. S. E., Wopat, M., and Wallace, J. P.: Pliocene-Recent rifting in SW Mexico and associated volcanism: an exotic terrain in the making, AAPG Memoir, 47, 425–445, 1991.

Alsina, D., Woodward, R. L., and Snieder, R. K.: Shear wave velocity structure in North America from large-scale waveform inversions of surface waves, J. Geophys. Res., 101, 15969–15986, doi:10.1029/96JB00809, 1996.

Ammon, C. J.: Notes on surface-wave tomography. Part I: Group velocity estimation, p. 44, Unpublished Notes, 1998.

Aragón-Arreola, M. and Martín-Barajas, A.: Westward migration of extension in the northern Gulf of California, Mexico, Geology, 35, 571, doi:10.1130/G23360A.1, 2007.

Aragón-Arreola, M., Morandi, M., Martín-Barajas, A., Delgado-Argote, L., and González-Fernández, A.: Structure of the rift basins in the central Gulf of California: Kinematic implications for oblique rifting, Tectonophysics, 409, 19–38, doi:10.1016/j.tecto.2005.08.002, 2005.

Auer, L., Boschi, L., Becker, T. W., Nissen-Meyer, T., and Giardini, D.: Savani: A variable resolution whole-mantle model of anisotropic shear velocity variations based on multiple data sets, J. Geophys. Res., 119, 3006–3034, doi:10.1002/2013JB010773, 2014.

Barmin, M., Ritzwoller, M. H., and Levshin, A. L.: A fast and reliable method for surface wave tomography, Pure Appl. Geophys., 158, 1351–1375, 2001.

Bedle, H. and van der Lee, S.: S velocity variations beneath North America, J. Geophys. Res., 114, B07308, doi:10.1029/2008JB005949, 2009.

Bensen, G. D., Ritzwoller, M. H., and Yang, Y.: A 3-D shear velocity model of the crust and uppermost mantle beneath the United States from ambient seismic noise, Geophys. J. Int., 177, 1177–1196, doi:10.1111/j.1365-246X.2009.04125.x, 2009.

Berteussen, K. A.: Moho depth determination based on spectral ratio analysis of NORSAR long-period P waves, Phys. Earth Planet. Inter., 31, 313–326, 1977.

Bijwaard, H. and Spakman, W.: Non-linear global P-wave tomography by iterated linearized inversion, Geophys. J. Int., 141, 71–82, 2000.

Bird, D. E., Burke, K., Hall, S. A., and Casey, J. F.: Gulf of Mexico tectonic history: Hotspot tracks, crustal boundaries, and early salt distribution, The American Association of Petroleum Geologists Bulletin, 89, 311–328, 2005.

Bird, P.: An updated digital model of plate boundaries, Geochem. Geophys. Geosyst., 4, 1027, doi:10.1029/2001GC000252, 2003.

Bozdağ, E. and Trampert, J.: On crustal corrections in surface wave tomography, Geophys. J. Int., 172, 1066–1082, doi:10.1111/j.1365-246X.2007.03690.x, 2008.

Burke, K.: Tectonic evolution of the Caribbean, Ann. Rev. Earth Planet. Sci., 16, 201–230, 1988.

Campillo, M., Singh, S. K., Shapiro, N. M., Pacheco, J. F., and Herrmann, R. B.: Crustal structure south of the Mexican volcanic belt, based on group velocity dispersion, Geofisica Internacional, 35, 361–370, 1996.

CGMW/UNESCO: Geological Map of the World at 1:25 million. Commission for the Geological Map of the World, UNESCO, second edition, September 2000.

Cordoba-Montiel F., Iglesias, A., Singh, S. K., Spicka, Z. and Legrand, D: Tomografía de Velocidad de Grupo para el Oriente de México y el Istmo de Tehuantepéc, Bol. Soc. Mex., 66, 3, 441–457, 2014.

Couch, R. W., Ness, G. E., Sanchez-Zamora, O., Calderon-Riverol, G., Doguin, P., Plawman, T., Coperude, S., Huehn, B., and Gumma, W.: Gravity anomalies and crustal structure of the Gulf and Peninsular Province of the Californias, in: The Gulf and Peninsular Province of the Californias: American Association of Petroleum Geologists, Memoir, edited by: Dauphin, J. P. and Simoneit, B. R. T., 47–70, 1991.

Dickinson, W. R.: The Gulf of Mexico and the southern margin of Laurentia, Geology, 37, 479–480, doi:10.1130/focus052009.1, 2009.

Divins, D. L.: Total sediment thickness of the world's oceans and marginal seas, NOAA National Geophysical Data Cente, Boulder, CO, 2003.

Ferrari, L., Valencia-Moreno, M., and Bryan, S.: Magmatism and tectonics of the Sierra Madre Occidental and its relation with the evolution of the western margin of North America., in: Geology of Mexico: Celebrating the Centenary of the Geological Society of México: Geological Society of America Special Paper, edited by: Alaniz-Álvarez, S. A. and Nieto-Samaniego, A. F., 1–39, 2007.

Fullea, J., Fernández, M., and Zeyen, H.: FA2BOUG-A FORTRAN90 code to compute Bouguer gravity anomalies from gridded free air anomalies: Application to the Atlantic-Mediterranean transition zone, Comput. Geosci., 34, 1665–1681, doi:10.1016/j.cageo.2008.02.018, 2008.

Gaite, B., Iglesias, A., Villaseñor, A., Herraiz, M., and Pacheco, J. F.: Crustal structure of Mexico and surrounding regions from seismic ambient noise tomography, Geophys. J. Int., 188, 1413–1424, doi:10.1111/j.1365-246X.2011.05339.x, 2012.

Godey, S., Snieder, R. K., Villaseñor, A., and Benz, H. M.: Surface wave tomography of North America and the Caribbean using global and regional broad-band networks: phase velocity maps and limitations of ray theory, Geophys. J. Int., 152, 620–632, doi:10.1046/j.1365-246X.2003.01866.x, 2003.

Goffe, W. L.: SIMANN: A Global Optimization Algorithm using Simulated Annealing, Studies in Nonlinear Dynamics & Econometrics, 1, 1558–3708, doi:10.2202/1558-3708.1020, 1996.

Goffe, W. L., Ferrier, G. D., and Rogers, J.: Global optimization of statistical functions with simulated annealing, Journal of Econometrics, 60, 65–99, doi:10.1016/0304-4076(94)90038-8, 1994.

Gomberg, J., Prietsley, K., and Brune, J.: The compressional velocity structure of the crust and upper-mantle of northern Mexico and the border region, Bull. Seism. Soc. Am., 79, 1496–1519, 1989.

González-Fernández, A., Dañobeitia, J. J., Delgado-Argote, L. A., Michaud, F., Córdoba, D., and Bartolomé, R.: Mode of extension and rifting history of upper Tiburón and upper Delfín basins, northern Gulf of California, J. Geophys. Res., 110, B01313, doi:10.1029/2003JB002941, 2005.

Grand, S. P.: Mantle shear structure beneath the Americas and surrounding oceans, J. Geophys. Res., 99, 11591–11621, doi:10.1029/94JB00042, 1994.

Handschy, J. W., Keller, G. R., and Smith, K. J.: The Oauachita system in northern Mexico, Tectonics, 6, 323–330, 1987.

Herrmann, R. B.: Computer programs in seismology, available at: http://www.eas.slu.edu/People/RBHerrmann/ComputerPrograms.html (last access: 5 January 2014), St. Luis University, St. Luis, Missouri, 1987.

Iglesias, A., Cruz-Atienza, V. M., Shapiro, N. M., Singh, S. K., and Pacheco, J. F.: Crustal structure of south-central Mexico estimated from the inversion of surface-wave dispersion curves using genetic and simulated annealing algorithms, Geofísica Internacional, 40, 181–190, 2001.

Iglesias, A., Clayton, R. W., Pérez-Campos, X., Singh, S. K., Pacheco, J. F., García, D., and Valdés-González, C.: S wave velocity structure below central Mexico using high-resolution surface wave tomography, J. Geophys. Res., 115, B06307, doi:10.1029/2009JB006332, 2010.

Laske, G., Masters, G., Ma, Z., and Pasyanos, M.: Update on CRUST1.0 – A 1-degree Global Model of Earth's Crust, Geophys. Res. Abstracts, 15, Abstract EGU2013-2658, 2013.15, 2013.

Lekić, V., Panning, M., and Romanowicz, B. A.: A simple method for improving crustal corrections in waveform tomography, Geophys. J. Int., 182, 265–278, doi:10.1111/j.1365-246X.2010.04602.x, 2010.

Levshin, A. L., Barmin, M. P., Ritzwoller, M., and Trampert, J.: Minor-arc and major-arc global surface wave diffraction tomography, Phys. Earth Planet. Inter., 149, 205–223, 2005.

Lin, F.-C., Moschetti, M. P., and Ritzwoller, M. H.: Surface wave tomography of the western United States from ambient seismic noise: Rayleigh and Love wave phase velocity maps, Geophys. J. Int., 173, 281–298, doi:10.1111/j.1365-246X.2008.03720.x, 2008.

Lizarralde, D., Axen, G. J., Brown, H. E., Fletcher, J. M., González-Fernández, A., Harding, A. J., Holbrook, W. S., Kent, G. M., Paramo, P., Sutherland, F. H., and Umhoefer, P. J.: Variation in styles of rifting in the Gulf of California, Nature, 448, 466–469, doi:10.1038/nature06035, 2007.

Londslade, P.: Geology and tectonic history of the Gulf of California, The Eastern Pacific Ocean and Hawaii (The Geology of North America, N), edited by: Husson, D., Winterer, E. L., and Decker, R. W., Geol. Soc. Am., Boulder, CO, 1989.

Luhr, J. F., Nelson, S. A., Allan, J. F., and Charmichael, I. S. E.: Active rifting in south-western Mexico: manifestations of an incipient eastward spreading ridge jump, Geology, 13, 54–57, doi:10.1130/0091-7613(1985)13<54:ARISMM>2.0.CO;2, 1985.

Marshall, J. S.: Geomorphology and physiographic provinces of Central America, in: Central America: Geology, resources and hazards, edited by: Bundschuh, J. and Alvarado, G., Taylor and Francis, London, 75–122, 2007.

McNamara, D. E., McCarthy J., and Benz H.: Improving earthquake and tsunami warning for the Caribbean Sea, Gulf of Mexico and the Atlantic coast, U.S. Geological Survey Fact Sheet, NO. 2006–3012, 4pp., 2006.

Melgar, D. and Pérez-Campos, X.: Imaging the Moho and subducted oceanic crust at the isthmus of Tehuantepec, Mexico, from receiver functions, Pure Appl. Geophys. 168, 1449–1460, doi:10.1007/s00024-010-0199-5, 2011.

Moreno, F. A., Mickus, K. L., and Keller, G. R.: Crustal structure and location of the Ouachita orogenic belt in northern Mexico, Geofísica Internacional, 39, 229–246, 2000.

Müller, R. D., Sdrolias, M., Gaina, C., and Roest, W. R.: Age, spreading rates, and spreading asymmetry of the world's ocean crust, Geochem. Geophys. Geosyst., 9, Q04006, doi:10.1029/2007GC001743, 2008.

Nieto-Samaniego, A. F., Alaniz-Álvarez, S. A., and Camprubí í Cano, A.: La Mesa central de México: estratigrafía, estructura y evolución tectónica cenozoica, Boletín de la Sociedad Geológica Mexicana, Boletín Conmemorativo del Centenario, Temas Selectos de la Geología Mexicana, LVII, 285–318, 2005.

Oskin, M. and Stock, J. M.: Marine incursion synchronous with plate-boundary localization un the Gulf of California, Geology, 31, 23–26, doi:10.1130/0091-7613(2003)031<0023:MISWPB>2.0.CO;2, 2003.

Panning, M. P., Lekić, V., and Romanowicz, B. A.: Importance of crustal corrections in the development of a new global model of radial anisotropy, J. Geophys. Res., 115, 1–8, doi:10.1029/2010JB007520, 2010.

Pasyanos, M. E., Masters, G. T., Laske, G., and Ma, Z.: LITHO1.0: An updated crust and lithospheric model of the Earth, J. Geophys. Res.-Solid Earth, 119, 2153–2173, doi:10.1002/2013JB010626, 2013.

Persaud, P., Pérez-Campos, X., and Clayton, R. W.: Crustal thickness variations in the margins of the Gulf of California from receiver functions, Geophys. J. Int., 170, 687–699, doi:10.1111/j.1365-246X.2007.03412.x, 2007.

Pindell, J. and Kennan, L.: Tectonic evolution of the Gulf of Mexico, Caribbean and northern South America in the mantle reference frame: an update. The origin and evolution of the region between North and South America, Geological Society London Special Publication, 328, 1–55, 2009.

Poole, F. G., Perry Jr., W. J., Madrid, R. J., and Amaya-Martínez, R.: Tectonic synthesis of the Ouachita-Marathon-Sonora orogenic margin of southern Laurentia: Stratigraphic and structural implications for timing of deformational events and plate-tectonic model, Geological Society of America Special Paper, 393, 543–596, doi:10.1130/0-8137-2393-0.543, 2005.

Ritsema, J., Van Heijst, H., and Woodhouse, J.: Global transition zone tomography, J. Geophys. Res., 109, 1–14, doi:10.1029/2003JB002610, 2004.

Ritzwoller, M. H., Shapiro, N. M., Barmin, M. P., and Levshin, A. L.: Global surface wave diffraction tomography, J. Geophys. Res., 107, 2335, doi:10.1029/2002JB001777, 2002.

Ritzwoller, M. H., Lin, F. C., and Shen, W.: Ambient noise tomography with a large seismic array, C. R. Geoscience, 343, 558–570, doi:10.1016/j.crte.2011.03.007, 2011.

Rogers, R. D, Mann, P., and Emmet, P. A.: Tectonic terraines of the Chortis block based on integration of regional aeromegnetic and geologic data, The Geol. Soc. of Am., Special Paper, 428, 65–88, doi:10.1130/2007.2428(04), 2007.

Sabra, K. G., Gerstoft, P., Roux, P., Kuperman, W. A., and Fehler, M. C.: Surface wave tomography from microseisms in Southern California, Geophys. Res. Lett., 32, L14311, doi:10.1029/2005GL023155, 2005.

Sandwell, D. T. and Smith, W. H. F.: Marine gravity anomalies from GEOSAT and ERS-1 satelite altimetry, J. Geophys. Res., 102, 10039–10054, doi:10.1029/96JB03223, 1997.

Sandwell, D. T., Müller, R. D., Smith, W. H. F., Garcia, E., and Francis, R.: New global marine gravity model from CryoSat-2

and Jason-1 reveals buried tectonic structure, Science, 346, 65–67, doi:10.1126/science.1258213, 2014.

Schaeffer, A. J. and Lebedev, S.: Global shear-speed structure of the upper mantle and transition zone, Geophys. J. Int., 194, 417–449, doi:10.1093/gji/ggt095, 2013.

Schaeffer, A. J. and Lebedev, S.: Imaging the North American continent using waveform inversion of global and USArray data, Earth Planet. Sci. Lett., 402, 26–41, doi:10.1016/j.epsl.2014.05.014, 2014.

Sedlock, R. L.: Tectonoestratigraphic terranes and tectonic evolution of Mexico, Geological Society of America Special Paper, 278, 153 pp., 1993.

Shapiro, N. M. and Ritzwoller, M. H.: Monte-Carlo inversion for a global shear-velocity model of the crust and upper mantle, Geophys. J. Int., 151, 88–105, 2002.

Shapiro, N. M., Campillo, M., Paul, A., Singh, S. K., Jongmans, D., and Sánchez-Sesma, F. J.: Surface-wave propagation accross the Mexican Volcanic Belt and the origin of the long-period seismic-wave amplification in the valley of Mexico, Geophys. J. Int., 128, 151–166, doi:10.1111/j.1365-246X.1997.tb04076.x, 1997.

Shapiro, N. M., Campillo M., Stehly L., and Ritzwoller, M. H.: High-resolution surface-wave tomography from ambient seismic noise, Science, 307, 1615–1618, doi:10.1126/science.1108339, 2005.

Siebert, L. and Simkin, T.: Volcanoes of the world: an illustrated catalog of Holocene volcanoes and their eruptions, http://www.volcano.si.edu (last access: 17 December 2014), 2002.

Swayer, D. S., Buffler, R. T., and Pilger, R. H. J.: Ther crust under the gulf of Mexico basin, in: The Geology of North America, edited by: Salvador, A., The Geological Society of America, 53–72, 1991.

Trampert, J., Paulsen, H., van Wettum, A., Ritsema, j., Clayton, R. W., Castro, R., Rebollar, C. J., and Pérez-Verti, A.: New array monitors seismic activity near the Gulf of California in Mexico, Eos Trans. AGU, 84, 29–32, doi:10.1029/2003EO040002, 2003.

Van der Lee, S. and Nolet, G.: Upper mantle S velocity structure of North America, J. Geophys. Res., 102, 22815–22838, doi:0148-0227/97/97JB-01168, 1997.

Vdovin, O., Rial, J., Levshin, A. L., and Ritzwoller, M. H.: Group-velocity tomography of South America and the surrounding oceans, Geophys. J. Int., 136, 324–340, doi:10.1046/j.1365-246X.1999.00727.x, 1999.

Wang, Y., Forsyth, D. W., and Savage, B.: Convective upwelling in the mantle beneath the Gulf of California, Nature, 462, 499–501, doi:10.1038/nature08552, 2009.

Wessel, P. and W. H. F. Smith: New, improved version of Generic Mapping Tools released, Eos Trans. AGU, 79, 579, doi:10.1029/98EO00426, 1998.

Yang, Y. and Ritzwoller, M. H.: Teleseismic surface wave tomography in the western U.S. using the Transportable Array component of USArray, Geophys. Res. Lett., 35, L04308, doi:10.1029/2007GL032278, 2008.

Yao, H., Beghein, C., and van der Hilst, R. D.: Surface wave array tomography in SE Tibet from ambient seismic noise and two-station analysis – II. Crustal and upper-mantle structure, Geophys. J. Int., 173, 205–219, doi:10.1111/j.1365-246X.2007.03696.x, 2008.

Zamora-Camacho, A., Espindola, V. H., Pacheco, J. F., Espindola, J. M., and Godinez, M. L.: Crustal thickness at the Tuxtla Volcanic

Field (Veracruz, Mexico) from receiver functions, Phys. Earth Planet. Inter., 182, 1–9, doi:10.1016/j.pepi.2010.05.009, 2010.

Zhang, X. and Paulssen, H.: Geodynamics of the Gulf of California from surface wave tomography, Phys. Earth Planet. Inter., 192–193, 59–67, doi:10.1016/j.pepi.2011.12.001, 2012.

Zhang, X., Paulssen, H., Lebedev, S., and Meier, T.: Surface wave tomography of the Gulf of California, Geophys. Res. Lett., 34, L15305, doi:10.1029/2007GL030631, 2007.

Zheng, Y., Shen, W., Zhou, L., Yang, Y., Xie, Z., and Ritzwoller, M.: Crust and upermost mantle beneath the north China Craton, northeast China, and the Sea of Japan from ambient noise tomography, J. Geophys. Res., 116, 1–25, doi:10.1029/2011JB008637, 2011.

Zhou, L., Xie, J., Shen, W., Zheng, Y., Yang, Y., Shi, H., and Ritzwoller, M. H.: The structure of the crust and uppermost mantle beneath South China from ambient noise and earthquake tomography, Geophys. J. Int., 189, 1565–1583, doi:10.1111/j.1365-246X.2012.05423.x, 2012.

4

Lithospheric-scale structures in New Guinea and their control on the location of gold and copper deposits

L. T. White[1,2], M. P. Morse[2,*], and G. S. Lister[2]

[1]Southeast Asia Research Group, Department of Earth Sciences, Royal Holloway University of London, Egham, Surrey, UK
[2]Research School of Earth Sciences, The Australian National University, Canberra, ACT, Australia
[*]current address: Geoscience Australia, Canberra, Australia

Correspondence to: L. T. White (lloyd.white@rhul.ac.uk)

Abstract. The locations of major gold and copper deposits on the island of New Guinea are considered by many to be controlled by a series of transfer faults that strike N–S to NE–SW, perpendicular to the long axis of the island. The premise is that these faults dilate perpendicular to the regional stress field, forming conduits for metalliferous gases and fluids to drop out of solution. However, the data on which this idea was first proposed were often not presented or, when the data were presented, were of poor quality or low resolution. We therefore present a review of the existing structural interpretations and compare these with several recently published geophysical data sets to determine if the mineralization controlling transfer faults could be observed. These data were used to produce a new lineament map of New Guinea. A comparison of the lineaments with the location of major gold and copper deposits indicates there is a link between the arc-normal structures and mineralization. However, it is only those deposits that are less than 4.5 million years old that could be associated with these structures. Gravity and seismic tomography data indicate that some of these structures could penetrate deep levels of the lithosphere, providing some support to the earlier idea that the arc-normal structures act as conduits for the younger mineral deposits of New Guinea. The gravity data can also be used to infer the location of igneous intrusions at depth, which could have brought metal-bearing fluids and gases closer to the Earth's surface. These regions might be of interest for future exploration campaigns, particularly those areas that are crosscut by deep, vertical faults. However, new exploration models are needed to explain the location of the deposits that are older than 5 Ma.

1 Introduction

The exploration for metalliferous deposits around the world presents a challenge as there is still much to learn about the processes that concentrate metals in particular areas. Many of the largest mineral deposits around the world were found from detailed exploration campaigns around areas where the "old timers" panned the streams and sunk shafts, particularly in the search for gold. There is a major geological challenge, however, in understanding the bigger picture, and why there are regions of the globe that are particularly well endowed with concentrations of precious metals. In the pursuit of this goal, geologists typically devise models that attempt to explain why some metals are more concentrated in particular regions. One such model is that gold and copper deposits are often found along strike of other deposits (e.g. along a line or a lineament) (cf. Richards et al., 2000). However, the problem here is that several deposits must be discovered before a line can be drawn between the points. So, lineaments are only proposed retrospectively, after several deposits have been identified. The effectiveness of this technique is questionable. Yet, these lineament style analyses are commonly used by the minerals exploration industry, particularly on the island of New Guinea because it is home to some of the world's richest gold and/or copper deposits (e.g. Grasberg, Ok Tedi, Frieda River, Porgera and Wafi-Golpu) (Fig. 1a).

The idea that part of the Papuan Fold Belt in central New Guinea was dissected by arc-normal transfer faults was proposed in Hill (1991). This work focussed on the structural geology of the region from a hydrocarbon perspective rather than copper and gold exploration. It presented data from

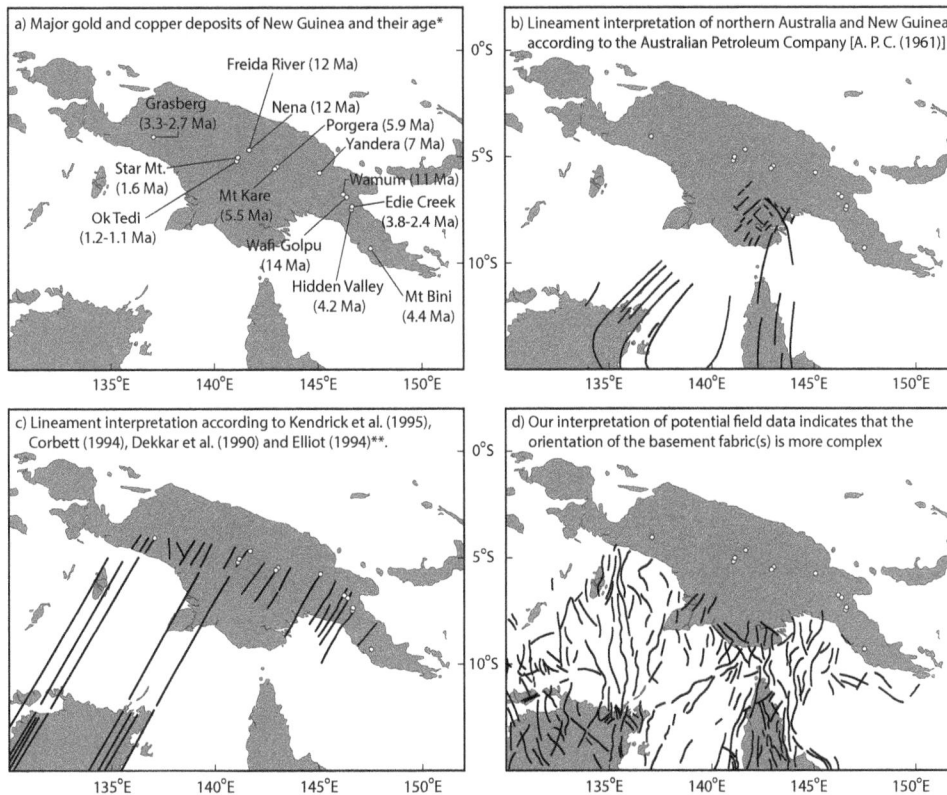

Fig. 1. Maps of northern Australia and New Guinea showing **(a)** the location of major gold and copper deposits and the age of mineralization; **(b)** the location of lineaments according to the Australian Petroleum Company (A. P. C., 1961), who suggested that the basement fabric of northern Australia persisted across the Arafura Sea and into southern Papua New Guinea; **(c)** the location of lineaments that were identified by different studies conducted along the length of New Guinea (Hill et al., 1991; Corbett et al., 1994; Dekker et al., 1994 and Kendrick et al., 1995). These interpretations assumed that NNE striking structures of the Australian mainland could be extrapolated across the Arafura Sea and into New Guinea (in this case the structures that are shown here for northern Australia were proposed by Elliot (1994)). **(d)** However, our interpretation of magnetic data (Milligan (2010): Magnetic Map of Australia, 5th Edition) and gravity data (Pavlis et al., 2008, 2013; this paper) indicates that the orientation of the basement fabric is much more complex than previous workers proposed. * Mineralization ages were taking from the compilation of Garwin et al. (2005). ** Image was adapted from a figure presented in Hill et al. (1996).

geological mapping campaigns and structural analyses that showed that four NNE-trending lineaments could be drawn through the SE limit of the Muller Anticline, the NW limit of the Tari Basin, the NW limit of the deformed Mio–Pliocene strata in the Andabare Plataeu, the SE limit of the Om metamorphic terrane, the NE limit of the Miocene Maramuni igneous province and offset of the NE–SW-trending Papuan Ultramafic Belt. This model drew on the idea that northern Australia had a dominant NNE–SSW-oriented crustal fabric that ran across the Arafura Sea and into southern New Guinea (Fig. 1b) (A. P. C., 1961), where the faults were said to be reactivated lateral ramps that were displaced during a period of shortening in the Neogene. This structural interpretation was also proposed to explain results from apatite fission-track analyses that indicated that parts of Papua New Guinea (PNG) had experienced varying rates of uplift (Hill and Gleadow, 1989). Further support for the model was provided by Davies (1990, 1991), who showed that NNE–SSW striking clusters of mantle-derived intrusives and volcanics

occurred near Bosavi (Fig. 2a). This chain of igneous rocks is referred to as the Bosavi Transfer Zone (Smith, 1990) or the "Bosavi Lineament" (e.g. Hill et al., 2004, 2010). A second magmatic lineament was also identified to the west, drawn through Anju (south) to Ananadi (north) (Fig. 2b). As the magmatic centres along these lineaments young to the south, Davies (1990) calculated that the magmatic front migrated at a rate of 5–8 mm per year (Fig. 2c–d). This rate of movement is an order of magnitude lower than that of the Australian plate (e.g. DeMets et al., 1990), so this magmatism was said to be controlled by a local decrease in pressure in the mantle beneath a deep crustal tensional structure oriented parallel to the axis of maximum horizontal stress (i.e. a NNE–SSW striking transfer fault) (Davies et al., 1990).

Additional arc-normal transfer structures were proposed to exist in New Guinea, east and west of those that were identified by Hill (1991) and Davies et al. (1990, 1991). These arc-normal structures were defined on the basis of the interpretation of gravity data and the orientation of geomorphological

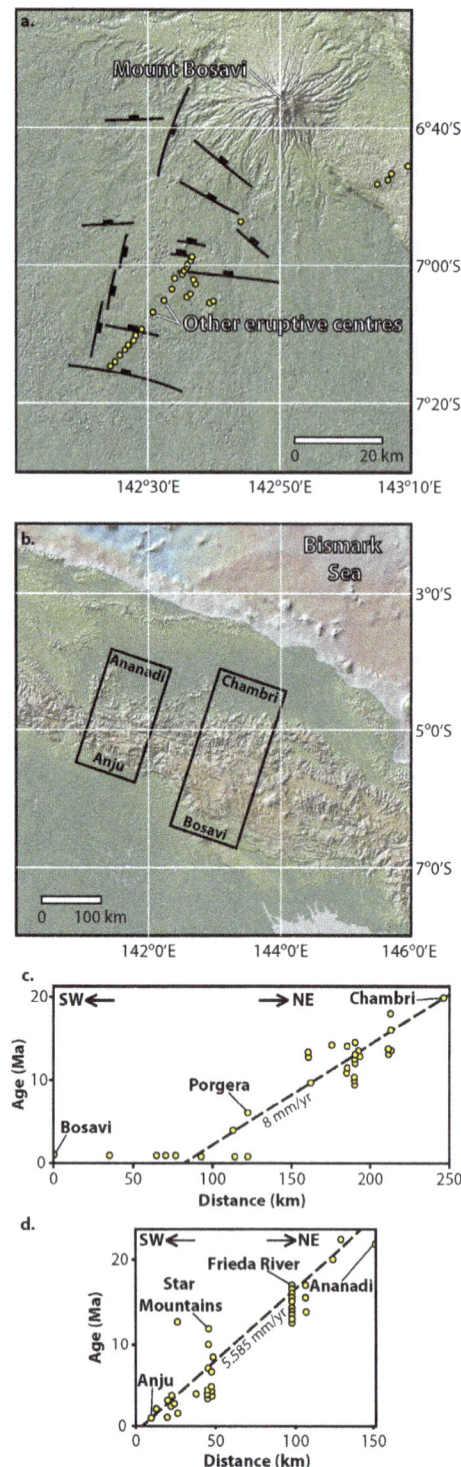

Fig. 2. A chain of volcanics near Bosavi was identified by Davies (1990, 1991) who showed that magmatism in this region progressed from north to south across Papua New Guinea. This pattern is observed (**a**) locally where active volcanism occurs at Mount Bosavi and in smaller eruptive centres to the south, and (**b**) regionally. Davies (1990) drew to cross sections from Ananadi to Anju and Chambri to Bosavi to show that (**c–d**) the volcanic front migrated at a rate of 6–8 mm yr^{-1}. Figure was modified from Davies (1990). The basemap in (**a**) and (**b**) is from the Global Multi-Resolution Topography (GMRT) data set (Ryan et al., 2009).

features such as drainage lines identified from Landsat imagery and aerial photography (Fig. 1c) (e.g. Dekker et al., 1990; Corbett, 1994; Kendrick et al., 1995; Gow and Walshe, 2005; Hill et al., 2008; 2010). Several workers then drew on the link between the structural and magmatic lineaments (e.g. Hill 1991; Davies et al., 1990, 1991) and the location of some of New Guinea's copper and gold deposits. The NNE–SSW striking lineaments were considered to dilate perpendicular/obliquely to the regional maximum principal stress direction associated with convergence between the Australian and Pacific plates. These dilationary zones were said to focus the ascent of evolved magmas and magmatic and metalliferous fluids from deep-crustal reservoirs to nearer the Earth's surface (e.g. Corbett, 1994; Kendrick et al., 1995; Hill et al., 1996, 2002; Richards, 2000). It did not take long before a NNE–SSW striking lineament was identified to intersect each copper and/or gold deposit in New Guinea (Corbett, 1994; Kendrick et al., 1995; Hill et al., 2002; Gow and Walshe, 2005) (Fig. 1c). These were proposed according to the interpretation of various forms of geophysical data.

Yet, while there has been much interest in mapping the arc-normal structures, it is difficult to evaluate if they actually exist. This is because the geophysical data that were used to interpret the structures is often not presented or, if data were shown, the resolution/quality was too low to clearly interpret the location of a geological structure. While there is some geological evidence that supports the existence of NNE–SSW-trending lineaments (e.g. Hill, 1991; Hill et al., 2008, 2010), we were skeptical as to whether all of the proposed transfer structures exist in New Guinea. So we were curious as to whether these structures were mappable using higher-resolution geophysical data that were not available when the lineament concept was originally proposed. We therefore discuss the background behind the concept of the lineaments in New Guinea and evaluate the earlier lineament interpretations against modern geophysical data sets.

2 Geophysical data that were used in this study

In this section we discuss the data that were used, including information about the source and resolution of the data, as well as how the data were processed and interpreted. Please note that we were unable to access the same gravity and magnetic data that were used in the earlier studies, so we instead used publicly accessible data, or data that we obtained permission to use.

2.1 Topography/bathymetry

Several workers proposed lineaments on the basis of interpretations of topographic data (Corbett, 1994; Kendrick et al., 1995; Gow and Walshe, 2005). We obtained topographic data for New Guinea from the Global Multi-Resolution Topography (GMRT) model (Ryan et al., 2009) (Fig. 3) as well as the Consultative Group on International Agricultural Research Consortium for Spatial Information (CGIAR-CSI), who reprocessed NASA's Shuttle Radar Topographic Mission (SRTM) global digital elevation data set (Version 4.0) (Jarvis et al., 2008) (Fig. 4).

The spatial resolution of the GMRT is variable as it ties together multiple sources of data. The resolution of the onshore topography data ranges between 30 and 90 m, whilst the resolution of bathymetric data is generally ∼ 100 m (and up to ∼ 50 m in some coastal areas). In comparison, the CGIAR-CSI SRTM data have a uniform resolution of 90 m, but the coverage is limited to land above sea level. The variable ∼ 30–100 m resolution of these data indicate that structural features can be mapped to a sufficient level of detail for a regional interpretation such as presented here, that is, provided that these geomorphological features actually represent faults.

The GMRT data were manipulated online using GeoMapApp (www.geomapapp.org). The CGIAR-CSI SRTM data are served as a series of tiles to the community from CGIAR-CSI's website (http://srtm.csi.cgiar.org). The relevant tiles were downloaded and were merged using ArcGIS 10.0.

We used the GMRT data set (Ryan et al., 2009) to assess whether lineaments could be observed in regions such as the Arafura Sea between Australia and New Guinea (e.g. Fig. 3). We used the SRTM data to evaluate the interpretation of Gow and Walshe (2005), who proposed that some NE-trending lineaments could be identified in PNG if only the 10, 100 and 1000 m topographic contours were drawn on a map (Fig. 4a). To do this we orthorectified the map that was shown in Gow and Walshe (2005) and compared this with the same selection of topographic contours generated from the SRTM data and displayed in ArcGIS (Fig. 4b).

2.2 Magnetics

To our knowledge there are no publicly available regional onshore magnetic data sets that cover New Guinea, so we could not produce a structural interpretation for the island. The EMAG2, Earth Magnetic Anomaly Grid (Maus et al., 2009), covers much of the offshore region that surrounds New Guinea (see Supplement file 1), and Geoscience Australia continue to publish maps of their compiled magnetic data sets (e.g. Milligan, 2010), but these do not extend far enough to the north to be useful for a structural interpretation of onshore New Guinea. These data sets do provide useful information about the orientation of fabric(s) along northern Australia, so these were used in conjunction with gravity data (see Sect. 2.3) to assess the orientation of lineaments along northern Australia, the Arafura Sea and southern New Guinea (Fig. 1d).

We accessed two smaller-scale high-resolution aeromagnetic data sets that cover parts of the PNG Central Highlands and the Papuan Peninsula (Figs. 3 and 5) that were collected by Fugro Ltd for the World Bank between 2006 and 2008. These data were collected in north–south-oriented lines with 400–500 m line spacing and a terrain clearance of 100 m during a helicopter-borne survey. The aero-magnetic data grids have a cell size of 100 m by 100 m such that these data can effectively show magnetic anomalies with wavelengths of 200 m. A raster was produced from the magnetic data using a minimum curvature, exploration industry standard interpolation. The Central Highlands data set covers an area that encompasses the Porgera (Au), Mt Kare (Au) and Yandera (Cu-Mo) deposits (Fig. 5), and therefore the northern segment of the Porgera–Mt Kare lineament (Hill, 1991). So, these data were used to determine if these lineaments could be identified with regional magnetic data. This involved an examination of the reduced-to-pole magnetic grid. We also examined the same grid after applying an upward continuation residual (UCR) filter (Jacobsen, 1987; Morse, 2010).

The UCR filter is a method of frequency separation that generates a high-frequency residual data set by subtracting a higher upward continued (UC) potential field data set from a lower UC or original (zero level) of the same data set. Upward continuation simulates the effect of measuring the potential field at a higher (larger) distance from the source (ground level), so this calculation effectively filters out the anomalies of smaller and shallower sources, producing a smoother, longer wavelength data set (a higher UC produces a smoother data set relative to a lower UC). The high-frequency signal in an UC or original data set can be accessed by subtracting the higher, smoother (longer wavelength) upward continued signal. This arguably produces a data set that shows significantly enhanced geological features (Morse, 2010). In this case, we generated new grids by subtracting (1) the 1 km UC grid from the original magnetic data grid, (2) the 5 km UC grid from the 1 km UC magnetic data grid, and (3) the 5 km UC grid from a 2 km UC magnetic data grid (Fig. 5).

2.3 Gravity

In addition to the examination of topographic and magnetic data sets, we also interpreted free-air and Bouguer gravity grids to determine if these data could be used to delineate arc-normal structures (Fig. 6). The grids used were spatial subsets of the global Earth Gravitational Model (EGM2008) (Pavlis et al., 2008), covering the greater New Guinea region. These data were downloaded from the Bureau Gravimétrique International's (BGI) data portal (http://bgi.omp.obs-mip.fr/index.php/eng/Data-Products/ Grids-and-models/Regional-gravity-anomaly-grids). The Bouguer data that were downloaded had been corrected using the FA2BOUG code that was proposed by Fullea et al. (2008). This included a topographic correction that was applied up to a distance of 167 km using the 1.0 × 1.0 arc-minute ETOPO1 digital elevation model (Amante and Eakins, 2009). The gravity data have a spatial resolution of

Table 1. Details of the UCR grids that were generated from the BGI Bouguer gravity data where one upward continued grid (grid(X)) (that was generated by simulating the measurement being made X km above the original surface) was subtracted from another upward continued grid (grid(Y)) (that was generated by simulating the measurement being made Y km above the original surface).

Upward continued height (X km)	Upward continued height (Y km)
30	1
40	5
50	10
80	20
100	40

2.5×2.5 arc minutes (~ 4–5 km). While this is quite coarse, they are the best publicly available data (to our knowledge) that cover onshore New Guinea. These data were used to produce a grid with a minimum curvature interpolation.

The UCR filter (see Sect. 2.2) (Jacobsen, 1987; Morse, 2010) was also applied to the Bouguer data set. In the case of the UCR Bouguer gravity data, new UCR grids were generated by subtracting the original Bouguer gravity grid upward continued X kilometres above the surface, from the same grid upward continued Y kilometres above the surface (i.e. grid(Y) minus grid(X)) (Figs. 7–8). The X and Y heights for the UCR grids that were generated are shown in Table 1.

2.4 Additional image processing of magnetic and gravity data prior to structural interpretation

The magnetic and gravity data sets were loaded in ER Mapper 7.1 before any structural interpretation began. Three duplicate gravity or magnetic images were draped on top of one another. The base layer was used to display the "intensity" of the magnetic or gravity signal. The intensity layer was illuminated with a light source from the north (000°), northeast (045°), east (090°) and southeast (135°) with an inclination angle of 30–40° so that we could highlight different orientations of geological structures, where a new image file was generated for each sun angle. The middle and highest layers are duplicate images shaded with a gradient fill. The only difference between the two layers is that the highest layer was made partially transparent to better enhance the data in shadowed areas.

2.5 Seismic Tomography

Seismic tomography has proven a useful tool to map geological structures in the lower crust and mantle. We used the global P wave seismic tomography data set of Li et al. (2008) and the global S wave seismic tomographic data set of Debayle et al. (2005) to determine if any arc-normal lineaments were imaged with tomography data and therefore major crustal structures since these structures reportedly focus metal-bearing fluids from such depths (Corbett, 1994; Kendrick et al., 1995; Hill et al., 1996, 2002). The tomographic data were imported as a grid of points into ArcGIS 10.0. Tomograms were produced by applying a spline interpolation to the point data for different depth slices (68, 113 and 203 km) (Fig. 11). The data were also contoured at discrete intervals to aid the identification of any linear structural features (Fig. 11). The spatial resolution of this data set is estimated to be ~ 100 km in the best sampled regions of the upper mantle (Li et al., 2008). The station coverage is relatively good across New Guinea and northern Australia (cf. Figure 1 of Li et al., 2008), but this coverage is less dense than other parts of the world, so the spatial resolution is likely to be greater than 100 km. This means that the tomography data will not be useful in identifying a particular fault or lineament in New Guinea, but they should still be useful in identifying if there are any major discontinuities in the lower crust and upper mantle.

3 Results

3.1 Topographic data

Our examination of the GMRT bathymetric and topographic data did not identify any regional NE to N striking lineaments in northern Australia, the Arafura Sea or in New Guinea (Fig. 3). We were also unable to identify any clear NE to N striking lineaments using SRTM data when using all of the data, or with selected contours as advocated by Gow and Walshe (2005). Our regional comparison is shown Fig. 4. Gow and Walsh (2005) suggested that NNE-trending lineaments could be readily identified when the 10, 100 and 1000 m topographic contours were drawn (Fig. 4a), but this is not evident when the same contours are drawn using the SRTM data (Fig. 4b). This means that the contours that were used by Gow and Walshe (2005) were smoothed or were derived from topographic data with a much lower spatial resolution. Considering these points, we concluded that topographic data are not the best data to use to discriminate arc-normal lineaments. While topographic data do have a use in identifying structural features, we did not use them any further in this study as topographic data cannot be used to determine if arc-normal structures are deep-level features that could focus metalliferous fluids.

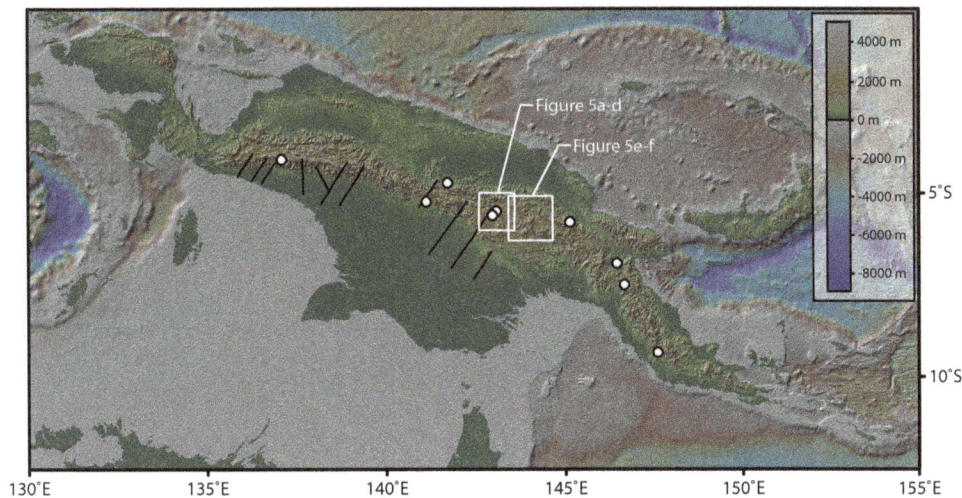

Fig. 3. A topographic and bathymetric map of New Guinea and the Arafura Sea using the Global Multi-Resolution Topography (GMRT) data set (Ryan et al., 2009) shows that there are no pronounced NNE striking lineaments. The location of New Guinea's copper and gold deposits (white circles) and the NNE-trending lineaments (thin black lines) that were proposed by Hill (1991) and Kendrick et al. (1995) are also shown.

3.2 Magnetics

The regional magnetic data UCR grid shows that the original regional magnetic data contain many subtle N–S- to NW–SE-oriented lineaments (Fig. 5a–b). We are unsure as to whether these lineaments represent orthogonal/polygonal faults (Fig. 5a–b), chevron-type folds (Fig. 5b), if they reflect a topographic feature, or if these do not represent geological features, and are a data artefact. As these lineaments are not visible in the lower-frequency UCR filtered data (Fig. 5c–d), it indicates that they do not occur deep in the crust.

We also observed several subtle NE striking magnetic features in the region directly east of Porgera (Fig. 5e). These lineaments were also not imaged in the lower-frequency UCR grids, indicating that they are not deep structures (Fig. 5f). We suspect that these lineaments could represent shallow dykes, veins or fractures given that they were only observed around volcanic centres (Fig. 5e).

Although there are possible NNE striking faults and dykes, these are subtle structural features. We envisaged that the lineament that was proposed to intersect the Mt Kare and Porgera gold deposits (Corbett, 1994; Kendrick et al., 1995) would be more pronounced in magnetic data. However, these data indicate the Mt Kare–Porgera lineament is not a magnetic feature, or that this particular lineament does not exist (Fig. 5a).

3.3 Gravity

The free-air and Bouguer gravity data indicate that NNE-trending lineaments occur in various regions of New Guinea (Fig. 6a–b). Yet these images also show that there are many other lineaments with different orientations. This is partic-ularly the case along the northern margin of Australia, the Arafura Sea and southern New Guinea where N–S-oriented lineaments are more pronounced than lineaments with other orientations (Figs. 1d, 6a–b).

The arc-normal faults that were proposed by Hill (1991) and Kendrick et al. (1995) were overlain on the free-air and Bouguer gravity data. This shows that some of the previously proposed NNE-trending lineaments correspond with sharp boundaries in the gravity data, yet many of the proposed lineaments do not (Fig. 6a–b).

While there are some lineaments that are apparent in all of the gravity grids – for example, the N–S- to NE–SW-oriented structures that cut across the Arafura Sea into southern New Guinea (Fig. 6) – we found that it was much easier to identify lineaments in the UCR filtered gravity grids relative to the free-air and Bouguer gravity data. The UCR filtered images highlight distinct edges of gravity anomalies, many of which could be interpreted as lineaments, and indeed arc-normal faults (Figs. 7–8). Some of these distinct breaks in gravity anomalies correspond with the lineaments proposed by earlier workers (e.g. Hill, 1991; Kendrick et al., 1995). Some of the distinct changes in gravity could also be interpreted as faults, as they show an apparent sense of displacement (Fig. 9).

3.4 A revised regional lineament map from the interpretation of gravity data

An updated lineament map of New Guinea was produced by examining each of the UCR filtered magnetic and gravity grids (Fig. 10). These grids indicate that NNE-trending lineaments do occur in New Guinea, particularly in central and northern parts of the island. However, these structures are not

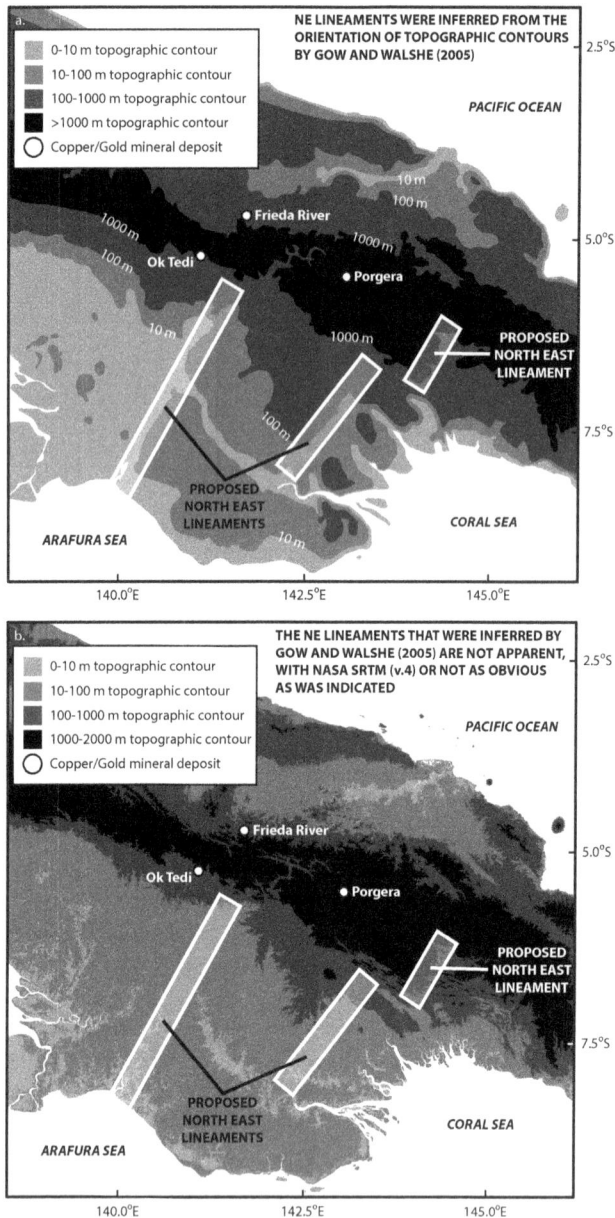

Fig. 4. (a) Gow and Walshe (2005) examined the topography of Papua New Guinea at three elevation contours (10, 100 and 1000 m). From these, they saw that there were several pronounced NE striking trends. **(b)** However, these features are not observed when the same elevation contour intervals are generated using the NASA Shuttle Radar Topographic Mission (SRTM) (version 4) data.

Table 2. List of Cu-Au, Cu-Mo, Au and Au-Ag deposits in New Guinea, their age and whether a NNE-trending structure that was identifiable in gravity and magnetic data dissects the region of the deposit. The deposit ages were taken from Garwin et al. (2005).

Deposit	Intersecting lineament	Deposit age
Ok Tedi	Yes	1.2–1.1 Ma
Star Mountain	Yes	1.6 Ma
Grasberg	Possibly	3.3–2.7 Ma
Edie Creek	Yes	3.8–2.4 Ma
Hidden Valley	Yes	4.2 Ma
Mt Bini	Yes	4.4 Ma
Mt Kare	No	5.5 Ma
Porgera	No	5.9 Ma
Yandera	No	7.0 Ma
Wamum	No	11.0 Ma
Frieda River	No	12.0 Ma
Nena	No	12.0 Ma
Wafi-Golpu	No	14.0 Ma

structures of the Australian lithosphere have been folded or rotated into their current orientation. Our structural interpretation also indicates that only some of the mineral deposits overlie (or are at least proximal to) an arc-normal lineament (Fig. 10/Table 2). These include Star Mountain, Ok Tedi, Hidden Valley, Edie Creek, Mt Bini and possibly Grasberg. It is interesting to note that these deposits are relatively young (< 4.5 Ma) and the deposits that are older than 4.5 Ma do not correspond with arc-normal structures (Table 2).

3.5   Seismic tomography

The P wave seismic tomography grids show that there is a NE–SW- to N–S-oriented boundary between what we interpret to represent cold Australian lithosphere (Fig. 11). This boundary is most pronounced in the 113 and 203 km depth tomograms (Fig. 11b–f), but it is also visible in the 68 km depth slice. The tomographic feature is also imaged at the same depths in the global S wave tomographic data set of Debayle et al. (2005) (Supplement file 2). We are confident that this tomographic anomaly represents a lithological/mantle structure as it is imaged in multiple tomographic data sets.

When our revised lineament interpretation (Fig. 10) is overlain on the 68, 113 and 203 km tomograms (Fig. 11b, d, f), it clearly shows that several NNE-trending lineaments follow the western edge of the tomographic anomaly (Fig. 11b, d, f). It is therefore possible that the geometry of this mantle structure could control the orientation of faults/lineaments observed at, or near, the surface.

as laterally extensive as earlier workers proposed. The geophysical data also indicate that there are multiple orientations of fabrics in the northern Australian lithosphere, and this conflicts with the idea that there is a pervasive NNE fabric between northern Australia and New Guinea (e.g. Fig. 1c). This indicates that the lineaments observed in New Guinea are not related to those in northern Australia, or that the older

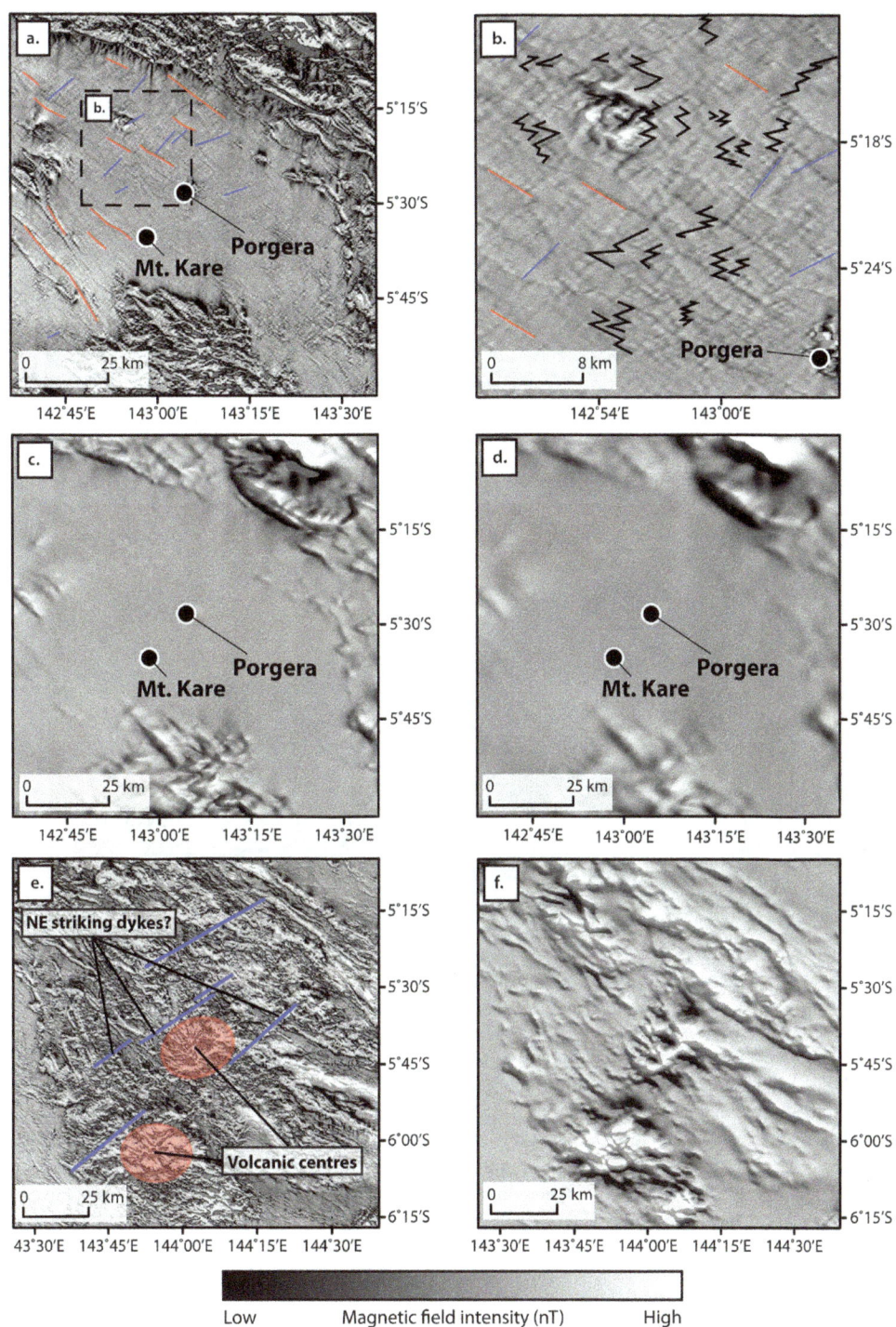

Fig. 5. An upward continuation residual (UCR) filter was applied to regional aeromagnetic data to determine if northeast-trending lineaments could be observed near the Mt Kare and Porgera Au deposits. While NE–SW and NW–SE striking lineaments are observed (**a, b, e**) in the images where the 1 km upward continuation grid was subtracted from the original data, the same structures were not observed in the longer-frequency ("deeper") UCR grids, where the 5 km upward continuation grid was subtracted from the 1 km (**c** and **f**) and 2 km (**d**) upward continuation grids. This indicates that the northeast-trending lineaments are shallow structures. Images (**a–d**) were generated with an illumination angle from the east. Images (**e–f**) were generated with an illumination angle from the northeast.

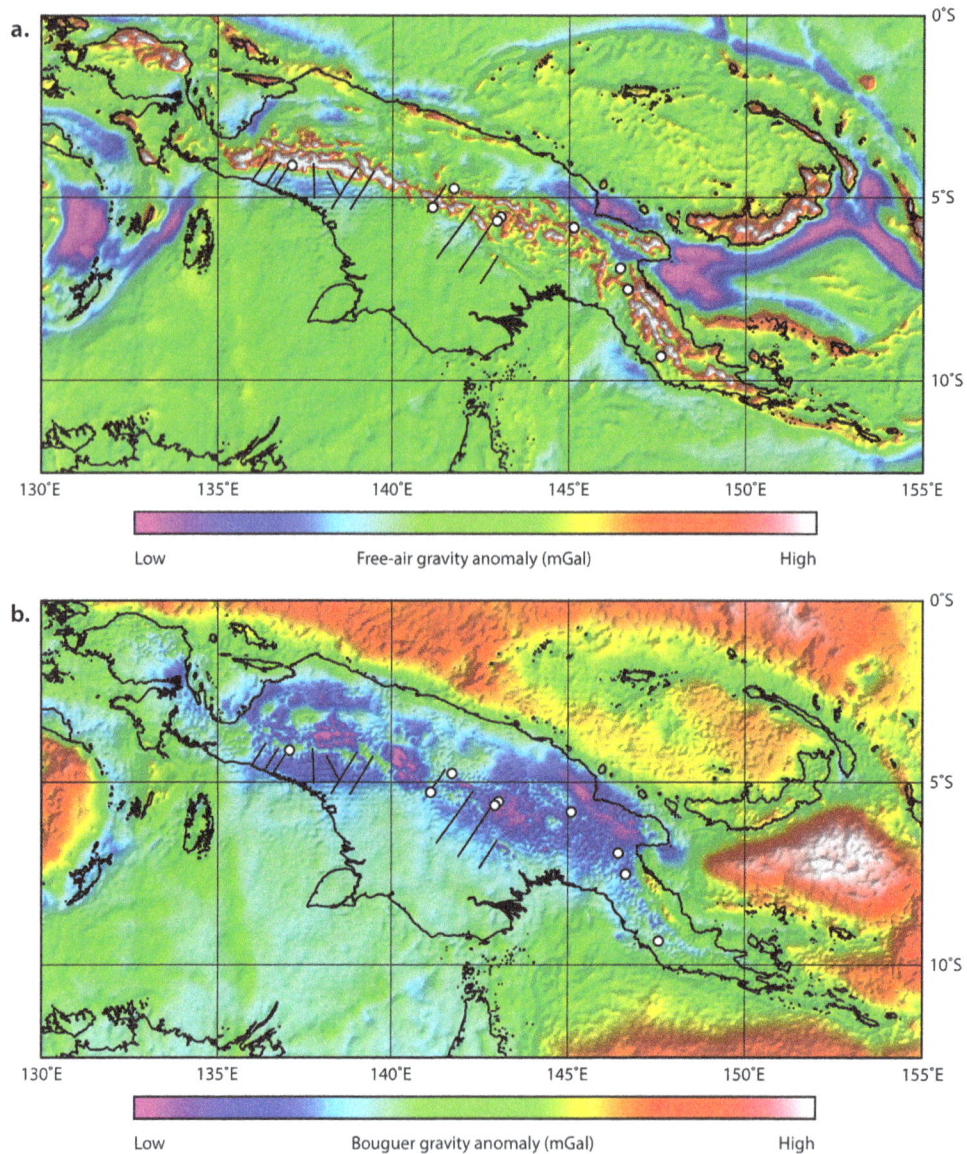

Fig. 6. (a) Free-air and **(b)** Bouguer gravity data sets obtained from the Bureau Gravimetrique Internationale (BGI) show that there are several major N–S- and NE–SW-trending structures cutting across the Arafura Sea. These data also indicate that the majority of structures in southern New Guinea trend N–S. However, there are some NNE–SSW-trending structures in central and northern New Guinea. This indicates that if any NNE–SSW striking faults occur in New Guinea, they are perhaps unrelated to the inherited fabric of the Australian lithosphere. The light source in these images is from the southeast to highlight NE-trending structures. The same images were generated with different illumination angles to interpret lineaments with other orientations. The location of major Au/Cu/Ag/Mo deposits is shown as white circles (See Fig. 1a for more details). The NNE-trending lineaments that were proposed by Hill (1991) and Kendrick et al. (1995) are shown as thin black lines.

4   Discussion

4.1   Structural interpretation of geophysical data

The application of a UCR filter to gravity and magnetic data were useful for interpreting lineaments when compared to using the original free-air gravity, Bouguer gravity and reduced-to-pole magnetic data. The use of different sun angle orientations to highlight lineaments of different orientations was an essential part of the process. This is because regional structural interpretations of geophysical images where only one sun angle orientation is used will bias which structures are drawn/interpreted.

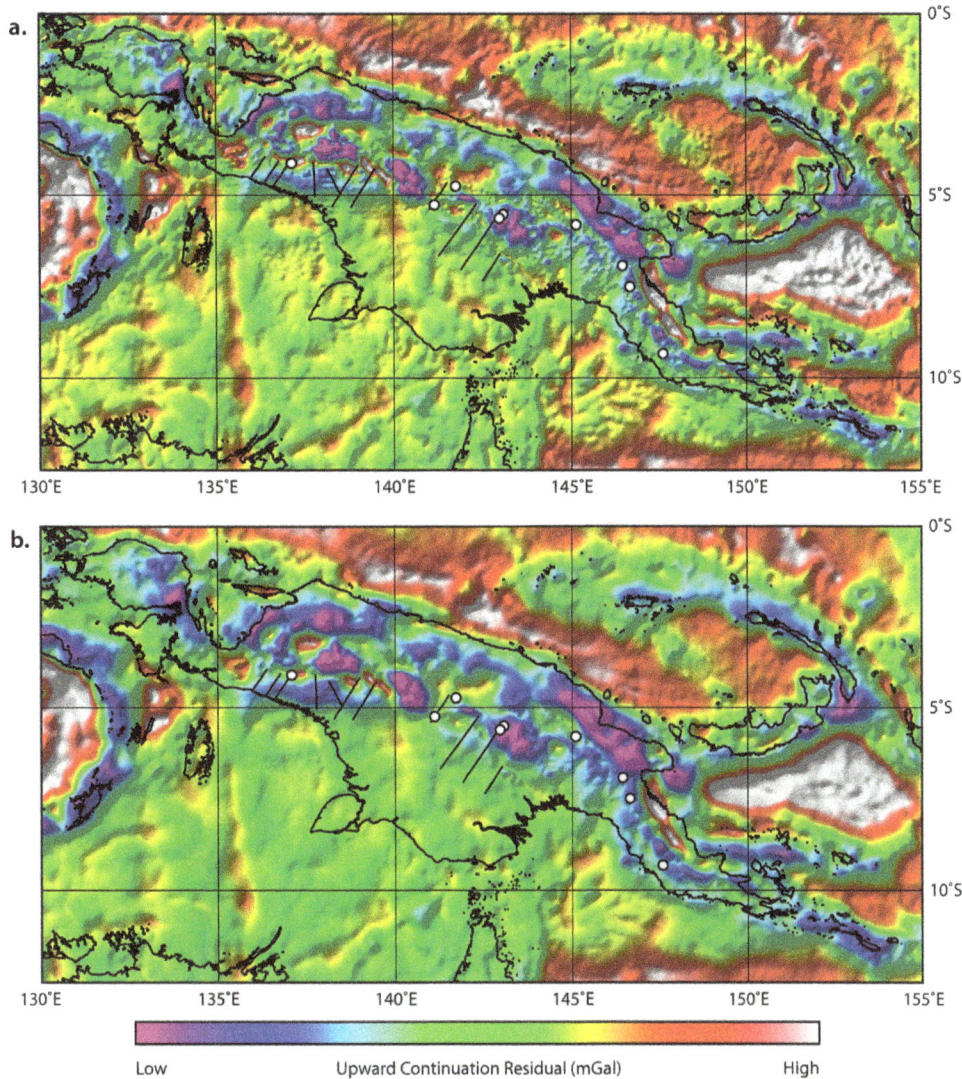

Fig. 7. Upwards continuation residual (UCR) filter applied to the BGI Bouguer gravity data. The Bouguer gravity data were calculated as if they were collected by a measurement at 5, 10, 40 and 50 km a.s.l., and new grids were produced by (**a**) subtracting the 40 km grid from the 5 km grid, and (**b**) subtracting the 50 km grid from the 10 km grid. The illumination direction that was used in (**a**) and (**b**) is from the southeast. This was used to highlight any NNE-trending structures. The same grids were generated with different illumination angles to interpret lineaments with other orientations. The location of major Au/Cu/Ag/Mo deposits is shown as white circles (See Fig. 1a for more details). The NNE-trending lineaments that were proposed by Hill (1991) and Kendrick et al. (1995) are shown as thin black lines.

While the geophysical data do not prove that a lineament is a geological structure (e.g. a fault), we are confident that many of the interpreted lineaments are geological structures as some of the gravity anomalies are offset (e.g. Fig. 9). However, it is also likely that some of the lineaments that were interpreted could reflect features that are not geological structures. Such issues will be refined through time as new, higher-resolution data become available.

4.2 Arc-normal faults and their control on mineralization

Our aim was to present a methodical and systematic method of mapping lineaments in a way that the reader could clearly evaluate our work and come to their own judgement as to whether arc-normal faults exist in New Guinea. We were initially skeptical about the existence of any arc-normal transfer structures in New Guinea as the evidence that was proposed in support of these structures was often not shown, or was not clearly presented. However, our analysis of various geophysical data do indeed support that some arc-normal lineaments occur along the length of the island (Fig. 10).

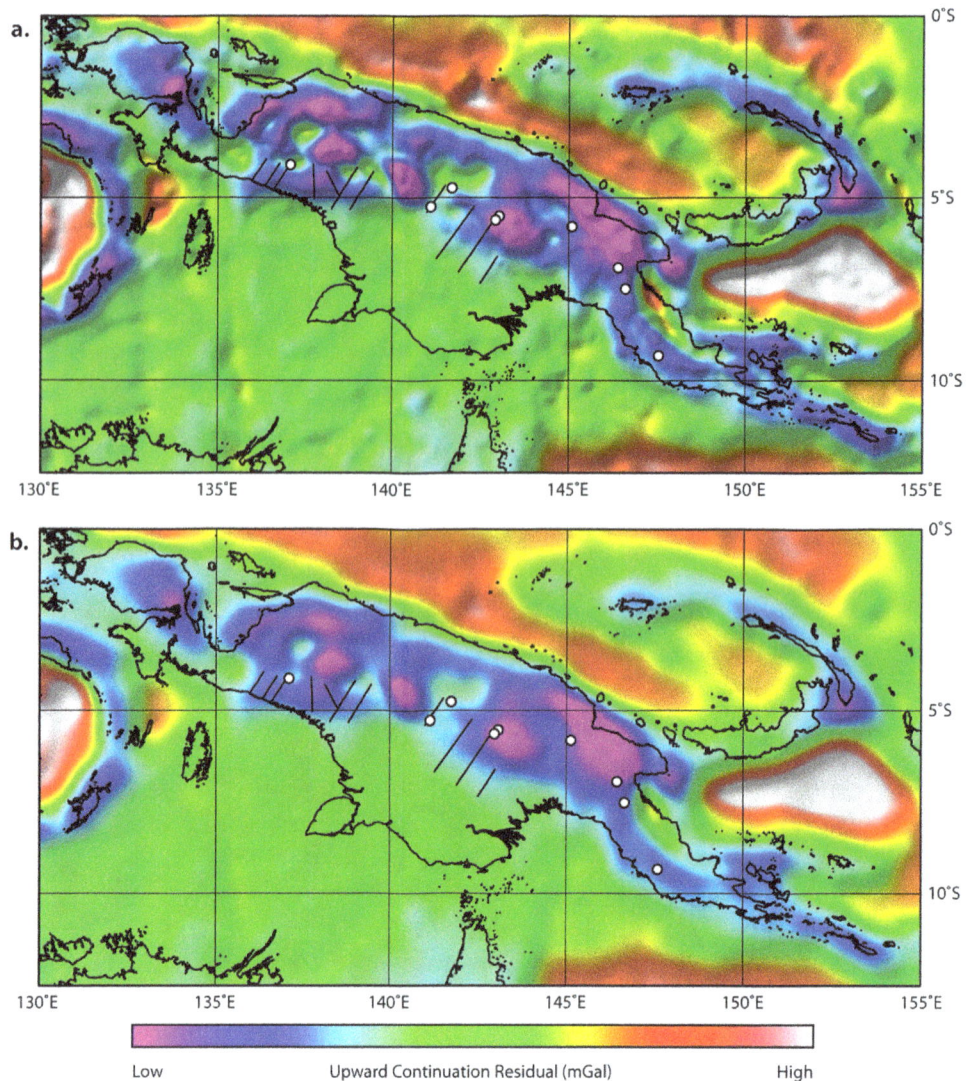

Fig. 8. Upwards continuation residual (UCR) filter applied to the BGI Bouguer gravity data. The Bouguer gravity data were calculated as if they were collected by a measurement at 20, 40, 80 and 100 km a.s.l., and new grids were produced by: **(a)** subtracting the 80 km grid from the 20 km grid, and **(b)** subtracting the 100 km grid from the 40 km grid. The illumination direction that was used in **(a)** and **(b)** is from the southeast. This was used to highlight any NNE-trending structures. The same grids were generated with different illumination angles to interpret lineaments with other orientations. The location of major Au/Cu/Ag/Mo deposits is shown as white circles (See Fig. 1a for more details). The NNE-trending lineaments that were proposed by Hill (1991) and Kendrick et al. (1995) are shown as thin black lines.

In regards to timing, Hill (1991) proposed that the arc-normal transfer faults first formed as lateral ramps along with arc-parallel normal faults during a phase of extension during the Late Cretaceous–Palaeocene (100–56 Ma). The arc-parallel normal faults were later inverted during a phase of shortening in the Late Miocene to Pliocene (∼ 11–2.5 Ma), and the arc-normal lateral ramps were reactivated as strike-slip faults/transfer zones at the same time. This deformation is marked by the offset of the Eocene-middle Miocene Darai Limestone (Hill 1991; Davies 2012). This time limit on the movement of the transfer faults is interesting considering that we found that the arc-normal structures only intersect the de-posits that are less than 5 million years old (Fig. 10 and Table 2).

4.3 Inherited basement fabric

The orientation of the arc-normal lateral ramps (otherwise known as accommodation, or transfer, zones; cf. Gawthorpe and Hurst, 1993) was said to be controlled by a Proterozoic–Palaeozoic fabric that was the same orientation along northern Australia, the Arafura Sea and New Guinea (Hill 1991) (Fig. 1b–c). Accommodation zones are common in various extensional basins around the world (e.g. Gawthorpe and Hurst 1993; Longley et al., 2002; McClay et al., 2002);

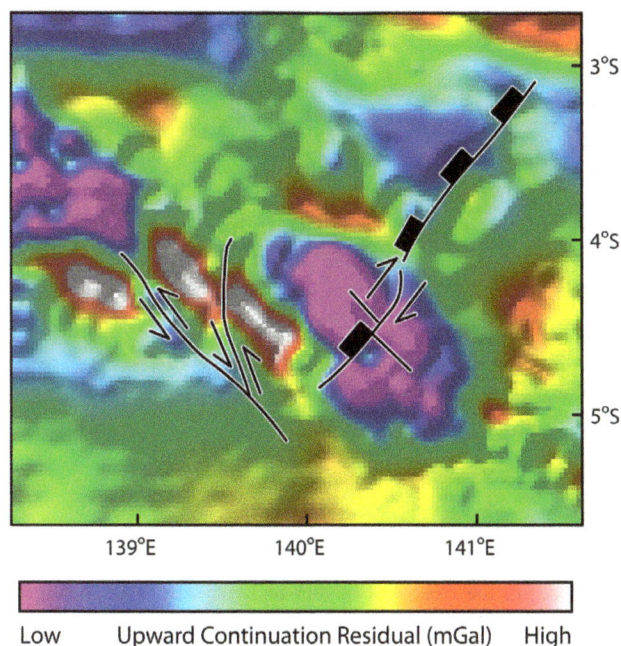

Fig. 9. There were several cases where we could identify lineaments where we could infer an apparent sense of displacement. For example, we show that there may have been some left-lateral displacement of gravity highs (red–white colours), as well as a possible NE-trending extensional fault with a right-lateral sense of movement. The grid that is shown here was generated by subtracting the 40 km upward continuation (UC) from the 5 km UC Bouguer grid. The grid is illuminated from the southeast.

however, the mechanism for their formation is disputed. One model for their formation proposes that these are associated with strike-slip or oblique-slip transfer faults (cf. Gibbs 1983, 1984; Lister et al., 1986); the other model proposes that these are zones of accommodation of distributed faulting (cf. Bosworth 1985; Rosendahl et al., 1986; Morley et al., 1990; McClay et al., 2002). Considering (1) that we found no link between what we could identify as deeper-level basement structures in New Guinea and (2) that analogue modelling of rift systems has shown that accommodation zones oriented obliquely to perpendicular to the rift axis are not necessarily related to hard-linked strike-slip basement faults (McClay et al., 2002), it is thus worth considering that the lateral ramps proposed by Hill (1991) for the central Papuan Fold Belt (and beneath the Porgera and Mt Kare gold deposits) are not associated with deeper-level faults and/or an inherited basement fabric. However, if this is the case, it is likely that there is much more to understand about the structural history of the region as there is some apparent NE–SW strike-slip offset of units in this area.

In addition, our analysis of the trend of the basement fabric(s) in this region shows that there is much more variation in the orientation of these structures (Fig. 1d and 10) than what others had proposed earlier (Fig. 1b–c). This means (1)

that the arc-normal structures are not related to the dominant structural grain of the Australian lithosphere; (2) that these structures are related to the dominant structural grain of the Australian lithosphere, but have been folded into their current orientation in New Guinea; (3) that these faults are inherited from the Australian lithosphere, but the orientation of the structural grain was always different in New Guinea and Australia; or (4) that the arc-normal structures are unrelated to any existing fabric and are associated with more recent orogenic events. Further work is required to resolve these points.

4.4 Penetration depth of arc-normal faults

As the arc-normal lineaments were said to direct metalliferous fluids from deep in the crust, it was important to determine if these were deep faults. This is why we used various UCR filters on magnetic and gravity data sets, and why we looked at seismic tomography data. NNE-trending lineaments were only observed in the higher-frequency UCR filtered magnetic data, indicating that these particular structures are shallow features, or are not magnetic. The UCR filtered gravity data indicate that arc-normal lineaments are also mappable in the high- and low-frequency grids, suggesting that many of the arc-normal lineaments are significant structures that cut through deeper levels of the crust. This idea is supported by seismic tomography data that show that the arc-normal structures might sole into deeper levels of the lithosphere and controlled by deeper lithospheric faults (Fig. 11). This indicates that the arc-normal faults could potentially tap mantle melts and fluids, and this is supported further by the field observations and geochemical data from chains of mantle-derived melts and volcanoes (e.g. Davies 1990, 1991). It is therefore possible that the arc-normal faults control the distribution of metals in New Guinea, but considering our other findings, perhaps only those that are less than 5 million years old.

4.5 Exploration under cover – mapping the intersection of intrusions and faults

One of the biggest challenges faced by mineral explorers today is the concept of finding new deposits that do not crop out at the surface. This is especially the case in regions that have very thick soil profiles or within heavily forested regions, such as in New Guinea. Gravity data have long been used as a tool for mineral explorers to infer information about the composition of the basement, where zones of low gravity typically reflect sedimentary basins or felsic intrusions. So we produced a map of the low-gravity zones that were identified in the UCR filtered Bouguer gravity data in the hope that these zones represent zones of felsic to intermediate intrusions – the idea being that explorers might be able to focus their efforts on these regions as they could represent zones where hot metal-bearing fluids and gases associated

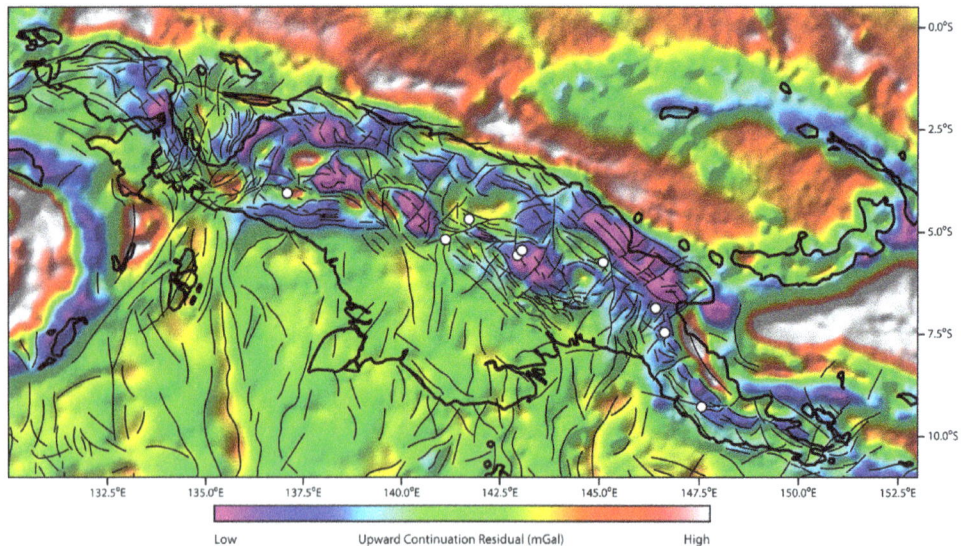

Fig. 10. A new lineament map was generated by interpreting structural features in the free-air, Bouguer and the various UCR filtered gravity data sets. We also compared these interpretations with structural information that was interpreted from the Global Multi-Resolution Topographic data set (Ryan et al., 2009). Here we show the lineament map and the location of major Au/Cu/Ag/Mo deposits draped over the 10 km-minus-50 km UCR filtered gravity grid (illuminated from the southeast). This figure shows not only that the location of some of the Au/Cu/Ag/Mo deposits (white circles) correlated with identifiable lineaments (transfer faults?), but it also shows that not every deposit is linked with an arc-normal lineament.

with magmatism were once deposited. To examine this, we used the 10–50 km UCR grid to ensure that we were effectively looking at the gravity signal from a depth of approximately 5 km, in the hope that the zones of lower gravity reflected igneous, rather than sedimentary, rocks (Fig. 12a). This indicates that there are several areas that might represent future exploration targets (Fig. 12a). Explorers might be able to further concentrate their efforts on the low-gravity zones that are intersected by faults, or where faults occur above the gravity anomalies as the structures could act as conduits for fluids and gases, and thus zones where precious metals might come out of solution.

However, there are several problems associated with this approach: (1) the resolution of the gravity data that were available to us is quite coarse (4–5 km), and (2) we do not have much in the way of geological control that could tell us more about the composition of the gravity anomalies. The first problem will only be resolved once more onshore gravity surveys are conducted on New Guinea. We addressed the second problem by comparing the position of low-gravity zones with the Coordinating Committee for Geoscience Programmes in East and Southeast Asia (CCOP) 1:2 000 000 geological map of SE Asia (downloaded from http://www. orrbodies.com/resources/item/orr0052), noting the location of felsic intrusives and extrusives and other lithologies at the surface. We then highlighted the areas where the felsic intrusives and extrusives intersected, or were near to the regions of low gravity. While we understand that this cannot be ad-

equately tested here, we considered it useful to include this approach, as well as a map so that this might be tested further in the future.

5 Conclusions

Our interpretation of the several different geophysical data types indicates that arc-normal lineaments do occur in New Guinea, but these structures are not as prevalent or as simple as they have previously been portrayed. Our approach in using these different data sets means that our interpretation covered some three orders of magnitude in terms of scale (topographic and magnetic data at ~ 100 m (10^2); gravity data at 5 to 10 km ($\sim 10^3$–10^4); and tomography data at 100 km (10^5)). The lineaments were not identified in the highest-resolution data and data representing the shallowest levels of the crust (e.g. topographic and magnetic data); however, there was evidence of arc-normal lineaments in the lower-resolution data that imaged deeper levels of the crust (gravity and tomography). Our results show that these lineaments do not intersect every New Guinea gold and copper deposit. Instead, it is only the deposits that are less than 5 million years old that might be associated with arc-normal structural features. These findings mean that exploration geologists might be able to better target which host rocks to investigate for mineralization and that new models are required to explain the genesis of the > 5 Ma deposits.

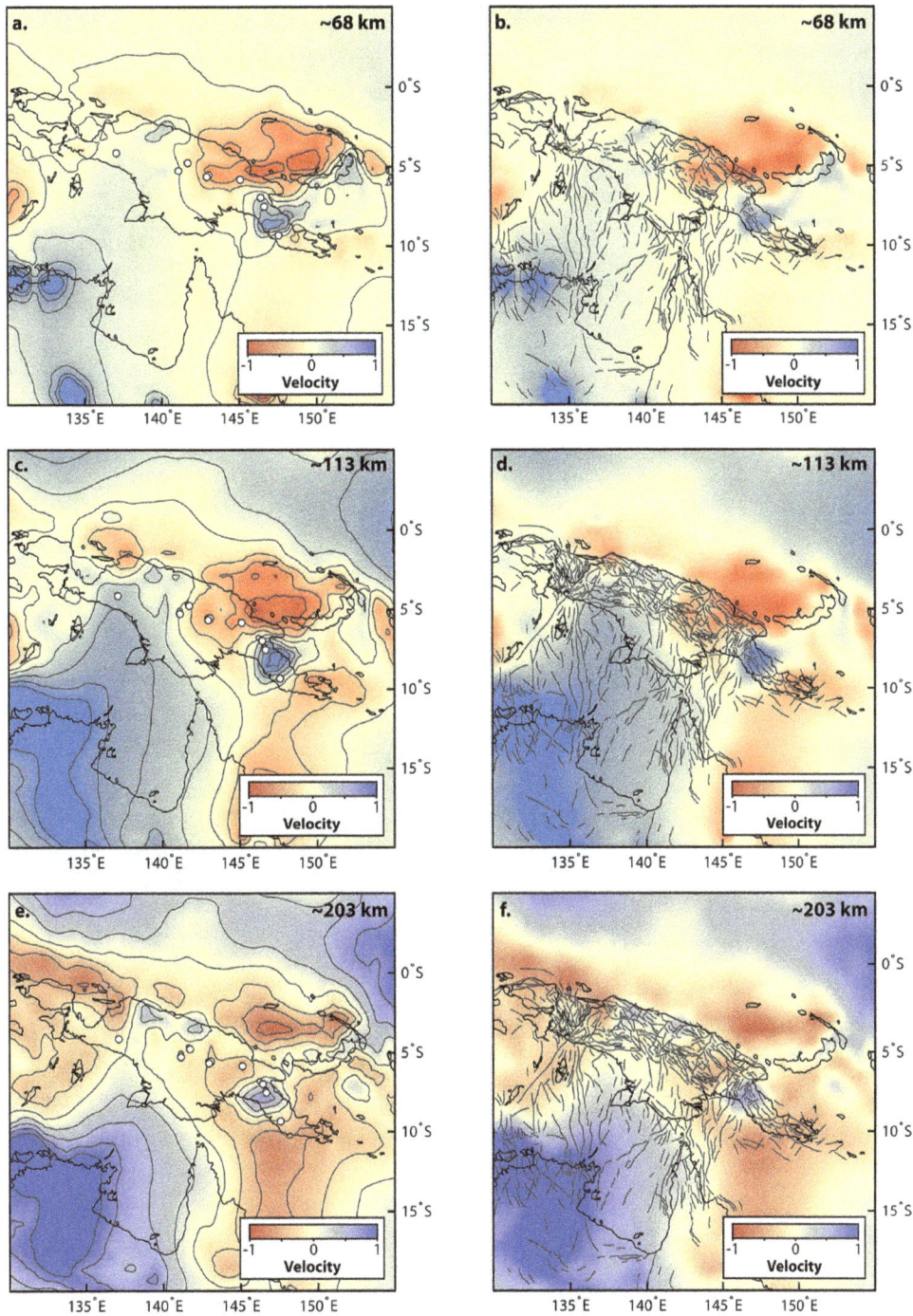

Fig. 11. Seismic tomograms were generated for the New Guinea region from a global P wave tomography data set (Li et al., 2008). Here we show slices at different depths: (**a–b**) 68 km; (**c–d**) 113 km; and (**e–f**) 203 km. The images that are shown in the left-hand column (**a, c, e**) have been contoured and also show the location of the Au/Cu/Ag/Mo deposits (white circles) that are described in Fig. 1a. The images that are shown in the right-hand column (**b, d, f**) compare the tomography data with the lineament interpretation that is shown in Fig. 9. There is some similarity between the orientation of some lineaments and the tomographic "anomalies". These tomography data also indicate that there is a keel of cold Australian lithosphere below much of western New Guinea. This possibly represents a rigid indentor. The Grasberg, Frieda River, Nena, Star Mountain and Ok Tedi deposits are located at the margins of this tomographic feature. This might indicate that the edges of the keel have influenced the location of faults in crust, and therefore helped to concentrate the deposition of metalliferous fluids/magmas.

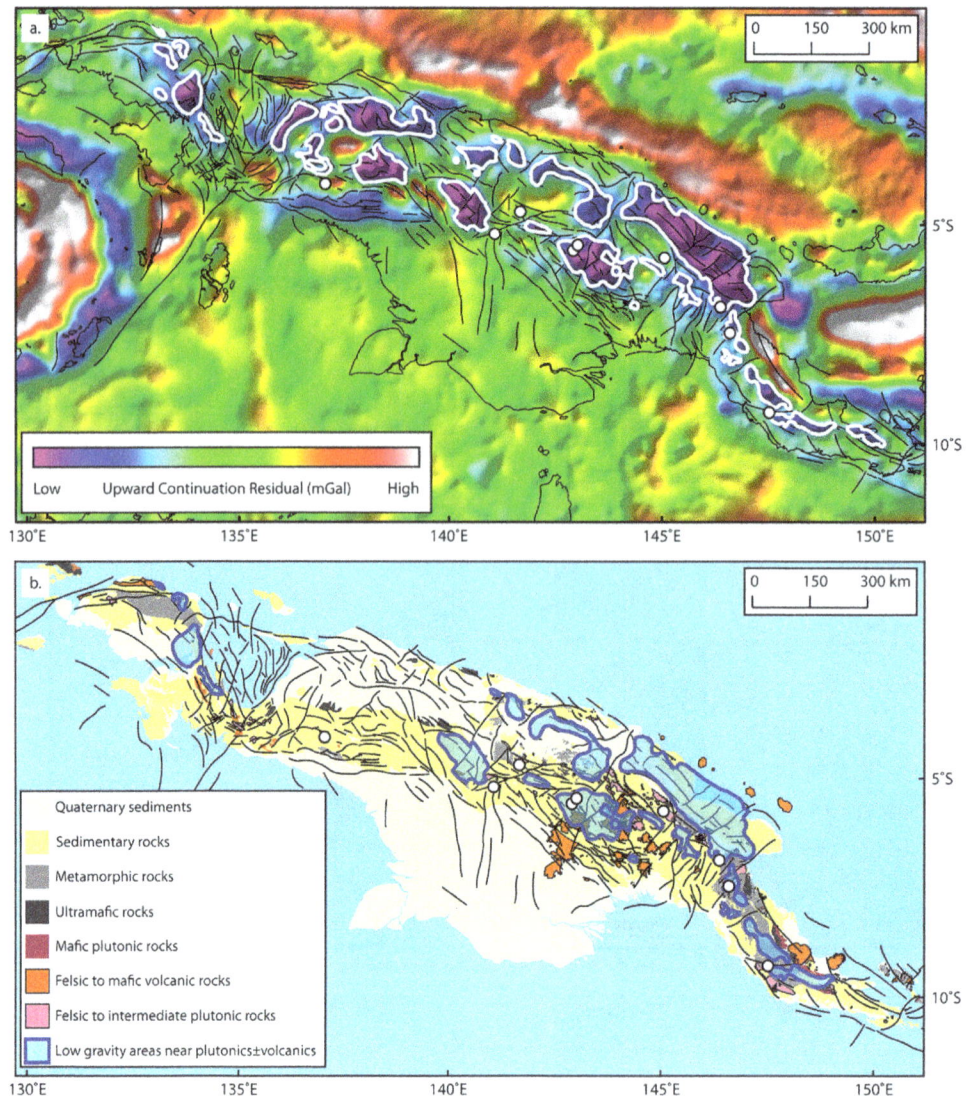

Fig. 12. (**a**) The 10 km-minus-50 km UCR filtered grid was used to delineate zones of low gravity (white rimmed polygons) in the hope that these regions represented plutons or volcanic feeder zones at depth. These zones of low gravity were compared to (**b**) the lithologies that are shown on the CCOP 1:2 000 000 geological map of SE Asia. The low-gravity zones that were highlighted in (**a**) that did not correspond with regions of volcanic rocks or felsic to intermediate plutons rocks were deleted to better highlight the regions where plutonic/sub-volcanic rocks might be found at depth, and therefore areas where metalliferous fluids may have been deposited above. The regions that are intersected by arc-normal faults might add further control to where any metalliferous fluids are concentrated.

Acknowledgements. This work was funded by a consortium of mineral exploration companies. L. T. White is also grateful for financial support provided by a consortium of hydrocarbon companies that sponsor the Southeast Asia Research Group at Royal Holloway University of London. The authors thank S. Occhipinti, D. Sewall, C. Swager, G. Cant, M. Roberts, M. Lindsay, N. Hayward, M. Armstrong, M. Cawood, T. Craske, J. Sinclair, S. Cox, D. Tanner and various people at Geoscience Australia for discussing and commenting on various aspects of this work. The authors would also like to thank Hugh Davies and Kevin Hill for reviewing the paper and the Papua New Guinea Mineral Resources Authority for allowing us to publish images that were generated from aero-magnetic data sets of PNG that were collected by Fugro on behalf of the World Bank (http://www.mra.gov.pg/Investors/AeromagneticDatasetsArea12,PNG.aspx).

Edited by: C. Gaina

References

Amante, C. and Eakins, B. W.: ETOPO1 1 Arc-Minute Global Relief Model: Procedures, Data Sources and Analysis, NOAA Technical Memorandum NESDIS NGDC NGDC-24, 19 pp., 2009.

A. P. C. (The Australasian Petroleum Company Proprietary): The geological results of petroleum exploration in western Papua, Journal of the Geological Society of Australia, 8, 1–133, 1961.

Bosworth, W.: Geometry of propagating continental rifts, Nature, 316, 625–627, 1985.

Corbett, G. J.: Regional structural control of selected Cu/Au occurrences in Papua New Guinea, in: Proceedings of the Papua New Guinea Geology, Exploration and Mining Conference: Melbourne, edited by: Rogerson, R., Australasian Institute of Mining and Metallurgy, 57–70, 1994.

Davies, H. L.: Structure and evolution of the border region of New Guinea, in: Petroleum Exploration in Papua New Guinea, Proceedings of the First PNG Petroleum Convention, Port Moresby, edited by: Carmen, G. J. and Carmen, Z., Papua New Guinea Chamber of Mines and Petroleum, Port Moresby, 245–269, 1990.

Davies, H. L.: Tectonic Setting of Some Mineral Deposits of the New Guinea Region, AusIMM PNG Geology, Exploration and Mining Conference, 49–57, 1991.

Davies, H. L.: The geology of New Guinea – the cordilleran margin of the Australian continent, Episodes, 35, 87–102, 2012.

Debayle, E., Kennett, B., and Priestley, K.: Global azimuthal seismic anisotropy and the unique plate motion deformation of Australia, Nature, 433, 509–512, 2005.

Dekker, F., Balkwill, H., Slater, A., Herner, R., and Klampschuur, W. A.: Structural interpretation of the eastern Papuan Fold Belt, based on remote sensing and fieldwork, in: Petroleum Exploration in Papua New Guinea, Proceedings of the First PNG Petroleum Convention, Port Moresby, edited by: Carmen, G. J. and Carmen, Z., Papua New Guinea Chamber of Mines and Petroleum, Port Moresby, 319–336, 1990.

DeMets, C., Gordon, R., Argus, D., and Stein, S.: Current plate motions, Geophys. J. Int., 101, 425–478, 1990.

Elliot, C. I.: Australian Lineament Tectonics, unpublished Ph.D. thesis, University of Melbourne, 262 pp., 1994.

Fullea, J., Fernandez, M., and Zeyen, H.: FA2BOUG – a FORTRAN 90 code to compute Bouguer gravity anomalies from gridded free-air anomalies: application to the Atlantic Mediterranean transition zone, Comput. Geosci., 34, 1665–1681, 2008.

Garwin, S., Hall, R., and Watanabe, Y.: Tectonic setting, geology and gold and copper mineralization in Cenozoic magmatic arcs of Southeast Asia and the West Pacific, Econ. Geol., 100th Anniversary Volume, 891–930, 2005.

Gawthorpe, R. L. and Hurst J. M.: Transfer zones in extensional basins: their structural style and influence on drainage development and stratigraphy, J. Geol. Soc. Lond., 150, 1137–1152, 1993.

Gibbs, A. D.: Balanced section constructions from seismic sections in areas of extensional tectonics, J. Struct. Geol., 5, 153–160, 1983.

Gibbs, A. D.: Development of extension and mixed-mode sedimentary basins, in Coward, M. P., Dewey, J. F., and Hancock, P. L. (Eds.): Continental extensional tectonics, Geol. Soc. Special Publication, 28, 19–33, 1984.

Gow, P. and Walshe, J.: The role of pre-existing geologic architecture in the formation of giant porphyry-related Cu ± Au deposits: examples from New Guinea and Chile, Econ. Geol., 100, 819–833, 2005.

Hill, K. C.: Structure of the Papuan fold belt, Papua New Guinea, AAPG Bull., 75, 857–872, 1991.

Hill, K. C., Simpson, R. J., Kendrick, R. D., Crowhurst, P. V., O'Sullivan, P. B., and Saefudin, I.: Hydrocarbons in New Guinea, controlled by basement fabric, Mesozoic extension and Tertiary convergent margin tectonics, in: Petroleum Exploration, Development and Production in Papua New Guinea: Proceedings of the Third PNG Petroleum Convention, Port Moresby, 9–11 September 1996, edited by: Buchanan, P. G., 63–76, 1996.

Hill, K. C.: Tectonics, timing and economic deposits in Papua New Guinea (abstract), Australasian Institute of Mining and Metallurgy, Geology, Exploration and Mining Conference, Madang, 10–12 October, Abstracts, 233–234, 1997.

Hill, K., Kendrick, R., Crowhurst, P., and Gow, P.: Copper-gold mineralization in New Guinea: tectonics, lineaments, thermochronology and structure, Aust. J. Earth Sci., 49, 737–752, 2002.

Hill, K. C., Keetley, J. T., Kendrick, R. D., and Sutriyono, E.: Structure and hydrocarbon potential of the New Guinea Fold Belt, in: Thrust tectonics and hydrocarbon systems, edited by: K. R. McClay, AAPG Memoir, 82, 494–514, 2004.

Hill, K. C., Bradey, K., Iwanec, J., Wilson, N. and Lucas, K.: Structural exploration in the Papua New Guinea Fold Belt, in: Eastern Australian Basins Symposium III, edited by: Blevin, J. E., Bradshaw, B. E., Uruski, C., Petroleum Exploration Society of Australia, Special Publication, 225–238, 2008.

Hill, K. C., Lucas, K., and Bradley, K.: Structural styles in the Papuan Fold Belt, Papua New Guinea: constraints from analogue modelling, Geol. Soc. Lond., Special Publications, 348, 33–56, 2010.

Jacobsen, B. H.: A case for upward continuation as a standard separation filter for potential field maps, Geophysics, 52, 1138–1148, 1987.

Kendrick, R., Hill, K., Parris, K., Saefudin, I., and O'Sullivan, P.: Timing and style of Neogene regional deformation in the Irian Jaya Fold Belt, Indonesia, Proceedings of the Annual Convention – Indonesian Petroleum Association, 249–262, 1995.

Li, C., van der Hilst, R. D., Engdahl, E. R., and Burdick, S.: A new global model for P wave speed variations in Earth's mantle, Geochem. Geophy. Geosy., 9, Q05018, doi:10.1029/2007GC001806, 2008.

Lister, G. S., Etheridge, M. A., and Symonds, P. A.: Detachment faulting and the evolution of passive continental margins, Geology, 14, 246–250, 1986.

McClay, K. R., Dooley, T., Whitehouse, P., and Mills, M.: 4-D evolution of rift systems: Insights from scaled physical models, AAPG Bulletin, 86, 935–959, 2002.

Milligan, P.: New magnetic datasets to identify energy, geothermal and mineral resources, AusGeo News, September 2010, Issue 99, available at: http://www.ga.gov.au/ausgeonews/ausgeonews201009/magnetic.jsp (last access: 16 October 2013), 2010.

Morley, C.K., Nelson, R. A., Patton, T. L., and Munn, S. G.: Transfer zones in the East African rift system and their relevance to hydrocarbon exploration in rifts, AAPG Bulletin, 74, 1234–1253, 1990.

Morse, M.: Potential field methods prove effective for continental margin studies, AusGeo News, June 2010, Issue 98, http://www.ga.gov.au/ausgeonews/ausgeonews201006/geology.jsp (last access: 16 October 2013), 2010.

Pavlis, N. K., Holmes, S. A., Kenyon, S. C., and Factor, J. K.: An earth gravitational model to degree 2160: EGM2008, EGU General Assembly 2008, Vienna, Austria, 13–18 April, available at: http://earth-info.nga.mil/GandG/wgs84/gravitymod/egm2008 (last access: 16 October 2013), 2008.

Pavlis, N. K., Holmes, S. A., Kenyon, S. C., and Factor, J. K.: The development and evaluation of the Earth Gravitational Model 2008 (EGM2008), J. Geophys. Res., 117, B04406, doi:10.1029/2011JB008916, 2012.

Richards, J.: Lineaments revisited, Society of Economic Geologists Newsletter, 42, 14–20, 2000.

Rosendahl, B. R., Reynolds, D., Lorber, P., Burgess, C., McGill, J., Scott, D., Lambiase, J., and Derksen, S.: Structural expressions of rifting: lessons from Lake Tanganyika, in: Sedimentation in the East African rifts, edited by: Frostick, L. E., Renaut, R. W., Reid, I., and Tiercelin, J. J., Geol. Soc. Special Publication, 25, 29–43, 1986.

Ryan, W. B. F., Carbotte, S. M., Coplan, J. O., O'Hara, S., Melkonian, A., Arko, R., Weissel, R. A., Ferrini, V., Goodwillie, A., Nitsche, F., Bonczkowiski, J., and Zemsky, R.: Global multi-resolution topography synthesis, Geochem. Geophy. Geosy., 10, Q03014, doi:10.1029/2008GC002332, 2009.

Smith, R. I.: Tertiary Plate Tectonic Setting and Evolution of Papua New Guinea, in: Petroleum Exploration in Papua New Guinea, p. 229–244, edited by: Carmen, G. J. and Carmen, Z., Proceedings of the First PNG Petroleum Convention, Port Moresby, Papua New Guinea Chamber of Mines and Petroleum, Port Moresby, 1990.

5

The ring-shaped thermal field of Stefanos crater, Nisyros Island: a conceptual model

M. Pantaleo and T. R. Walter

Department 2, Physics of the Earth, Helmholtz Centre Potsdam, GFZ German Research Centre for Geoscience, Potsdam 14473, Germany

Correspondence to: M. Pantaleo (pantal@gfz-potsdam.de)

Abstract. Fumarole fields related to hydrothermal processes release the heat of the underground through permeable pathways. Thermal changes, therefore, are likely to depend also on the size and permeability variation of these pathways. There may be different explanations for the observed permeability changes, such as fault control, lithology, weathering/alteration, heterogeneous sediment accumulation/erosion and physical changes of the fluids (e.g., temperature and viscosity). A common difficulty, however, in surface temperature field studies at active volcanoes is that the parameters controlling the ascending routes of fluids are poorly constrained in general. Here we analyze the crater of Stefanos, Nisyros (Greece), and highlight complexities in the spatial pattern of the fumarole field related to permeability conditions. We combine high-resolution infrared mosaics and grain-size analysis of soils, aiming to elaborate parameters controlling the appearance of the fumarole field. We find a ring-shaped thermal field located within the explosion crater, which we interpret to reflect near-surface contrasts of the soil granulometry and volcanotectonic history at depth. We develop a conceptual model of how the ring-shaped thermal field formed at the Stefanos crater and similarly at other volcanic edifices, highlighting the importance of local permeability contrast that may increase or decrease the thermal fluid flux.

1   Introduction

Thermal anomalies in volcanic areas may be detected before, during and long after eruptions, allowing assessment of precursors, of fluid flux and degassing intensity levels, and quantification of the volcanic heat discharge at the surface (Sekioka and Yuhara, 1974; Stevenson, 1993). Besides the magmatic/hydrothermal source itself, different factors may affect the expression and intensity of fumaroles. These may be, for instance, the stress field, the presence of faults and fractures, or the lithology (Mongillo and Wood, 1995; Dobson et al., 2003; Finizola et al., 2003; Revil et al., 2008; Schöpa et al., 2011; Peltier et al., 2012). Thermal anomalies can be detected at the surface by direct measurements and by satellite-based or hand-held infrared camera measurements (Bukumirovic et al., 1997; Harris and Maciejewski, 2000; Chiodini et al., 2007; Harris et al., 2009). These measurements alone cannot however explain which factors control the permeability complexities and the thermal expression. Therefore the question of how the thermal expression at volcanoes is affected by permeability complexities remains to be studied. Few case studies have investigated the effect of permeability on fumaroles (Mongillo and Wood, 1995; Schöpa et al., 2011; Peltier et al., 2012), highlighting the entanglement between the stress, the faults and the lithologies. Mongillo et al. (1995) showed that, at White Island, New Zealand, the tectonic structures control the site permeability at an edifice scale, whereas the lithology has a local influence on it. Schöpa et al. (2011) indicated that at Vulcano Island, Italy, the topography-induced stress field focuses the permeable pathways toward the morphological crests. At these sites both the position (e.g., inner or outer flank) and geometry (e.g., radial or concentric) are controlled by the lithology and the shallow fractures. Peltier et al. (2012) suggested that at the Yasur–Yenkahe complex, Vanuatu, the stratigraphic layering dictates the permeability setting together with faults and fractures. At a larger scale, the relation between porosity, permeability and fluid flow was studied at the Yellowstone caldera (Dobson et al., 2003). Here, results showed that sediments and non-welded tuffs have high permeability thanks to primary porosity. Other welded lithologies have even higher permeability because of fractures and veins, which represent the secondary porosity.

The reconnaissance of the permeability background becomes relevant to decipher the temporal and spatial variability of thermal fields as it also relates to unrest. Faults and fractures, in particular, control the permeability of the rock masses according to Darcy's cubic law (Caine et al., 1996; Faulkner et al., 2010), whereas the permeability of soils relates to the grain-size distribution as well as compaction, cementation and alteration (Shepherd, 1989; Benson et al., 1995); consequently they all accomplish convective heat flow (Hardee, 1982). Variations in volcanic and geothermal activity have been frequently observed at sites such as Vulcano Island, Italy (Bukumirovic et al., 1997; Harris and Maciejewski, 2000), at Iwodake Volcano, Japan (Matsushima et al., 2003), at the Solfatara of Pozzuoli, Italy (Chiodini et al., 2007), and at Colima, Mexico (Stevenson and Varley; 2008), but they were alternatively attributed to changes in the magmatic or hydrothermal source (Stevenson, 1993), to permeability changes due to conduit sealing by deposition or tectonic activity (Harris and Maciejewski, 2000), or a combination thereof. Examples of chemical and thermal changes were also recently documented for the 2011–2012 unrest episode at Santorini (Parks et al., 2013; Tassi et al., 2013).

In this work we test the influence of permeability by analyzing the stratigraphic and volcanotectonic setting in controlling the degassing sites at Nisyros Island. We first explore if the results from thermal mapping and soil analysis correlate, both possibly reflecting the local geology. On the one hand, we make use of a portable infrared (IR) camera, which is efficient in imaging volcanic regions at metric to sub-metric resolutions, overcoming the low spatial resolution of satellites and the cost- and time-consuming thermometer-based measurements. On the other hand we collect and sieve soil samples at the fumarole field to define soil types and related permeabilities. Finally we compare the spatial permeability contrasts with the spatial distribution of the thermal anomalies. Our testing location was the Stefanos crater on Nisyros (Fig. 1). This volcanic island has a long history of phreatic eruptions, the latest in 1871–1873 and 1887 (Marini et al., 1993), experiencing an episode of unrest in 1996–2001 (Papadopoulos et al., 1998; Chiodini et al., 2002; Sachpazi et al., 2002). The Stefanos crater is one of several phreatic craters on Nisyros, and is also the major contributor to the total heat budget of the island (Lagios et al., 2007; Ganas et al., 2010). Below we first introduce the study area, the infrared and soil analysis methods, followed by our results that allow an interpretation of how the permeability might control the appearance of thermal anomalies.

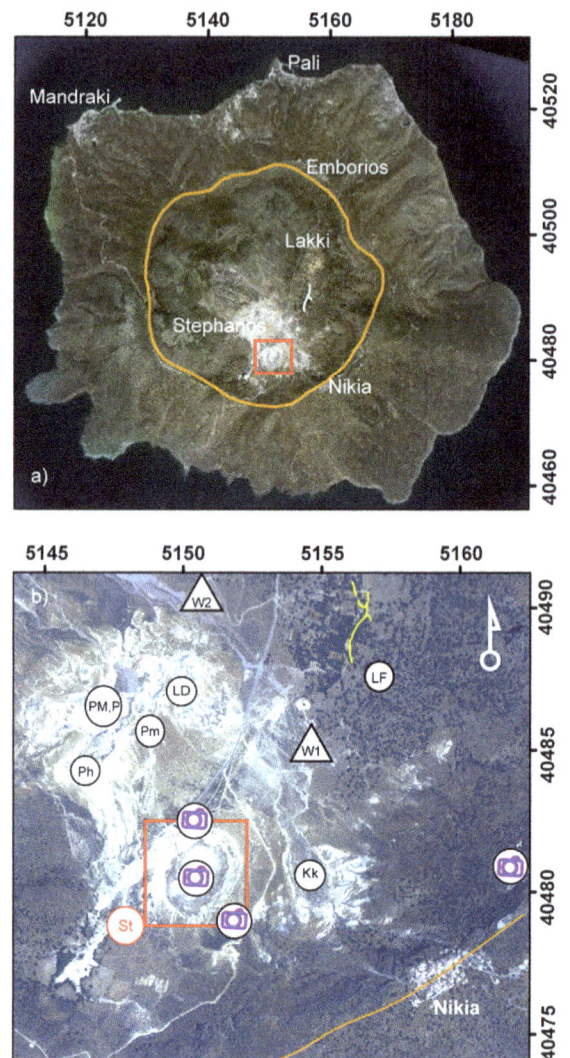

Fig. 1. (a) Nisyros Island. Coordinates are in UTM, zone 35, grid ticks are at 2 km. The brown line marks the border of the caldera. Inside the caldera: the red square shows the position of the Stefanos crater; the white line highlights the 2001–2002 fissure at the Lakki plain. Other toponyms outside the caldera indicate the villages on the island. **(b)** Satellite image (WV02) showing the main volcanic features inside the caldera; grid ticks are at 500 m. The red square indicates the Stefanos crater labeled St; other labeled sites are the Kaminakia crater (Kk), the Lofos dome (LD), the nested Polybotes Megalos and Polybotes craters (PM,P), the Polybotes Micro crater (Pm) and the Phlegeton crater (Ph). LF is the fissure in the Lakki plain. The camera icons point out the position where IR and OP images were collected, inside/outside of the Stefanos crater and at the caldera border (brown line). W1 and W2 indicate the positions of two geothermal wells.

2 Study area

2.1 Geological background

Nisyros is a volcanic island in the South Aegean active volcanic arc related to the northward subduction of the African plate below the Aegean plate. The island is sub-circular in plan view with a diameter of ~ 7 km and morphologically appears like a truncated cone. The volcanic edifice developed through five distinguished stages (Marini et al., 1993; Tibaldi et al., 2008) that led to the formation of a ~ 4 km-wide caldera, hosting rhyodacitic domes in the west and an alluvial plain in the eastern part, respectively. Superheated geothermal fluids have triggered hydrothermal explosions forming several phreatic craters, most recently in 1887 (Marini et al., 1993). The largest of these craters is the Stefanos crater, with a diameter of ~ 300 m (Fig. 1), which is in the focus of our study.

2.2 Hydrothermal activity

The remarkable hydrothermal activity, hot springs and fumaroles motivated site studies and the drilling of two deep wells for geothermal exploitation (Geotermica Italiana 1983, 1984; Marini et al., 1993). These gave a direct view into the hydrogeological and hydrothermal system. Geochemical analyses (Chiodini, 1993; Lagios et al., 2007) helped to characterize the system, and the fumaroles on Nisyros were investigated in detail (Chiodini, 1993; Chiodini et al., 2002; Teschner et al., 2007).

The hydrogeological system of the caldera consists of two separate permeable zones, as identified by the drilling of two geothermal wells (Geotermica Italiana, 1983, 1984; W1 and W2 in Fig. 1b). One permeable zone is found at a depth > 1000 m, the other is located at a depth of ~ 200 m. Both the permeable zones are deeper in the eastern part of the caldera. This vertical offset relates to a graben-like structure of the caldera basement (Caliro et al., 2005). The deep permeable zone is developed within the intrusive dioritic basement, but ends at the overlying carbonate and volcanic sequence. This permeability contrast is attributed to diverse fracture density between the lithologies (Ambrosio et al., 2010). The deep permeable zone has a temperature range of 300–350 °C and the heat is provided by magmatic fluids (Caliro et al., 2005; Lagios et al., 2007). The deep permeable zone provides vapor to the shallow permeable zone, which has a temperature range of 150–260 °C (Chiodini et al., 2002, Lagios et al., 2007). Finally, Lagios et al. (2007) proposed that a third still shallower reservoir exists and is fed by condensates.

At the surface, fumaroles occur mainly at phreatic craters such as at Stefanos, Phlegethon, Polybotes Micros (St, Ph, Pm in Fig. 1b), and at the eastern base of the Lofos dome (LD in Fig. 1b). Temperature measurements at the phreatic craters have been conducted for decades, recording fumarole outlet temperatures of mostly 96–100 °C (Chiodini et al.,

2002; Teschner et al., 2007). Smaller fumaroles occur along the flanks and on top of the Lofos dome and at the Kaminakia crater flank (LD, Kk in Fig. 1b). Some degassing vents occur also along the southern and western internal flank of the caldera following the NE–SW trends, which represents one of the main fault strikes recognized by Caliro et al. (2005), Lagios et al. (2005) and Tibaldi et al. (2008).

A 100-year period of quiescence ended with the 1996–2001 unrest episode. During this period, field observations report increased fumarole activity in 1997 (Vougioukalakis and Fytikas, 2005). Other phenomena attributed to the unrest include the progressive opening of a fissure in the Lakki plain since 2001 (Vougioukalakis and Fytikas, 2005; Fig. 1). Monitoring of the fumaroles at Nisyros since 2003 (Teschner et al., 2007) indicates that the outlet temperature only fluctuates by a few degrees due to variations in meteorological conditions. Satellite TIR data from 2000–2005 indicate that the Stefanos crater hosts a high temperature anomaly and contributes the most to the total heat flux budget of the caldera (Lagios et al., 2007; Ganas et al., 2010). In the same period, gravity surveys reveal short-term changes (Gottsmann et al., 2005) occasionally associated with height changes, inferred to reflect instabilities of the hydrothermal system (Gottsmann et al., 2007). However, these variations do not appear in the temperature data of Teschner et al. (2007). More recent ground-based InSAR measurements (2010) could not detect a significant displacement signal, implying a general decrease in activity and suggesting that the hydrothermal system is close to rest conditions (Pantaleo, 2014).

While the previous studies suggest that hydrothermal activity at depth and at the surface are related, a direct study on how the fumarole field depends on the near-surface permeability has not been elaborated. In this paper, we investigate the relationship between the thermal activity and the permeability contrast. Thermal activity was recorded by means of infrared measurements, whereas the permeability was analyzed indirectly by granulometry. A better understanding of the thermal field allows important implications in the framework of hazard assessments for the hydrothermal activity at the site.

3 Method

3.1 Infrared (IR) survey

A forward-looking infrared camera (FLIR P620) was used to collect images of the 200×300 m-large Stefanos crater in normal and panorama mode (Fig. 1). The camera operates at 7.5–13 μm bandwidth. The image size is 640×480 pixels and the resolution is 0.33 or 0.65 mrad (different lenses were used at different monitoring distances). The dimension of the pixel size is given by the resolution (mrad) times the distance of the target (D). The camera also hosts a digital optical sensor, allowing joint acquisitions of target regions. In

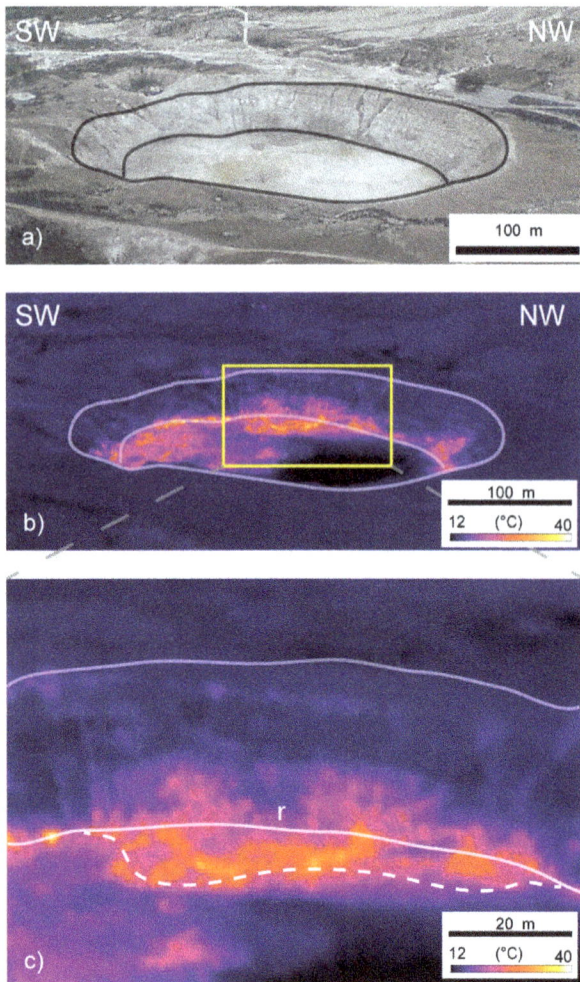

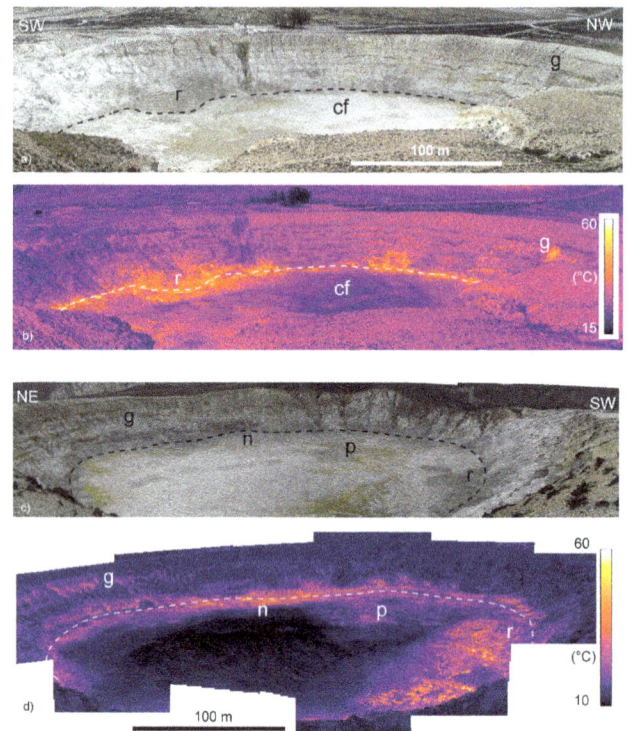

Fig. 2. (a, b) OP and IR images collected from long distance (caldera rim, Fig. 1b). Solid lines mark the upper and lower crater rim. Both the OP and IR images display part of the crater floor and the east-facing flank; the IR images show thermal anomalies along the bottom rim and a weaker anomaly toward the center of the crater but also propagating to the south. The yellow rectangle in (b) defines the area shown by the close-up in (c). (c) Magnification of the wide thermal anomaly (r, dashed line) on the western side of the crater. The coarse spatial resolution (∼ 0.8 m) causes the smoothed appearance of the temperatures.

Fig. 3. (a, b) OP and IR daytime panorama viewing westward; the dashed line marks the crater's bottom rim. (b) The thermal anomaly is visible at the break in slope of the flank and is wider in correspondence to the mound (r). The crater floor (cf) and the flank are generally cold; the lateral thermal gradient (from SW to NW) on the flank appears because of the insolation. The thermal anomaly along the flank (g) is volcanic and is stronger than the insolated areas. The insolation and the fine spatial resolution allow the IR data to display morphological features (e.g., horizontal layering, gullies). Temperature values are saturated at 15 and 60 °C to optimize the view. (c, d) OP and IR nighttime panorama facing southeastward; the dashed line marks the crater bottom rim. (d) The thermal anomaly clearly shows its ring-shaped pattern at the break in slope of the flank; at the eastern (n) and western (r) sides the anomaly appears to be wider. Another anomaly (p) appears close to the center in the southern direction and represents the mud pits. Also visible is the thermal anomaly (g) along the flank, as observed in (Fig. a, b). Temperature values are saturated at 10 and 60 °C to optimize the view. (a) and (c) also show the sites with high sulfur content (yellowish).

total we recorded > 200 images to ensure ideal clear viewing conditions and the feasibility of image stitching, in order to investigate the thermal field at different scales (Figs. 2, 3 and 4). On average 4–10 images were necessary to create mosaic panoramas (Fig. 3) for both infrared and optical images, and ∼ 30 images to create the synoptic IR map of high spatial resolution (Fig. 4).

A data set of images was collected on the 7 April 2010 during a single day. During daytime the images from the caldera rim were collected at 08:30 (local time), and the images from outside and inside the crater were collected between 12:30 and 14:00 (local time). During nighttime the images were collected from outside and inside the crater between 22:15 and 23:00 LT. Images from long distances were collected from the caldera rim close to Nikia (Fig. 1); here we used a telezoom to obtain a spatial resolution of ∼ 0.8 m (Fig. 2). Images from shorter distances were recorded in panorama mode from two opposite vantage points along the crater rim. One subset was recorded from a position ∼ 200 m southeast of the crater (Fig. 1b) showing the center of the crater and the east-facing Stefanos crater flank (Fig. 3a and b). The other subset viewed the crater from ∼ 100 m northwest of the Stefanos crater rim (Fig. 1b) showing the bottom of the crater and its southeastern flank (Fig. 3c and d). For these subsets, the pixel dimension ranged between 0.05 and 0.15 m. Differences exist between the panoramas because the vantage points had different distances leading to a different field of view and incidence angles. A 360° panorama was recorded at the bottom of the crater (Fig. 1b), and these images were used to generate an IR mosaic of the entire crater (Fig. 4) and to detail small-scale features on the crater floor.

The IR images display temperatures on a color-coded scale, and these temperatures are considered apparent. That is, the temperature value represents the thermal energy distribution integrated over the pixel footprint (Dozier, 1981); accordingly,

$$T_{obj} = A_v \times T_v + (1 - A_v) \times T_b, \tag{1}$$

where (T_{obj}) is the pixel temperature, and A_v and $(1 - A_v)$ are the vent area and the vent-free area within the unitary pixel, respectively; T_v and T_b are the temperature (°C) of the vent and of the background, vent-free area, respectively. Otherwise T_{obj} depends on parameters like the target-to-sensor distance (D), the emissivity of the target, and the transmittance of the atmosphere as a function of atmospheric temperature (T_{atm}) and relative humidity (RH). Finally, the accuracy of the measurements depends on the orientation of the field of view, which should be as parallel as possible to the target (Ball and Pinkerton, 2006).

For each data set acquired in panorama mode, the images were sampled in fast sequence steering the IR camera and allowing a sufficient overlap between consecutive pictures. Because the distance was almost constant for each of the shooting positions and the time required between each acquisition is a few seconds only, the parameters D, T_{atm} and RH were assumed constants. Values of T_{atm} and RH are assumed to be suitable to site conditions in the range of 10–20 °C and 50 %, respectively. Also, ε is assumed to be constant and equal to 0.93 according to the literature (Lagios et al., 2007). All these assumptions were valid also for the other IR images collected as independent snapshots. We did not consider a pixel-by-pixel correction approach. Also, geometric complexities arising from the different viewing field and topography were not corrected for.

We processed the raw images by FLIR ThermaCAM software. The results are displayed (Figs. 2–4) with temperature scales saturated and clipped at the 10–60 °C interval for the night panorama, and at 15–60 °C and 15–90 °C for day panoramas, respectively, to enhance the thermal patterns. The stitching of infrared and digital images is finally executed using a combined perspective–cylindrical merging tool as embedded in common image software (Photoshop). The images taken from within the crater floor were also used to generate a crater-wide mosaic that is subsequently georeferenced in map view (Fig. 4) with GIS software (ArcGIS 9.3 by ESRI). This necessitates the application of a matching proce-

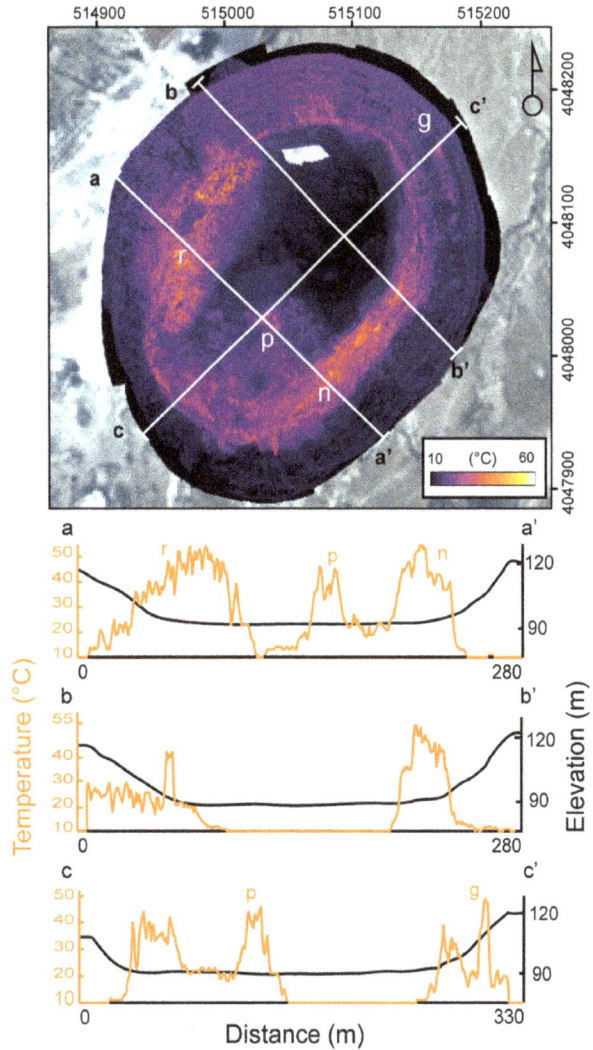

Fig. 4. (Above) Georeferenced IR mosaic showing the full extent of the ring-shaped thermal field along the bottom crater rim. Grid ticks are at 100 m. This map better highlights the NW–SE trending of the wider anomalies including the mounds (r, n) as well as the isolated position of the boiling ponds (p) and of the anomaly along the northern flank (g). Temperature values are saturated at 10 and 60 °C to optimize the view. Three orthogonal traces (a–a′, b–b′, c–c′) are shown intersecting those features. (Below) Temperature–topography profiles (a–a′, b–b′, c–c′); the distances along the trace and the topographic height (x–y axes are not scaled) are in black; the temperature axis and the values are in orange. These profiles highlight the fact that thermal anomalies occur mostly at breaks in the slope.

dure of ground control points recognized in both the satellite image (WorldView02, visible bands) and the IR mosaic.

A second field survey on 17 January 2013 allowed for the collection and verification of the IR images collected in 2010. The survey followed a similar procedure, except that the images were taken from the northern and southern borders of

Fig. 6. (upper row) Digital pictures of the pits where soil samples were collected and (lower row) corresponding material. The difference between granular and cohesive types can be distinguished visually. S01 is loose and shows few clasts ≥ 1 cm dispersed in an ∼ uniform fine sand matrix. S03 occurs in blocks of cohesive material and no particle can be distinguished. S06 is loose and shows few clasts ≥ 1 cm, but the matrix is graded. S08 has few clasts ≥ 1 cm and appears graded. S01 and S06 also have sulfur grains, as indicated by the yellowish color.

3.2   Soil analysis

Soil samples (12) were collected (Fig. 6) during our 2013 campaign on 18 January. Sampling locations were chosen at the crater floor mainly along an E–W profile (Fig. 7) crossing fumarole-bearing and fumarole-free areas. Two samples were collected along the eastern flank and two more samples in the southern and northern sectors, corresponding to sites of anomalous and normal temperature, respectively (Fig. 7). The sampling sites were selected according to the need to (i) represent the different thermal conditions and expressions (fumarole, diffusively heated ground, mud pools, boiling runoff) highlighted by the 2010 infrared survey, (ii) to have a sampling dense enough for spatial comparison to thermal data, and (iii) to limit the total weight. The E–W profile allowed us to intersect the thermal anomalies at the crater border, the center, and the interposed cool areas. A N–S profile instead would have neglected the large thermal anomalies observed at the western and eastern flank bases. Moreover, the N–S elongation of the flooded area would have caused a large gap in the spacing of soil samples. The representativeness of the bulk grain-size distribution was ensured by collecting 1–2 kg of a sample, depending on its cohesive or granular aspect defined in the field.

Each sample was collected at a depth up to 20 cm below the surface. We ensured that the sample corresponded to a single soil type. We dug down to 50 cm in an attempt to check for a possible vertical gradient in temperature. Each soil sample has been characterized by grain-size analysis in the laboratory. Samples were oven dried at 70 °C to avoid melting of sulphur crystals before weighing. Samples were wet sieved (ASTM-D6913-04, 2009) to discern the relative percentage of granular fractions, particles with diameter (d) larger than 0.064 mm, and cohesive fraction, particles with d smaller than 0.064 mm. The cohesive component is washed away by running water filtered by sieves, while the granular

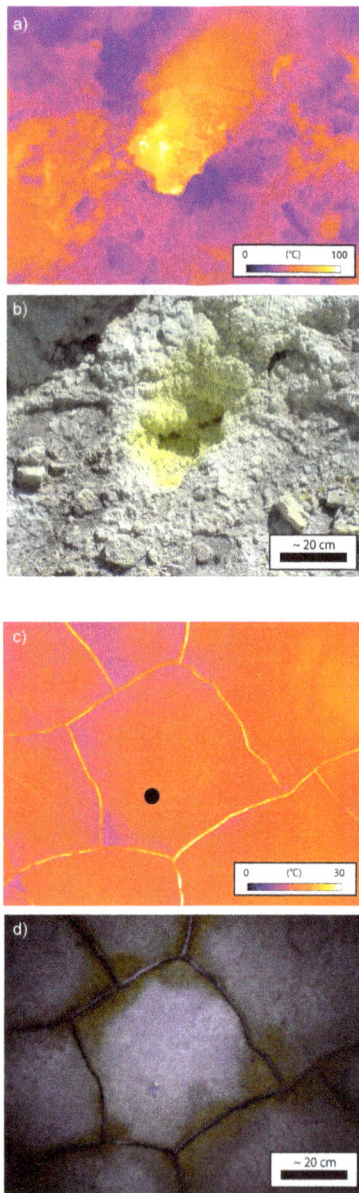

Fig. 5. (a, b) IR and OP close-ups of a fumarole on the western mound. The IR image shows that the temperature at the vent reaches 100 °C, according to direct measurements. This is possible as the pixel size is sub-mm. It also shows that temperature rapidly decreases sideways. The same can be deduced by the OP, where solid sulfur exists close to the degassing vent. **(c, d)** IR and OP close-up of the dessication polygon. The IR image shows the plate as cool, whereas the bounding fracture is ∼ 5 °C warmer. The black circle is a 1 Euro coin for scale; it appears cooler because it has an emissivity different from the emissivity of the soil. The OP image shows the sulfur crystallization close to the fracture.

the crater. The crater floor could not be imaged because it was partially flooded by recent rainfall.

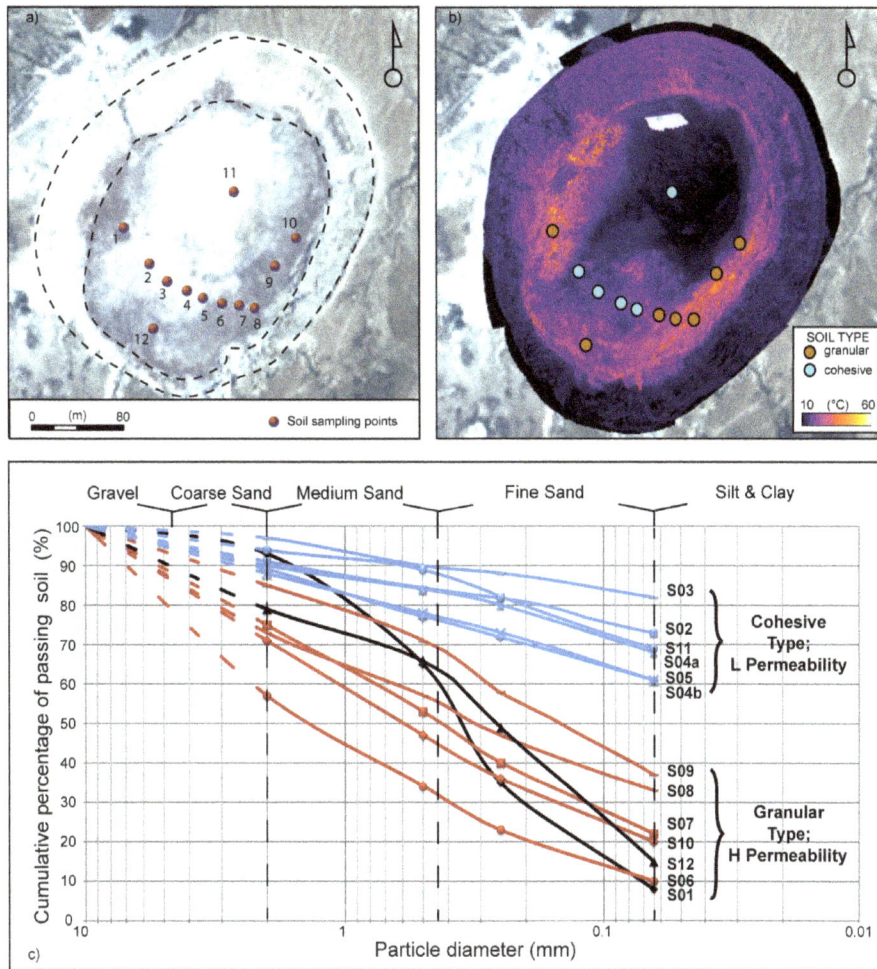

Fig. 7. (a) Optical satellite image (WV02) of the Stefanos crater: the numbered dots mark the sites where soil was sampled. **(b)** Georeferenced IR mosaic overlapping the satellite images; it shows the coupling of soil types with the thermal field. **(c)** Semi-logarithmic plot of grain-size curve. The vertical axis indicates the cumulative percentage (by weight) of material passing through the sieves; the lower horizontal axis indicates the mesh diameter of the sieves. The upper horizontal axis indicates the fraction names and intervals according to international standards. The blue lines are the grain-size curves of cohesive type soils, which have low (L) permeability. The red and black lines indicate the grain-size curves of granular type soils, which have high (H) permeability. The red curves are for sorted samples, the black for more uniform samples. Blue, red and black lines are solid within the measured intervals and dashed where uncertainty arises by assuming a maximum diameter of 10 mm and interpolating the curve (see text).

material is kept on them. Finally, the soil trapped by each sieve is dried and weighed again.

We used four sieves that allowed us to separate gravel and coarse sand fraction ($d > 2.0$), medium sand fraction ($2.0 < d < 0.5$), fine sand fraction ($0.5 < d < 0.064$), and silt and clay fraction ($d < 0.064$). The maximum d for each sample was noted visually; few clasts exceeded 10 mm. We did not perform a settling analysis to compute the relative percentage of silt and clay.

Following geotechnical practice, the results are presented as distribution curves in a semi-logarithmic plot (Fig. 7); the vertical axis indicates the cumulated percentage (by weight) of soil passing through the sieve whose mesh size is labeled on the horizontal axis. In this plot, we adopt $d = 10$ mm

as the maximum value and assume that all the material is smaller, so the grain-size curve results are interpolated from $d = 2$ mm to $d = 10$ mm (dashed curves in Fig. 7).

Although our main analysis and result is based on the granulometry, we also estimate the permeability (k) of our samples by applying the empirical formula by Alyamani et al. (1993). Permeability is accordingly based on the knowledge of d_{10}, d_{50} (particle diameter values at the 10 and 50 percentage of the grain-size curve, respectively) and I_0 (the intercept of the line formed by d_{50} and d_{10} with the diameter axis). The parameters d_{50} and d_{10} were extracted from linear interpolation of the grain-size curves. Both particle diameter values could be directly obtained from samples S01 and S06. For the other samples, however, the measured grain-size curve does not intercept the 10 % of passing soil. To provide a first-order estimate for these other samples, we arbitrarily constrain $d_1 = 0.002$ mm, which means that samples have 1 % of clay content in the cumulate curve. Furthermore, we constrain $I_0 = 0.001$ mm. The aforementioned constraint cannot be applied to the cohesive samples, because they are mainly silt–clay material and so, given our grain-size curves, any prediction of d_{50}, d_{10} and I_0 would be unrealistic. For the cohesive samples, we hence refer to the related literature (Bowles, 1988).

4 Results

4.1 Far field view of large-scale thermal architecture

The IR images collected from the caldera rim, about 1000–1200 m away from the Stefanos crater (Fig. 1), have a coarse resolution with pixel size of about 0.60–0.80 m. They show the western flank of the crater and part of the bottom, the remaining parts are not imaged because of the topography-related shadowing effect (Fig. 2a and b). The upper rim of the crater does not express any thermal anomaly, whilst the bottom rim of the crater does; the temperature field is shown by a ring not perfectly concentric bordering the crater floor (Fig. 2b). The temperature distribution is smoothed and homogenized because of the coarse resolution, therefore single or grouped vents cannot be recognized (Fig. 2c). On the western and southern sides, more elevated temperatures and wider thermal fields can be observed. At the southern side the apparent temperature (T) is in the range 22–30 °C and rises to 30–40 °C westward (r in Fig. 2c) before decreasing to 24–30 °C northward. The background cool temperature inside the crater is ≤ 18 °C. Further clustered anomalies are observed toward the center of the crater; these display an apparent T of 20–25 °C and mark the transition between a relatively warm southern sector and relatively cold northern sector.

4.2 Close field view of medium-scale structures

The mosaics produced with images collected close to the Stefanos crater have a high resolution of about 0.05 and 0.15 m in pixel size (Fig. 3). These mosaics show that the thermal anomalies are distributed in a roughly circular-shaped pattern at the margin of the crater floor (Fig. 3b and d). The anomalies are wider along the west, through the south, to the eastern side and narrower at the opposite side. Hotter spots corresponding to clustered vents are now distinguished as sparsely distributed within the geothermal field, showing apparent temperatures of $T \geq 60$ °C. The western side hosts a mound of material (r in Fig. 3a and b), 2–3 m high, ≤ 10 m wide and ~ 50 m long, covered by a hard crust dotted by centimeter-size vents; around the mound the crater floor is diffusively heated at apparent $T \sim 40$ °C. The southern side has a rough surface with an extensively developed hard crust

covered by pebbles/boulders fallen from the adjacent flank (Fig. 3c and d). Here there are few vents and the heat is pervasively distributed across the surface, so the apparent T is ~ 30 °C. The eastern side is flat except for a mound (n in Fig. 3c and d), 0.5–1 m high, 2–3 m wide and ~ 30 m long, hosting vents; the base of this flank also hosts surficial boiling runoff and small boiling ponds (up to tens of cm in size). The apparent T of the thermal anomaly at this side is ≥ 40 °C. Toward the center of the crater embedded in the cooler terrain, we identify thermal anomalies – apparent $T \sim 30$ °C – corresponding to sub-metric to metric wide boiling ponds fed by meteoric water and superficial runoff. Another localized anomaly is isolated along the northeastern flank on the slope (g in Fig. 3b and d); its apparent temperature is 25–30 °C or higher, depending on the viewing geometry. Both optical panoramas show the aforementioned morphologies (mounds r, n and ponds) as well as the sites with high sulfur content (yellowish) at the crater floor and along the flanks (Fig. 3a and c), which are not detectable in the infrared panoramas.

Temperature values from IR images are generally lower than inferred from pointwise measurements because the vent usually occupies a small part of the pixel and contributes proportionally to the pixel thermal budget. Only the IR close-up of the fumarole clearly displays T values of ~ 100 °C.

4.3 Close-up of small-scale structure

Using the images collected within the crater, we can further explore particular features that are not revealed from larger distances. Indeed, the resolution approach millimeters scale when standing above the targets. The IR measurements taken directly above the fumaroles show that the vent temperature decreases rapidly from ~ 100 °C to 30–40 °C sideways (Fig. 5). Such a vent temperature is in agreement with the temperature measured inside the vent by the K-type thermocouple during our survey, and confirms findings by previous authors (Chiodini et al., 2002; Teschner et al., 2007). Similarly, the close-up of the mud pits shows the temperature decreasing rapidly from ~ 60 °C at the deep-seated boiling surface to almost to 20 °C at the surface. The ground surrounding the ponds is cold and a network of cracks separate desiccation polygons (Fig. 5), usually 10–20 cm a side, whose surfaces have sulfur films developing from the crack inside the polygon. The cracks are warmer than the plate and we also observed the soil temperature increasing ~ 5–10°C a few centimeters below the crust. Such a pattern resembles on a small scale the one observed at the crater scale.

4.4 Mosaic of large- to small-scale structure

The thermal features are visualized in the georeferenced IR mosaic giving a synoptic view of the crater (Fig. 4). A total of > 30 images have been stitched and ≥ 5 ground control points have been selected to reference each image. The

Table 1. Permeability values for granular soil samples computed according to Alyamani et al. (1993).

Sample	S01	S06	S07	S08	S09	S10	S12
Permeability (ms^{-1})	1.79E-06	2.17E-05	1.94E-06	1.12E-06	3.82E-07	4.56E-06	6.50E-07

IR map shows most of the crater except for a small patch in the cold region, which was not covered by any image. The geothermal field, expressed by the ring-shaped thermal zone, is located along the border of the crater floor. The ring is not perfectly symmetric, but appears wider on the eastern and western sides (n and r in Fig. 4). Here, the two opposing flanks are expressed by the highest temperatures recorded. This geometric symmetry reveals alignments of elevated thermal fields trending SW–NE. We note that the same SW–NE trend is already identified by the long axis of the ellipticity of the crater and by one of the main structural trends seen elsewhere on the island (Caliro et al., 2005; Tibaldi et al., 2008). The mosaic also shows the thermal anomaly associated with the boiling ponds, in contrast with the cold zone close to the crater center and the thermal anomaly along the NE flank (p and g in Fig. 4). The spatial analysis reveals that the extent of the thermally anomalous area along the western flank is $\sim 2400\,m^2$, and on the opposite flank it is $\sim 2700\,m^2$, whereas the anomalies at the southern and northern sites cover $\sim 500\,m^2$ each. The anomaly at the center is smaller, being only $\sim 70\,m^2$. The overall heated ($> 30\,°C$) area is therefore $\sim 6200\,m^2$ and corresponds to 17 % of the $\sim 35\,300\,m^2$-wide crater floor. We selected $30\,°C$ as the threshold because it effectively separates the geothermal effect from effects of insolation. We show three cross sections to highlight the spatial relationship between the thermal anomalies and the topography (Fig. 4). All the sections show that temperatures augment toward the crater floor and that the increase is gradual or sharp, depending on whether the ground is insolated or shadowed, respectively. For example, the profile $b-b'$ shows temperature fluctuations along the western flank that reflect the stratigraphic layering. The highest temperatures are reached at the breaks in slope, but the extents are different, as $a-a'$ crosses the wide anomalies (r, n), whereas $b-b'$ and $c-c'$ do not. The temperatures have a tendency to diminish toward the crater center and to flatten at $10\,°C$ (clipped value imposed by our elaboration) in the cold areas; the only exceptions are the temperature peaks in correspondence to the boiling ponds (p). Profile $c-c'$ also shows the thermal anomaly (g) that occurs along the flank and reaches values comparable to the anomalies at the flank base.

4.5 Soil analysis

We present the results from the soil analysis in the form of a semi-logarithmic grain-size curve and in an ArcGIS framework on the IR map to provide a combined view of temperature–grain-size distributions (Fig. 7). The grain-size plot shows two main groups: the cohesive type, which has ≥ 60 % of cohesive fraction, and the granular type, which has ≤ 40 % of cohesive fraction (Fig. 7). Both soil types are generally well graded when considering the sand fraction, with the exception of S01 and S12, which have a dominant medium and fine sand fraction. In a first-order approximation, we describe the granular type as medium-to-fine sand and the cohesive type as a silt–clay. Spatially we observe that the sample S01 (Fig. 6) collected at the base of the western mound has 93 % granular content, with cohesive material being almost absent. The sample S12 collected at the southern border and below a hard crust has 85 % granular content, again with a very low cohesive fraction. At the eastern flank, the samples S07, S08 (Fig. 6), S09 and S10 are granular, with percentages ranging between 64 and 80. The samples S04 and S05 are collected as close as possible to the ponds, just a few meters, but out of the area temporarily flooded by the rain. They have granular content below 40 % and significant cohesive content of about 60–70 %. A similar value is found in a deeper sample at the same location (S04 at 50 cm), collected because the IR data showed a $\sim 35\,°C$ temperature increase, thus suggesting a reduction in the vertical efficiency in heat transfer. The samples S02 and S03 (Fig. 6) collected between the western flank and the mud pools have cohesive fractions ranging between 70 and 80 %. The sample S06 (Fig. 6) collected between the eastern flank and the mud pools is ~ 90 % granular. The soil sample S11, at the north of the mud pools where there is no thermal anomaly, has only ~ 68 % of cohesive particles.

Granular soils have higher permeability values than cohesive soils, because granular soils allow higher effective porosity (Graton et al., 1935; Shepherd 1989). The empirical formulation used to calculate the permeability of our granular samples indicates that permeability values range between 10^{-5}–$10^{-7}\,ms^{-1}$, with sample locations S06 and S09 being the most and least permeable, respectively (Table 1). The permeability value of the cohesive soils taken from the literature corresponds to $k \sim 10^{-11}\,ms^{-1}$ (Bowles, 1988).

4.6 Comparison of IR and soil analysis

Our results show that the locations of thermal anomalies match with the locations of granular soils. Lower temperatures match with the locations of more cohesive soils (Fig. 7). We notice that S06 diverges from the general behavior by having a high granular fraction but being located in a relatively cold area. This particular sample location was close to the flooded area, associated with a possible cooling effect. We also note an apparent correlation between changes in the granular fraction with the spatial thermal gradient. Looking at the granular content along the E–W profile (S01–S08), we note that it decreases by over 60 % from S01 to S02, which are ∼ 40 m apart. A further 10 % decrease is found from S02 to S03, and a 10 % increase from S03 to S06; these locations are ∼ 20 m apart. Higher rates (20–30 % each 20 m) occur between S06-S07-S08. Similarly we observe strong horizontal thermal gradients from S01 to S02 and from S06 to S07 (30–40 °C), whilst from S02 to S06, the gradient is ∼ 15 °C. The granular content along the N–S profile (S08–S10) appears more stable, as indicated by the horizontal thermal gradient too.

5 Discussion

This work demonstrates a relationship between the thermal anomaly and the grain-size distribution of the soil present in the largest explosive crater on Nisyros Island. The Stefanos crater is known for the significance of the geothermal activity, the fluid emissions and the short-term episodes of uplift and gravity changes. We analyzed the thermal field by means of a multi-scale infrared study, retrieving spatial resolutions (pixel size) of 0.05 to 0.8 m. Results show a complex, accurate, high-temperature field bordering the center of the crater floor that is expressed more strongly on the western and eastern sides. Here the temperature is not only higher but the thermal field is also wider. To test this temperature distribution in more detail, we collected soil samples and identified the granulometry. We find a first-order correlation between the grain-size distribution of the soil, which we consider as a proxy for soil permeability, and the temperature distribution.

Understanding the dynamics of the degassing close to the surface at Nisyros is of relevance, as the volcano is currently ranked in the "Very High Threat" class (Kinvig et al., 2010), considering the vulnerability of the population (∼ 1000) and seasonal tourists (∼ 60 000). The Stefanos crater was chosen for our study because of a recent increase in fumarole activity and outflow of melted sulfur and hot mud (Papadopoulos et al., 1998; Chiodini et al., 2002; Sachpazi et al., 2002; Vougioukalakis and Fytikas, 2005; Lagios et al., 2007), as well as its strong thermal signature in comparison to other phreatic craters (Lagios et al., 2007; Ganas et al., 2010). Moreover, our results may have implications relevant for other sites. The Stefanos crater is a rather typical explosive phreatic crater with a morphometry similar to those found on other volcanoes. The Stefanos crater also displays a ring-shaped thermal pattern of fumaroles similarly observed at other volcanic craters, such as at Vulcano (Bukumirovic et al., 1997), Satsuma-Iwojima (Shinohara et al., 2002), Colima (Varley and Taran, 2003), and Kudrayavy (Yudovskaya et al., 2008). Vulcano and Kudrayavy in particular could be excellent case studies to validate or challenge our findings because their summit craters are collectors of sediments and because they are safely accessible. At the other active volcanoes, direct soil sampling, however, is dangerous, making similar studies difficult.

5.1 Limitations

We surveyed the site during a single day in 2010 and for 2 days in 2013, in both cases collecting images at day and night time. During this interval Nisyros Island was quiescent and we decided to test whether some long-term changes might have occurred in the hydrothermal system. Seasonal changes affecting the air temperature and the water table level may occur, and affect the magnitude and pattern of the thermal anomalies. However, Teschner et al. (2007) reported seasonal changes on the order of ∼ 5°C at nearby fumarole. This suggests that the changes are negligible for our purpose. Moreover, at Nisyros the fumarole temperature outlet does not change on a day-to-day basis (Teschner et al., 2007), at least when the volcano is quiescent. Consequently we consider that the IR data, based on single day recording, are representative of the site and of the activity during the encompassed period. Besides, we observe that the geothermal fields appeared (in the IR data) almost identical in 2010 and 2013. Therefore we are confident to relate IR data collected in 2010 to the soil analysis of samples collected in 2013. Indeed, considering the deposition processes, we do not observe geological/geomorphological evidence indicating clear changes in sedimentation, neither at fast rates (e.g., recent slumpings) nor at slow rates (e.g., upward grain-size changes) (Fig. 6). Also, we do not expect changes in sedimentation near mud pools, because the pits drain the surface runoff rather than acting as springs. Despite the flooding in this area preventing us from imaging parts of the crater floor in 2013, we expect temporary cooling of the water and the soil at the pits because of the mixing of rising hot gasses with large volumes of cold rain water. This phenomenon possibly explains the disagreement between the granular aspect of S06 and the low temperature at that location. Limitations may also arise from the infrared technique, the environmental conditions, the image processing and mosaicking, or a combination thereof. Moreover, soil sampling and laboratory analysis may affect the conclusions drawn, as detailed below.

5.1.1 IR imaging

One of the main limitations affecting IR imaging relates to the changes in emissivity and transmittance, which depend on geometric conditions (i.e., on distance and viewing angle, and on the physical condition of the target and of the atmosphere). Here we detail the limitations encountered at Nisyros and their influence on the results. A detailed description of general limitations is provided by Spampinato et al. (2011).

The IR mosaics we present are computed using a constant camera position. However, D changes within a single image and across each set of images from foreground to background, and T_{obj} is expected to vary consequently. In our data set we found that, doubling D, the T_{obj} varies in the range of 1–3 % according to the occurrence of T_{obj} close to or higher than T_{atm}. For the processing, we use the mean distances in the field of view. Therefore the temperature is expected to be overestimated at shorter distances. We note that these differences are small and therefore negligible for the purpose of this study. Moreover, D influences the geometric and thermal resolutions. Increasing D has the effect of widening the pixel size and lessening the apparent temperature, because the pixel will include, comparatively, more cold surface than degassing vents. This also leads to a lower sensitivity in the detection of apparent temperature contrasts. Further errors arise from assuming a single emissivity value, because it changes with the material and the viewing angle (Ball and Pinkerton, 2006). We use $\varepsilon = 0.93$ for the average site conditions following previous works (Lagios et al., 2007). Laboratory tests showed that emissivity is expected to decrease as the viewing geometry deviates from the perpendicular to the surface (Ball and Pinkerton, 2006). Such a phenomenon should affect the crater floor due to the oblique viewing, but it should not affect the flanks, as the viewing geometry is almost perpendicular. Nevertheless we do not observe any temperature gradient that may relate to the viewing angle. Another issue related to the site conditions is the presence of steam in the line of sight, which dampens the propagation of the thermal signal (Sawyer and Burton, 2006). As shown by the digital panoramas (Figs. 2a, 3a and c), we managed to collect our IR data during low-to-absent (visible) steam output, although we cannot rule out its contribution.

Considering the atmospheric conditions, we corrected the transmittance using values of temperature and relative humidity suitable for the site conditions. This does not affect our results because the purpose of this work is to investigate the spatial distribution of thermal anomalies. Indeed, we tested that changes in T_{atm} of $\pm 10\,^{\circ}$C cause variations of T_{obj} ranging between 3 and 1 $^{\circ}$C for cool ground and $< 1\,^{\circ}$C for heated ground.

Finally, in terms of data processing, the georeferencing appears accurate along the flank because there are matching features (layering) between hand-held IR and satellite OP images, as well as only a small amount of geometric distortion. We did not further analyze the accuracy of georefencing; at the crater floor this accuracy may be lower because of the lack of recognizable features. Moreover, the IR images taken from the crater floor may suffer from large geometric distortion due to the viewing geometry. However, since our data are actually collected from a short distance (a few hundred meters) with a lens providing a field of view of $24^{\circ} \times 18^{\circ}$, which is somewhat narrow, large distortions are consistent.

5.1.2 Soil analysis

Soil samples were collected within the top 50 cm only. Sampling at a different depth might provide different information because we commonly observe an increase in alteration and crystallization at the interface of lithological beddings. However, we are confident that the correlation of grain size with the IR data is beyond coincidence.

Grain-size distribution of volcanic deposits is generally investigated to understand eruptive processes and energy (Brazier et al., 1983) or to address the permeability of volcanic deposits (Peltier et al., 2012). A major limitation during our sieving operation was in the artificial modification of the grain-size distribution. That is, in our samples, native sulfur crystals and chemically weathered volcanic clasts (Fig. 7) were fragile and could break into smaller granular particles. Conversely, the cohesive material appears as aggregates simulating granular clasts. We attempted to minimize these issues with a careful washing to avoid strong particle collision and by using a brush to disperse the cohesive aggregates. The substantial difference in sand contents between the granular and cohesive type (Fig. 6) suggests that our sieving procedure prevented the convergence to a single sand class. Another source of error affecting the fraction distribution derives from having volcanic silicate particles with a density of 2.4–2.6 g cm^{-3} mixed with native sulfur crystal of 1.9 g cm^{-3} density. The density contrast could cause a 7 % difference in weight on equal volumes of silicate and sulfur, but as the quantity of silicate in our samples is much larger, the bias becomes negligible.

A systematic bias in the grain-size description of the soil would cause a different or even misleading estimation of permeability values and consequently affect the interpretation of the thermal field. We might have overestimated the amount of gravel and coarse sand by having the largest sieve mesh at 2.0 mm and by fixing the maximum $d < 10$ mm based on visual description. We evaluate this effect to be ~ 10 % for the cohesive type and up to ~ 20 % for the granular type (~ 40 % for S01). This error affects only the relative percentage of granular fractions, not the relative percentage of granular vs. cohesive, though we disregarded the occurrence of gravel to estimate the soil permeability. The permeability can be measured by laboratory or in situ tests, but both the procedures are technically demanding for our purpose (Lambe and Whitman, 2008). Alternatively, empirical formulae based on particle size distribution can be used. We applied the formulation of Alyamani et al. (1993) to the granular samples. The permeability values of samples S01 and S06 are based on real grain-size data (Fig. 7). The permeability values of samples S07, S08, S09, S10, and S12 required a strong assumption about the minimum grain diameter and its mass percentage. Nevertheless, all the estimated permeability range corresponds to tabulated values (Bowles, 1988) and thus are reasonable. We also tested that the permeability remains in the same order of magnitude when the mass percentage of the

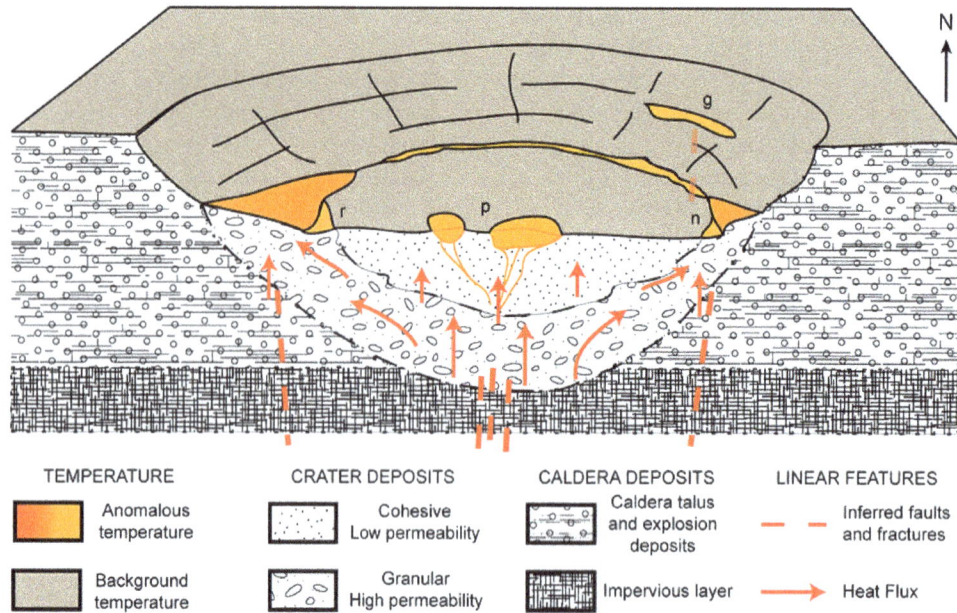

Fig. 8. Sketch (not to scale) of the conceptual model. At the bottom an impervious layer is broken by faults that drive the phreatic explosion through the caldera talus and the deposits of an older eruption. On top of the impervious layer is the crater. The heat rising from the depths enters the crater through the faults and propagates upward. According to our hypothesis, the crater is initially filled by granular and permeable deposits that progressively appear only at the sides. Later cohesive and impermeable material deposits on top of and in heteropy to the coarse ones. The heat flux rises with different efficiency: high in granular deposits (longer arrows), low in cohesive deposits (shorter arrows). Because of the low permeability at the crater center, fluids are forced to move sideways, increasing the degassing at the border. Possibly there are also other sideways faults that enhance the heat release at the NW and SE sites (r, n) and funnel hot fluids higher along the NE flank (g). Fault(s) at the crater center are also responsible for drainage and erosion of the crater filling, by which finally mud pits generate (p). At the surface, the thermal anomalies are presented in light and dark orange to indicate temperature differences. Brown is the cool terrain corresponding to the background temperature.

minimum grain diameter increases ten times. This suggests that permeabilities in the range of 10^{-5}–10^{-7} ms^{-1} are realistic for the granular soils in the Stefanos crater. Note that the values refer to water permeability, whereas air permeability, which is generally higher (Springer et al., 1998), should be considered. This approximation is nevertheless considered representative of the different behavior of the soil types identified at Stefanos.

5.2   Conceptual model

The Stefanos crater has an explosive phreatic genesis and is elongated NE–SW, consistent with a major fault trend observed elsewhere on the island (Caliro et al., 2005; Tibaldi et al., 2008). Marini (1993) suggested that a fault may have initiated the cratering by (i) connecting two aquifers and causing the flashing of the fluids, or (ii) weakening/opening the sealing of a deep, over-pressured, hot aquifer that flashed. One may assume that hot fluids continue to migrate upward through the fault into the crater (Fig. 8). In this scenario, we would expect to observe a linear thermal anomaly consistent with the trace of the fault trend. Our data, instead, show a near-circular thermal anomaly mainly at the border of the

crater. Therefore faults are not the major control on the near-surface expression of the heat flux.

Our soil analysis suggests that (i) cohesive and impervious material seals the center of the crater and thus inhibits the convective fluid ascent, which is then diverted and forced laterally toward the crater border. (ii) There, granular and permeable materials effectively transfer the heat along the crater border. Indeed, we found a permeability contrast of three orders of magnitude in the granular material only. Combined with cohesive material, our study suggests a contrast of up to six orders of magnitude. Consequently, the heat release appears to be controlled by the soil texture distribution. The soil type, in turn, is related to depositional processes (Fig. 8). The depositional processes are initiated with rockfalls and landslides. The remobilized material corresponds to the original talus of the caldera escarpment and to the deposits of the Kaminakia and Stefanos explosions. Afterwards, mixed granular and cohesive sediments drained by the superficial runoff into the crater from the surrounding relief and gradually filled the crater (Fig. 8). We observe that the soil particles range from pebbles to silt/clay and that there is a selective deposition based on the energy of the runoff; the particles travel either short or long distances from the crater rim according to their dimension.

The IR observations indicate that the western, southern and eastern borders have a wider extent of the thermal anomaly, which might relate to a more widespread deposition of granular material. We conjecture that the sediment input at those sites is larger because of the adjacent highs in relief, the Lofos dome and the caldera rim (Fig. 1b), whereas in the north the topography is flat (Lakki plain in Fig. 1b).

Only the local mud pools and the diffuse heating along the northern flank (p and g in Figs. 3–5) contrast the lithological control and can be speculatively attributed to the volcanotectonic control. Buried and inactive fractures can drain the finer fraction of soils opening large or small permeable channels.

5.3 Comparison with other results

5.3.1 Thermal data

The comparison of our ground-based survey with previous satellite-based observations (Lagios et al., 2007) can only address the spatial distribution of the thermal anomalies and not the amplitude. This is because the geometric resolution differs by two orders of magnitude and therefore the thermal resolution is different. Moreover, the observation angle from our view point favors accurate thermographic imaging of the flank, whereas satellites have a better view of the crater floor. Nevertheless, both data sets reveal that the southern sector of the crater shows a more pronounced thermal flux than the northern sector. Prior to the present study, only a single ground-based IR survey had been performed at Nisyros (immediately after an episode of unrest in 2002) with the aim of validating satellite IR data (Lagios et al., 2007). Images were collected from the caldera rim at Nikia (Fig. 1) and showed all of the Stefanos crater except for the eastern flank and the surrounding floor. The spatial resolution of the previous survey is one order of magnitude lower than the resolution of our images. Considering this difference, the amplitude of the anomalies still appears in the same range of values, and the spatial pattern has remained similar. Only small anomalies within the current cold region and in the proximity of the upper crater rim have disappeared since 2002. That might indicate a small decrease in activity following the unrest and simultaneously exclude a significant decline in heat flux since 2002. The general pattern of temperature distribution remains mostly constant. The anomaly at the crater floor occurred where we expect the deposition of cohesive sediment, thus confirming our hypothesis on the depositional sealing. The comparison of our 2010 and 2013 surveys, where possible, also does not show significant change in the extent of the thermal field, confirming the stability of the degassing system and suggesting that depositional sealing acts at a low rate.

5.3.2 Lithological control

At Nisyros we describe the thermal field of the Stefanos crater and the concurrence with the soil texture causing permeability contrasts on the order of 10^4–10^6 ms^{-1}. The role of soil texture in controlling the thermal permeability was also suggested at White Island, New Zealand (Mongillo and Wood, 1995), where lacustrine sediment and fractured lavas control the distribution of fumarole or thermal anomalies. At Yenkahe, Vanuatu (Peltier et al., 2012), differences in soil permeability and thermal expression were observed involving scoria layers and ash layers. Whereas at these two sites the permeability contrasts are inherited mostly from primary eruptive products and processes, at Stefanos it is inherited from sedimentary depositional processes. A temporal development of sediment accumulation and, accordingly, a temperature field change at volcanic craters is open for speculation. In the first stage, we suspect the occurrence of crater inward sliding of peripheral blocks as observed at Anatahan, Northern Mariana Island (Nakada et al., 2005) and at Miyakejima, Japan (Geshi et al., 2012). In a later stage, the crater collects sediments through ephemeral fluvial-to-deltaic depositional processes, which activate with seasonal rainfalls. Our work hence suggests that permeability contrasts generated by depositional processes may be relevant for the evolution of the fumarole field elsewhere, possibly even being independent from the crater's genesis. Any process reducing the soil permeability (e.g., compaction, argillification) or increasing the conductive loss (e.g., increased sediment thickness) might be a factor of risk, because, while reducing the appearance of the heat signature, it also contributes to the seal the hydrothermal system. The sealing of hydrothermal sources may lead to overpressure in the hydrothermal system, facilitating phreatic explosions (Marini et al., 1993).

5.3.3 Structural control

While the role of soil texture has been generally underestimated, many previous studies at White Island, New Zealand (Mongillo and Wood, 1995), at Etna, Vulcano, and Stromboli, Italy (Finizola et al., 2003; Aubert et al., 2008), at Kudryavy, Kurili islands (Yudovskaya et al., 2008), and Yenkahe, Vanuatu (Peltier et al., 2012) identified and attributed the thermal anomalies to tectonic fault planes or crater border facilitating the migration and escape of fluids. These discontinuities may have a geomorphologic expression at the surface or may be buried.

At the Nisyros caldera the elongation of the crater's major axis, the structural data and CO_2 flux measurements indicate that tectonics exert a strong control on fluid pathways (Caliro et al., 2005). Fractures are identified crosscutting the southern flank of the Stefanos crater (Caliro et al., 2005). These fractures are expected to control the fluid flow at depth, whereas at a shallow level their directional permeability may be overprinted by the soil distribution, which pervasively distributes the thermal expression. Indeed, we observed that the wider thermal features lie parallel to the fracture direction, but we do not recognize a well-defined linear thermal pattern. Also, the gas flux measured during the unrest in 1999–2001 appeared almost homogeneously distributed across the crater (Caliro et al., 2005), which may result from the enhanced activity level. Nevertheless, there are two more pieces of evidence, the boiling ponds and the thermal anomaly along the northern flank (p and g in Figs. 3, 4, and 7, respectively), suggesting that fractures occur in the subsurface and locally influence the thermal field.

5.3.4 Stress control

Previous authors have also studied the influence of the gravitational stress field on fluid uprise (Schöpa et al., 2011). We address this topic at Nisyros only qualitatively because the topography of the Stefanos crater is minor. The crater is excavated ~ 30 m below the mean caldera topography, and little is known about the mechanical parameters of the sediments. A suitable analogue for this morphology is represented by open pit mines, so we can adopt results from such studies (Stacey et al., 2003) to our case study. Firstly, the stress field generated by opposite topographic reliefs does not interact when the distance of the reliefs is 0.8 times their height. At Nisyros, this condition is met both at the scales of the caldera and of the Stefanos crater. Secondly, close to the surface, the trajectories of the maximum principal stress ($\sigma 1$) follow the topography, being near-vertical at the crater walls and horizontal at the crater floor (Stacey et al., 2003). Similarly, another study about the stress field inside a crater morphology suggests that maximum horizontal stresses are compressive on the crater floor (Schöpa et al., 2011). This may explain the absence of thermal degassing and thermal anomalies in the crater floor at Vulcano Island (Schöpa et al., 2011). However, an important difference exists between the two sites; at the crater floor of Stefanos the thermal anomalies are clearly expressed, whereas at Vulcano thermal anomalies are seen in the crater floor.

5.4 Implications for future studies

We think that additional case studies might improve the description and quantification of the geological factors influencing fluid flow in geothermal areas. While the present study introduced additional complexities to be considered when inferring deeper geological settings from thermographic data, repeated surveys can witness the evolution of the fumarole field. The thermal field of some volcanoes, e.g., Vulcano Island, Iwodake, and Colima (Harris and Maciejewski, 2000; Matsushima et al., 2003; Stevenson and Varley, 2008) have experienced temporal and spatial changes. Such changes are common in active volcanic and geothermal settings and our results suggest that the observation and moni-

toring of lithologies, fault and fractures, and stresses can improve the understanding of those changes. We propose that thermal anomalies may be reduced by the additional deposition of sediments acting as an insulating layer; conversely, a removal of material should augment the appearance. The deposition or removal of large volumes may be associated with structural changes such as faulting, landslide or with eruptive processes, which have the potential to modify the degassing and thermal anomaly.

From a technical perspective, the interpretation of thermal fields benefits from the integration of IR imaging and geological information. GIS-based analysis finally allows us to generate historical databases that easily integrate different information. This would undoubtedly benefit the hazard assessment by improving the interpretation of infrared data in terms of increased activity or increased permeability.

6 Conclusions

We used IR measurements to map the spatial arrangement of the thermal field at the phreatic crater of Stefanos on Nisyros Island (Greece). The combination of IR images and soil grain-size analysis suggests that the temperature field is influenced by depositional processes as well as structural features such as faults. Warmer areas occur near the edge of the crater floor within permeable soil, while colder areas occur within the cohesive sediments at the center. Localized thermal anomalies might be associated with local or island-wide fractures, with increased erosional propensity. Consequently, three main factors control fumarole activity: lithology (Peltier et al., 2012, this study), volcanotectonic structures (Finizola et al., 2003), and the stress field (Schöpa et al., 2011). The relative importance of these three main players may change from site to site. Assuming that permeability changes control the thermal field at the surface, thermal mapping might allow us to assess the permeability setting of the near subsurface. We suggest that the parameters controlling the fumarole sites might be applicable elsewhere, though their hierarchy and efficiency may vary according to the genesis of the volcanic morphology and the surrounding geological setting.

Acknowledgements. We gratefully acknowledge A. Klatten, E. Günther, M. Ramatschi and K. Wiencke for help and support during the various field campaigns, and M. Zahariadis for logistic support on the island. We thank G. Frijia and C. Fisher for the provision of the laboratory facilities and materials at the University of Potsdam. Detailed reviews by D. Pyle and H. Kinvig, and by the Topical Editor, Y. Lavallee, are acknowledged. Financial support by GFZ Potsdam, the Helmholtz Initiating and Networking award, and the Helmholtz EOS-2 program are greatly appreciated.

Edited by: Y. Lavallee

References

Alyamani, M. S. and Zekâi, Ş.: Determination of Hydraulic Conductivity from Complete Grain-Size Distribution Curves, Groundwater, 31, 551–555, 1993.

Ambrosio, M., Doveri, M., Fagioli, M. T., Marini, L., Principe, C., and Raco, B.: Water–rock interaction in the magmatic-hydrothermal system of Nisyros Island (Greece), J. Volcanol. Geothermal Res., 192, 1, 57–68, 2010.

ASTM-D6913-04: Standard Test Methods for Particle-Size Distribution (Gradation) of Soils Using Sieve Analysis, Am. Soc. Test. Materials, 6913-04, doi:10.1520/D6913-04R09, 2009.

Aubert, M., Diliberto, S., Finizola, A., and Chébli, Y.: Double origin of hydrothermal convective flux variations in the Fossa of Vulcano (Italy), Bull. Volcanol., 70, 743–751, 2008.

Ball, M. and Pinkerton, H.: Factors affecting the accuracy of thermal imaging cameras in volcanology, J. Geophys. Res. Solid Earth, 111, B11203, doi:10.1029/2005JB003829, 2006.

Benson, C. H. and Trast J. M.: Hydraulic conductivity of thirteen compacted clays, Clays Clay Min., 43, 669–681, 1995.

Bowles, J. E.: Foundation analysis and design, McGraw-Hill Book Company Limited, England, 1988.

Brazier, S., Sparks, R. S. J., Carey, S. N., Sigurdsson, H., and Westgate, J. A.: Bimodal grain size distribution and secondary thickening inait-fall ash layer, Nature, 301, 115–119, 1983.

Bukumirovic, T., Italiano, F., and Nuccio, P.: The evolution of a dynamic geological system: the support of a GIS for geochemical measurements at the fumarole field of Vulcano, Italy, J. Volcanol. Geothermal Res., 79, 253–263, 1997.

Caine, J. S., Evans J. P., and Forster C. B.: Fault zone architecture and permeability structure, Geology, 24, 1025–1028, 1996.

Caliro, S., Chiodini, G., Galluzzo, D., Granieri, D., La Rocca, M., Saccorotti, G., and Ventura, G.: Recent activity of Nisyros volcano (Greece) inferred from structural, geochemical and seismological data, Bull. Volcanol., 67, 358–369, 2005.

Chiodini, G., Brombach, T., Caliro, S., Cardellini, C., Marini, L., and Dietrich, V.: Geochemical indicators of possible ongoing volcanic unrest at Nisyros Island (Greece), Geophys. Res. Lett., 29, p. 16, doi:10.1029/2001GL014355, 1759.

Chiodini, G., Cioni, R., Leonis, C., Marini, L., and Raco, B.: Fluid geochemistry of Nisyros island, Dodecanese, Greece, J. Volcanol. Geothermal Res., 56, 95–112, 1993.

Chiodini, G., Vilardo, G., Augusti, V., Granieri, D., Caliro, S., Minopoli, C., and Terranova, C.: Thermal monitoring of hydrothermal activity by permanent infrared automatic stations: Results obtained at Solfatara di Pozzuoli, Campi Flegrei (Italy), J. Geophys. Res. Solid Earth, 112, B12206, doi:10.1029/2007JB005140, 2007.

Dobson, P. F., Kneafsey, T. J., Hulen, J., and Simmons, A.: Porosity, permeability, and fluid flow in the Yellowstone geothermal system, Wyoming, J. Volcanol. Geothermal Res., 123, 313–324, 2003.

Dozier, J.: A method for satellite identification of surface temperature fields of subpixel resolution, Remote Sens. Environ., 11, 221–229, 1981.

Faulkner, D. R., Jackson C. A. L., Lunn R. J., Schlische R. W., Shipton Z. K., Wibberley C. A. J., and Withjack M. O.: A review of recent developments concerning the structure, mechanics and fluid flow properties of fault zones, J. Struct. Geol., 32, 1557–1575, 2010.

Finizola, A., Sortino, F., Lénat, J.-F., Aubert, M., Ripepe, M., and Valenza, M.: The summit hydrothermal system of Stromboli. New insights from self-potential, temperature, CO₂ and fumarolic fluid measurements, with structural and monitoring implications, Bull. Volcanol., 65, 486–504, 2003.

Ganas, A., Lagios, E., Petropoulos, G., and Psiloglou, B.: Thermal imaging of Nisyros volcano (Aegean Sea) using ASTER data: estimation of radiative heat flux, Internat. J. Remote Sens., 31, 4033–4047, 2010.

Geotermica Italiana: Nisyros 1 geothermal well, PPC-EEC report, p. 160, 1983.

Geotermica Italiana: Nisyros 2 geothermal well, PPC-EEC report, p. 44, 1984.

Geshi, N., Acocella, V., and Ruch, J.: From structure- to erosion-controlled subsiding calderas: evidence thresholds and mechanics, Bull. Volcanol., 74, 1553–1567, 2012.

Gottsmann, J., Carniel, R., Coppo, N., Wooller, L., Hautmann, S., and Rymer, H.: Oscillations in hydrothermal systems as a source of periodic unrest at caldera volcanoes: Multiparameter insights from Nisyros, Greece, Geophys. Res. Lett., 34, L07307, doi:10.1029/2007GL029594, 2007.

Gottsmann, J., Rymer, H., and Wooller, L.: On the interpretation of gravity variations in the presence of active hydrothermal systems: Insights from the Nisyros Caldera, Greece, Geophys. Res. Lett., 32, L23310, doi:10.1029/2005GL024061, 2005.

Graton, L. C. and Fraser, H. J.: Systematic packing of spheres: with particular relation to porosity and permeability, J. Geol., 43, 785–909, 1935.

Hardee, H.: Permeable convection above magma bodies, Tectonophysics, 84, 179–195, 1982.

Harris, A. and Maciejewski, A.: Thermal surveys of the Vulcano Fossa fumarole field 1994–1999: evidence for fumarole migration and sealing, J. Volcanol. Geothermal Res., 102, 119–147, 2000.

Harris, A. J., Lodato, L., Dehn, J., and Spampinato, L.: Thermal characterization of the Vulcano fumarole field, Bull. Volcanol., 71, 441–458, 2009.

Kinvig, H. S., Winson, A., and Gottsmann, J.: Analysis of volcanic threat from Nisyros Island, Greece, with implications for aviation and population exposure, Nat. Hazards Earth Syst. Sci., 10, 1101–1113, doi:10.5194/nhess-10-1101-2010, 2010.

Lagios, E., Vassilopoulou, S., Sakkas, V., Dietrich, V., Damiata, B., and Ganas, A.: Testing satellite and ground thermal imaging of low-temperature fumarolic fields: The dormant Nisyros Volcano (Greece), ISPRS J. Photogram. Remote Sens., 62, 447–460, 2007.

Lambe, T. W., and Whitman, R. V.: Soil mechanics SI version, John Wiley&Sons, 2008.

Marini, L., Principe, C., Chiodini, G., Cioni, R., Fytikas, M., and Marinelli, G.: Hydrothermal eruptions of Nisyros (Dodecanese, Greece), past events and present hazard, J. Volcanol. Geothermal Res., 56, 71–94, 1993.

Matsushima, N., Kazahaya, K., Saito, G., and Shinohara, H.: Mass and heat flux of volcanic gas discharging from the summit crater of Iwodake volcano, Satsuma-Iwojima, Japan, during 1996–1999, J. Volcanol. Geothermal Res., 126, 285–301, 2003.

Mongillo, M. and Wood, C.: Thermal infrared mapping of White Island volcano, New Zealand, J. Volcanol. Geothermal Res., 69, 59–71, 1995.

Nakada, S., Matsushima, T., Yoshimoto, M., Sugimoto, T., Kato, T., Watanabe, T., Chong, R., and Camacho, J. T.: Geological aspects of the 2003–2004 eruption of Anatahan Volcano, Northern Mariana Islands, J. Volcanol. Geothermal Res., 146, 226–240, 2005.

Pantaleo, M.: Geothermal and deformation activity observed at volcanoes by using high resolution imaging, PhD Thesis, Mathematisch-Naturwissenschaftlichen Fakultat der Universitat Potsdam, 2014.

Papadopoulos, G. A., Sachpazi, M., Panopoulou, G., and Stavrakakis, G.: The volcanoseismic crisis of 1996–1997 in Nisyros, SE Aegean Sea, Greece, Terra Nova, 10, 151–154, 1998.

Parks, M., Caliro, S., Chiodini, G., Pyle, D., Mather, T., Berlo, K., Edmonds, M., Biggs, J., Nomikou, P., and Raptakis, C.: Distinguishing contributions to diffuse CO_2 emissions in volcanic areas from magmatic degassing and thermal decarbonation using soil gas ^{222}Rn–δ^{13}C systematics: Application to Santorini volcano, Greece, Earth Planet. Sci. Lett., 377, 180–190, 2013.

Peltier, A., Finizola, A., Douillet, G. A., Brothelande, E., and Garaebiti, E.: Structure of an active volcano associated with a resurgent block inferred from thermal mapping: The Yasur–Yenkahe volcanic complex (Vanuatu), J. Volcanol. Geothermal Res., 243/244, 59–68, 2012.

Revil, A., Finizola, A., Piscitelli, S., Rizzo, E., Ricci, T., Crespy, A., Angeletti, B., Balasco, M., Barde Cabusson, S., Bennati, L., Bolève, A., Byrdina, S., Carzaniga, N., Di Gangi, F., Morin, J., Perrone, A., Rossi, M., Roulleau, E., and Suski, B.: Inner structure of La Fossa di Vulcano (Vulcano Island, southern Tyrrhenian Sea, Italy) revealed by high-resolution electric resistivity tomography coupled with self-potential, temperature, and CO_2 diffuse degassing measurements, J. Geophys. Res., 113, B07207, doi:10.1029/2007JB005394, 2008.

Sachpazi, M., Kontoes, C., Voulgaris, N., Laigle, M., Vougioukalakis, G., Sikioti, O., Stavrakakis, G., Baskoutas, J., Kalogeras, J., and Lepine, J. C.: Seismological and SAR signature of unrest at Nisyros caldera, Greece, J. Volcanol. Geothermal Res., 116, 19–33, 2002.

Sawyer, G. M. and Burton, M. R.: Effects of a volcanic plume on thermal imaging data, Geophys. Res. Lett., 33, L14311, doi:10.1029/2005GL025320, 2006.

Schöpa, A., Pantaleo, M., and Walter, T.: Scale-dependent location of hydrothermal vents: Stress field models and infrared field observations on the Fossa Cone, Vulcano Island, Italy, J. Volcanol. Geothermal Res., 203, 133–145, 2011.

Sekioka, M. and Yuhara, K.: Heat flux estimation in geothermal areas based on the heat balance of the ground surface, J. Geophys. Res., 79, 2053–2058, 1974.

Shepherd, R. G.: Correlations of permeability and grain size, Groundwater, 27, 633–638, 1989.

Shinohara, H., Kazahaya, K., Saito, G., Matsushima, N., and Kawanabe, Y.: Degassing activity from Iwodake rhyolitic cone, Satsuma-Iwojima volcano, Japan: Formation of a new degassing vent, 1990–1999, Earth Planets Space, 54, 175–186, 2002.

Spampinato, L., Calvari, S., Oppenheimer, C., and Boschi, E.: Volcano surveillance using infrared cameras, Earth-Sci. Rev., 106, 63-91, 2011.

Springer, D., Loaiciga, H., Cullen, S., and Everett, L.: Air permeability of porous materials under controlled laboratory conditions, Ground Water, 36, 558–565, 1998.

Stacey, T. R., Xianbin, Y., Armstrong, R., and Keyter, G. J.: New slope stability considerations for deep open pit mines, J. South African Inst. Mining Metall., 103, 373–390, 2003.

Stevenson, D. S.: Physical models of fumarolic flow, J. Volcanol. Geothermal Res., 57, 139–156, 1993.

Stevenson, J. A. and Varley, N.: Fumarole monitoring with a hand-held infrared camera: Volcán de Colima, Mexico, 2006–2007, J. Volcanol. Geothermal Res., 177, 911–924, 2008.

Tassi, F., Vaselli, O., Papazachos, C. B., Giannini, L., Chiodini, G., Vougioukalakis, G. E., Karagianni, E., Vamvakaris, D., and Panagiotopoulos, D.: Geochemical and isotopic changes in the fumarolic and submerged gas discharges during the 2011–2012 unrest at Santorini caldera (Greece), Bull. Volcanol., 75, 1–15, 2013.

Teschner, M., Faber, E., Poggenburg, J., Vougioukalakis, G. E., and Hatziyannis, G.: Continuous, direct gas-geochemical monitoring in hydrothermal vents: Installation and long-term operation on Nisyros Island (Greece), Pure Appl. Geophys., 164, 2549–2571, 2007.

Tibaldi, A., Pasquarè, F., Papanikolaou, D., and Nomikou, P.: Tectonics of Nisyros Island, Greece, by field and offshore data, and analogue modelling, J. Struct. Geol., 30, 1489–1506, 2008.

Varley, N. R. and Taran, Y.: Degassing processes of Popocatepetl and Volcan de Colima, Mexico: Geological Society, London, Special Publications, 213, 263–280, 2003.

Vougioukalakis, G. and Fytikas, M.: Volcanic hazards in the Aegean area, relative risk evaluation, monitoring and present state of the active volcanic centers, Develop. Volcanol., 7, 161–183, 2005.

Yudovskaya, M. A., Tessalina, S., Distler, V. V., Chaplygin, I. V., Chugaev, A. V., and Dikov, Y. P.: Behavior of highly-siderophile elements during magma degassing: A case study at the Kudryavy volcano, Chem. Geol., 248, 318–341, 2008.

6

Conventional tillage versus organic farming in relation to soil organic carbon stock in olive groves in Mediterranean rangelands (southern Spain)

L. Parras-Alcántara and B. Lozano-García

Department of Agricultural Chemistry and Soil Science, Faculty of Science, Agrifood Campus of International Excellence – ceiA3, University of Cordoba, 14071 Cordoba, Spain

Correspondence to: L. Parras-Alcántara (qe1paall@uco.es)

Abstract. Soil organic carbon (SOC) concentration is a soil variable subject to changes. The management system is a key factor that influences these changes. To determine the long-term effects of the management system on SOC stocks (SOCS) in olive groves, 114 soil profiles were studied in the Los Pedroches Valley (Mediterranean rangelands – southern Spain) for 20 years. The management practices were conventional tillage (CT) and organic farming (OF) in four soil types: Cambisols (CMs), Regosols (RGs), Luvisols (LVs) and Leptosols (LPs). Soil properties were statistically analysed by management techniques, soil types and horizons. Significant differences ($p < 0.05$) were found between soil types and management practices. It was equally observed that the management system affected SOCS. In addition, the total SOCS during the 20-year experiment increased in OF with respect to CT by 72 and 66 % in CMs and LVs respectively. SOC showed significant differences for horizons ($p < 0.05$) in relation to the management type. The stratification ratio (SR) was used as an indicator of soil quality based on the influence of surface SOC levels on erosion control, water infiltration and nutrient conservation with respect to deep layers. The SR of SOC from the surface to depth was greater in CT compared to OF with the exception of RGs. In all cases, the SR of SOC was > 2. These results indicate high soil quality and that management practices affect SOC storage in the Los Pedroches Valley.

1 Introduction

Over the centuries, olive groves (OGs) have been a relevant social and economic heritage of Mediterranean areas. In Spain, the olive growing surface area is 2.58 million ha, increasing on average by 1–1.5 % per year from 1995 to the present (ESYRCE, 2012). Subsidies and the rising price of olive oil have conditioned this growth (Louwagie et al., 2011). The olive oil production in Andalusia (Spain) is more than 1.3 million tonnes, constituting 85 % of total Spanish production (MAGRAMA, 2012) and 41 % of what is produced worldwide (IOC, 2012).

This activity has brought economic benefits to the region, but there have also been adverse effects. Olive production has traditionally been based on low tree density (100 trees ha^{-1}), weeds being controlled by tillage and tree size limited by pruning (Álvarez et al., 2007). Traditionally, the management strategy has been conventional tillage (CT), marked by the frequent use of mouldboard ploughs, mineral fertilisation and herbicides. CT in OGs has caused a loss of soil quality with significant economic and environmental implications. CT has contributed to alterations in the nitrogen cycle (Fernández-Escobar et al., 2009), water loss by evaporation (Cerdà and Doerr, 2007), destruction of the soil structure (Gómez et al., 2009), loss of soil organic matter (SOM) and nutrients (Paustian et al., 2000) and high erosion rates (Calatrava et al., 2011). Traditional OG management has been associated with soil erosion (Castro et al., 2008), river and water body pollution (Colombo et al., 2005), degradation of landscape (Parra-López et al., 2009), and climate change (Rodríguez-Entrena et al., 2012). Besides, CT reduces soil fertility and OG production, increasing production costs (Calatrava-Leyva et al., 2007). This is particularly aggravated in Mediterranean climatic conditions (Gómez et al., 2009).

Recent studies show that the restriction on tillage (Parras-Alcántara et al., 2013a), and/or the addition of organic residues to soils (Lozano-García et al., 2011; Lozano-García and Parras-Alcántara, 2013a) may improve soil quality. Accordingly, organic farming (OF) may be an attractive option for reducing the soil degradation processes (Cerdà et al., 2010; Aranda et al., 2011).

In Mediterranean areas, the elements that affect soil carbon (C) variability are mainly climate (Wang et al., 2010), land use (Lozano-García and Parras-Alcántara, 2013b), management (Parras-Alcántara et al., 2013b, 2014a), slope and altitude (Hontoria et al., 2004; Lozano-García and Parras-Alcántara, 2014), tillage intensity and no-till duration (Conant et al., 2007). Of all these factors, soil management is the best tool for climate change mitigation and adaptation (Lal et al., 2011). Over time, some researchers have studied the relationship between soil management effects in OGs and C capture and storage in soils as soil organic carbon (SOC) (Parras-Alcántara et al., 2013a, b; Romanya and Rovira, 2011). However, many of these studies have evaluated SOC content on the soil surface, whereas only a few studies have included deeper soil sections. In Mediterranean areas, SOC can be stored in deep layers (below 30 cm deep). This is important in OF, as SOC can be transported to deeper soil horizons, contributing to the subsoil C storage (Lorenz and Lal, 2005).

Climatic conditions in Mediterranean areas are limiting factors that affect accumulation of SOC. Thus, SOC determination might not be the best indicator of the improvement caused by management, as mineralisation rates vary with depth. Under these conditions, it may be more interesting to consider the stratification ratio (SR) of SOC (Corral-Fernández et al., 2013). The use of the SR as a soil quality indicator is based on the influence of surface SOC levels on erosion control, water infiltration and nutrient conservation (Franzluebbers, 2002). High SOC and nitrogen (N) SR values reflect undisturbed soil and high soil quality of the surface layer. The SR growth can be related to the rate and amount of SOC sequestration. However, ratios < 2 are frequent in degraded soils (Franzluebbers, 2002).

Very limited information is available regarding OF effects under semi-arid Mediterranean conditions in organic OGs. Consequently, the aims of this study were: (i) to determine the soil properties that affect soil development in the Los Pedroches Valley (OG in Mediterranean rangelands – southern Spain), (ii) to study the vertical distribution of SOC stock under two management practices (CT and OF) and (iii) to analyse the soil variables that are involved in the SR of SOC in OG using traditional and organic management systems in entire soil profiles.

2   Material and methods

2.1   Study site and management type

The study area is located in the Los Pedroches Valley (Cordoba, southern Spain), between 38.39 and 37.15° N, 4.50 and 4.15° W (Fig. 1) and comprises 10 600 ha where the dominant land use is OG. The average annual temperature is 16 °C and the annual thermal amplitude ranges from -2 °C in winter to 40 °C in summer. The study area is characterised by cold winters and warm summers, and the average annual rainfall is 600 mm. The climate is temperate semi-arid Mediterranean with continental influence. The moisture regime is dry Mediterranean. High temperatures and long drought periods cause water deficits up to 400 mm per year. The relief is smooth, with slopes ranging from 1 to 4 %, and the parent material is granite. According to IUSS Working Group WRB (2006), the most abundant soils are Cambisols (CMs), Luvisols (LVs), Regosols (RGs) and Leptosols (LPs).

Two soil management practices were selected for this study: OF and CT in four soil types (CMs, LVs, RGs and LPs). The OF were under this practice for 20 years (1989–2009), with no synthetic mineral fertilisation or pesticides and untilled soils. In these, the vegetative cover is kept under control by mowing from early to late spring with animal manure incorporation. CT is characterised by three ploughs per year at a depth of 15–20 cm from early spring (disk harrow and cultivator) to early autumn (tine harrow), weed control with residual herbicides, and annual application of ammonium sulphate (250 kg ha^{-1}). Moreover, two applications of foliar fertilisation per year were performed in CT, in spring (0.3 kg amino acids ha^{-1}, 0.7 kg N ha^{-1}, 0.4 kg P ha^{-1} and 0.3 kg K ha^{-1}) and in autumn (0.09 kg amino acids ha^{-1} and 1 kg K ha^{-1}).

In all cases (soil types and management practices), the average density of *Olea europaea* spp. *europaea* in OG is 100–110 trees ha^{-1}. Forty-year old olive trees, spaced 10×10 m, were selected for the study.

2.2   Soil sampling and analyses

Soil samples (entire profiles: 50 in CMs, 20 in LVs, 28 in LPs and 16 in RGs) were collected according to FAO (2006) in a random sample design (representative of the whole study zone): 70 samples in CT and 44 samples in OF (114 profiles $\times$ 3 or 2 horizons $\times$ 4 replicated) in the Los Pedroches Valley in 2009 in OG. We have selected fewer soil profiles in OF because the study area was smaller than the study area in CT. Soil entire profiles were collected in open areas.

Soil samples were air-dried at a constant room temperature (25 °C) and sieved (2 mm) to remove coarse soil particles. Four replicates of each sample were analysed in the laboratory to reduce the experimental error. Analytical methods and others parameters calculated are described in Table 1.

Table 1. Methods used in field measurements, laboratory analysis and calculated from field data.

Parameters	Method
Field measurements	
Bulk density (Mg m^{-3})	Cylindrical core sampler* (Blake and Hartge, 1986)
Laboratory analysis	
Particle size distribution	Robinson pipette method (USDA, 2004)**
pH–H$_2$O	1 : 2.5 suspension in water (Guitián and Carballas, 1976)
N (%)	Kjeldahl method (Bremner, 1996)
CO$_3$Ca equivalent	Volumetric with Bernard calcimeter (Duchaufour, 1975)
Organic C (%)	Walkley and Black method (Nelson and Sommers, 1982)
Ca^{2+}, Mg^{2+}, Na$^+$, K$^+$, C.E.C (cmol kg^{-1})	(Bower et al., 1952)
Base saturation (%)	(Duchaufour, 1975)
Hydraulic conductivity (mm h^{-1})	Mono-disc multiple potential process (Reynolds and Elrick, 1991)
Parameters calculated from field data	
C : N ratio	Ratio of organic C to organic N
SOC stock (Mg ha^{-1})	(SOC concentration × BD × d × $(1 - \delta_{2\,mm}$ %) × 0.1)*** (Wang and Dalal, 2006)
Total SOC stock (Mg ha^{-1})	$\Sigma_{horizons}$ SOC Stock$_{horizon}$ (IPCC, 2003)
SR	(SOC-Ap onto SOC-Bw/C)**** (Franzluebbers, 2002)

* 3 cm diameter, 10 cm length and 70.65 cm^3 volume.
** Prior to determining the particle size distribution, samples were treated with H$_2$O$_2$ (6 %) to remove organic matter (OM). Particles larger than 2 mm were determined by wet sieving and smaller particles were classified according to USDA standards (2004).
*** Where SOC is the organic carbon content (g kg^{-1}), d the thickness of the soil layer (cm), $\delta_{2\,mm}$ is the fractional percentage (%) of soil mineral particles > 2 mm in size in the soil, and BD the soil bulk density (Mg m^{-3}).
**** The SR is defined as a soil property on the soil surface divided by the same property at a lower depth. In this study, we defined two SRs for the CM and LV [SR1 (Ap/Bw-Bt) and SR2 (Ap/C)] and one SR for LP and RG [SR1 (Ah/C)].

Fig. 1. Study area.

2.3 Statistical analysis

A statistical analysis for each management system was applied to evaluate the physical and chemical horizon properties. Pearson's correlation coefficients were carried out to understand the relationships between different parameters (physical and chemical). An analysis of variance (ANOVA) was carried out to determine the importance of three sources of variability (main factors): soil type (CMs, LVs, LPs and RGs), soil management (CT and OF), and horizons (Ap-h, Bw-t and C) and their interactions. Significant differences were deemed statistically relevant from the Turkey test was used to compare the results, with significant differences considered at $p < 0.05$. A principal component analysis (PCA) was performed to minimise and explain the variability of the system; only the eigenvalues > 0.3 were considered for PCA interpretation. The Anderson–Darling normality test was used to check the normal distribution of SR. All calculations, including statistical analysis, were computed using the Minitab software package (Minitab, 2000).

3 Results and discussion

3.1 Soil properties and principal components analyses

The studied soils were CMs, LPs, RGs and LVs (IUSS-ISRIC-FAO, 2006). Data for the soil profiles are compiled in Table 2. The soils in the Los Pedroches Valley are conditioned by lithology (granite) and the formation processes of these soils take place with low slope (Nerger et al., 2007), thus they are shallow soils (Marañón, 1988).

LVs are fertile soils that are suitable for a wide range of Mediterranean crops such as cereals, fruit trees, olives and vineyards (Zdruli et al., 2011). The principal characteristic of these soils is a high clay content in the subsoil (CT-LVs-Bt: 39.6 %; OF-LVs-Bt: 30.2 %) compared to the topsoil

Table 2. Soil properties evaluated (average ± SD**) in the Los Pedroches Valley (Mediterranean rangeland) in olive groves.

(CT-LVs-Ap: 16.8 %; OF-LVs-Ap: 11.7 %) as a result of pedogenetic processes (clay migration – leading to an argillic subsoil horizon). An important feature of these soils is the low OM content (González and Candás, 2004). This is justified by climatic conditions (semi-arid Mediterranean) and soil texture (sandy soils), which contribute to a low organic matter content (Parras-Alcántara et al., 2013b, 2014b). With respect to CMs, RGs and LPs, these soils are characterised by low fertility, poor physical conditions and a marginal capacity for agricultural uses. CMs are then more developed than RGs and LPs. LPs are the less developed soils, influenced by topography and physiographic location (Recio et al., 1986).

Normally, the studied soils had an acid pH (5.3–6.7) and moderated or saturated base saturation (BS) (100–77.3 %), mainly calcium. With regard to N concentrations, these were lower in CMs and LVs (0.13 % A horizon in CMs-OF; 0.02 % B horizon in LVs-CT). Another important characteristic was a decrease in nutrient content with depth. Also, the sand proportion was higher in LVs, RGs and LPs under OF than in soils under CT. However, an opposite trend (more in CT than OF) was observed for clay content, pH, BS and hydraulic conductivity (HC). The CEC was high, ranging from 12.45 cmol kg^{-1} to 30.66 cmol kg^{-1} (limit proposed by Hazelton and Murphy, 2007 based on Metson, 1961). In this line, Ruiz et al. (2012) in Mediterranean rangelands obtained similar results, while Pulido-Fernández et al. (2013) in Iberian open woodland rangelands obtained low CEC values caused by nutrient scarcity. SOC content was generally low in all the studied soils, although the SOC decrease with depth was much more intense in CT soils. The higher SOC content in topsoil in OF could be attributed to the management type, which increases the soil vegetative cover and maximises the organic residue transfer. Also, it reduced the soil erosion risk. Similar results were obtained by Corral-Fernández et al. (2013) for evergreen oak woodland with OF in the Los Pedroches Valley. In that context, OF can be regarded as a key factor to take into account when considering environment-friendly management practices (Hathaway-Jenkins et al., 2011). The carbon–nitrogen (C : N) ratio suggested generally suitable conditions for active microbial development and humus recycling (Tables 2 and 3).

Factors affecting soil development in the study area were predominantly parameters related to the soil chemical properties and, to a lesser degree, chemical parameters. Numerous significant linear correlations were found among soil properties using Pearson's correlation matrix (Table 4). Some related variables were SOC and N, sand and clay content, and sand and silt content with an extremely strong correlation ($p < 0.001$; $r = 0.997$, $r = -0.810$, $r = -0.753$ respectively). Other significant relations were exchangeable Mg^{2+} and CEC, exchangeable Ca^{2+} and CEC, SOC and C : N ratio with a strong correlation ($p < 0.001$; $r = 0.662$, $r = 0.629$ and $r = 0.624$ respectively). Since extremely strong and strong correlations were found (Table 4), PCA was necessary to identify critical factors determining soil development in the

Table 3. Soil organic carbon, nitrogen and C:N ratio stock (average ± SD**) in the Los Pedroches Valley (Mediterranean rangeland) in olive groves.

Soil	Tillage	Horizon	SOC g kg⁻¹	T-SOC g kg⁻¹	N g kg⁻¹	TN g kg⁻¹	SOC-S Mg ha⁻¹	T-SOC-S Mg ha⁻¹	N-S Mg ha⁻¹	TN-S Mg ha⁻¹	C:N ratio
CM	CT n=32	Ap	7.63±3.09 a*		0.76±0.25 a*		24.84±5.46 a*		2.47±0.37 a*		10.06±1.02 a
		Bw	2.04±1.21 b*	11.67±1.99 a*	0.31±0.08 b*	1.33±0.17 a*	8.54±4.86 b*	43.75±5.11 a*	1.29±0.28 b*	5.11±0.81 a*	6.62±0.54 b*
		C	2.00±1.67 b*		0.26±0.17 c*		10.37±5.00 b*		1.35±0.48 b		7.68±0.57 c
	OF n=18	Ap	14.23±4.43 c		1.33±0.37 d		33.63±9.86 c		3.14±0.84 c		10.71±0.97 a
		Bw	4.88±3.09 d	23.31±3.23 b	0.53±0.26 e	2.34±0.27 b	24.52±5.15 a	74.7±6.67 b	2.66±0.39 a	7.69±0.71 b	9.21±0.84 a
		C	4.19±2.18 d		0.48±0.19 e		16.55±5.01 d		1.89±0.9 d		8.76±0.52 d
LV	CT n=12	Ap	7.74±1.61 a		0.77±0.12 a		23.22±1.29 a*		2.31±0.21 a*		10.05±0.95 a
		Bt	3.36±0.98 b*	14.31±1.15 a	0.23±0.10 b*	1.38±0.09 a*	17.75±4.06 b*	57.28±3.56 a*	1.22±0.25 b*	5.46±0.17 a*	14.55±1.14 b*
		C	3.21±0.87 b*		0.38±0.07 c*		16.31±5.32 b*		1.93±0.05 c*		8.45±0.69 c*
	OF n=8	Ap	7.74±1.93 a		0.77±0.16 a		30.59±9.71 c		3.04±0.16 d		10.06±0.89 a
		Bt	5.62±2.96 b	16.01±2.05 a	0.60±0.26 d	1.64±0.18 b	35.02±7.40 d	95.4±8.16 b	3.74±0.94 e	9.82±0.43 b	9.36±0.79 a
		C	2.65±1.26 c		0.27±0.12 e		29.79±7.38 c		3.04±0.20 d		9.80±0.74 a
LP	CT n=18	Ah	20.55±8.50 a		1.40±0.71 a		31.65±11.85 a		2.16±0.19 a*		14.65±1.12 a*
		C	5.34±1.62 b	25.89±5.06 a	0.41±0.13 b	1.81±0.42 a	19.99±7.43 b	51.64±9.64 a*	1.53±0.97 b*	3.69±0.58 a*	13.06±1.11 b*
	OF n=10	Ah	18.03±4.02 a		1.63±1.15 a		30.00±10.69 a		2.71±0.16 c		11.07±1.09 c
		C	4.88±3.36 b	22.91±3.69 b	0.49±0.31 b	2.12±0.73 a	12.77±3.60 c	42.77±7.15 b	1.28±0.77 b	3.99±0.47 b	9.98±0.78 c
RG	CT n=8	Ah	15.36±3.60 a*		1.19±0.17 a*		26.27±3.03 a*		2.03±0.88 a*		12.94±1.05 a*
		C	7.21±2.36 b*	22.57±2.98 a	0.58±0.17 b	1.77±0.17 a*	59.08±9.52 b*	85.35±6.27 a*	4.75±0.92 b*	6.78±0.90 a*	12.44±1.17 a*
	OF n=8	Ah	18.67±3.44 a		1.70±0.29 c		34.95±6.65 c		3.18±0.59 c		10.99±1.01 b
		C	5.60±3.01 b	24.27±3.23 a	0.60±0.25 b	2.31±0.27 b	18.23±5.23 d	53.18±5.94 b	1.95±0.81 a	5.13±0.70 b	9.35±0.89 b

SOC: Soil organic carbon; T-SOC: Total SOC; SOC-S: SOC stock; T-SOC-S: Total SOC stock; N: Nitrogen; TN: Total N; N-S: N stock; TN-S: TN stock. CM: Cambisol; LP: Leptosol; RG: Regosol; LV: Luvisol. CT: Conventional tillage; OF: Organic farming.

n = Sample size.

** Standard deviation;

* significant differences (P < 0.05) between CT and OF treatments (by horizons).

Numbers followed by different lower case letters within the same column have significant differences (P < 0.05) at different depths, considering the same soil type.

Los Pedroches Valley using PCA. Fifteen physical and chemical soil properties were included in the PCA, and four factors (eigenvalue > 3) were identified (Table 5). Considering average data for each soil properties, thirteen properties were found that account for 65 % of the variance. PCA explained 22.5, 19.9, 13.3 and 9.3 % of the variance from PC1, PC2, PC3 and PC4 respectively. Factor PC1 is positively correlated with exchangeable K^+ and Ca^{2+}, BD and SOC. This factor mostly groups parameters related to soil chemical conditions. PC2 received the greatest loading from sand content, clay content and thickness. All resulted as negative except sand content, grouping parameters related to the soil physical condition. PC3 grouped parameters related to the soil chemical condition (exchangeable Ca^{2+}, exchangeable Mg^{2+} and CEC); all with positive performance. Finally, PC4 grouped C:N ratio, exchangeable Na^+, BS and HC; all positive except C:N ratio (Table 5). As regards soil chemical properties, soils are defined by the dominance of basic cations, conditioned by lithology (Recio et al., 1986) and carbonates, especially in RGs and LPs. According to this, Nerger et al. (2007), in his study in the Los Pedroches Valley, justify the high influence of chemical properties caused by carbonate presence (high Ca^{2+}, basic pH and low OM in the top soil) that could affect soil development. In addition, the Ca^{2+} enrichment could be related to the continental influence (ion enrichment in the soil solution) due to climatic conditions (high rainfall in certain seasons). However, high tree density may affect SOC content (González et al., 2012). This is in agreement with our data (high nutrient contents and high CEC) compared with other Mediterranean rangelands. The PCA results showed that sand content (PC2) is a key factor that may affect soil development in the Los Pedroches Valley. In accordance with this, Hontoria et al. (2004) suggested that variables affecting soil development in the driest areas of Spain are those related to the specific texture more than those related to management or climate. Moreover, Castro et al. (2008) reported that soil texture is the first property that influences SOC in agricultural soils.

The ANOVA for PCA (horizons, soil type and management) showed that there were significant differences ($p < 0.001$) between soil types for PC2, related to physical and chemical properties. Also, when the management system was analysed, significant differences were found in PC1 ($p < 0.001$), PC2 ($p < 0.05$) and PC3 ($p < 0.05$), influenced principally by physical properties (Table 6). In the case of the horizon type, significant differences were found in PC1 ($p < 0.001$) and PC2 ($p < 0.05$) caused by thickness, SOC, N, BD, clay and K^+ (Table 6). However, when total SOC was analysed, no significant differences related to the management system were found in LVs, LPs and RGs. In this line, Parras-Alcántara et al. (2014b) indicates that OF has little effect on carbon stock in Mediterranean dehesa. Also, Bradford and Peterson's (2000) study assessing various land uses indicates that sometimes there is no difference between conventional, minimum, or reduced tillage.

Table 4. Pearson's correlation coefficients among soil properties in the Los Pedroches Valley. pH: pH (H_2O); SOC: soil organic carbon (g kg^{-1}); TN: total nitrogen (g kg^{-1}); C:N: C:N ratio; CO3Ca (%); exchangeable Ca^{2+}, Mg^{2+}, Na^+, K^+ (cmol kg^{-1}); CEC: cation exchange capacity (cmol kg^{-1}); BS: base saturation (%); BD: bulk density (Mg m^{-3}); HC: hydraulic conductivity (mm h^{-1}); sand, silt, clay (%).

	pH	SOC	TN	C/N	CO3Ca	Ca^{2+}	Mg^{2+}	Na$^+$	K+	CEC	BS	BD	HC	Sand	Silt	Clay
pH	1															
SOC	−0.045	1														
TN	−0.061	0.997**	1													
C/N	0.038	0.624**	0.581**	1												
CO3Ca	0.453**	0.064	0.067	0.025	1											
Ca^{2+}	0.546**	−0.128	−0.127	−0.189*	0.233*	1										
Mg^{2+}	0.141	−0.075	−0.093	−0.074	0.048	0.578**	1									
Na*	0.194*	0.002	−0.013	−0.009	−0.017	0.120	0.113	1								
K+	0.378**	0.042	0.041	−0.013	0.056	0.387**	0.010	0.297**	1							
CEC	0.272*	−0.147	−0.149	−0.159	−0.024	0.629**	0.662**	0.024	0.161	1						
BS	0.389**	0.033	0.022	0.098	0.226*	0.451**	0.231*	0.222*	0.308**	−0.231*	1					
BD	−0.087	−0.455**	−0.458**	−0.198*	0.102	−0.144	−0.044	−0.082	−0.459**	−0.043	−0.138	1				
HC	0.186*	−0.096	−0.092	−0.102	−0.034	0.183*	0.131	0.039	0.231*	0.090	0.129	−0.226*	1			
Sand	−0.034	−0.305**	−0.324**	0.178	0.012	−0.177	−0.083	−0.214*	−0.291**	−0.241*	0.063	0.036	0.035	1		
Silt	0.129	−0.122	−0.138	0.023	0.072	0.090	0.016	0.077	0.229*	0.108	−0.031	−0.001	−0.046	−0.753**	1	
Clay	−0.065	−0.343**	−0.358**	−0.285*	−0.081	0.181*	0.108	0.248*	0.226*	0.260*	−0.066	−0.053	−0.011	−0.810**	0.224*	1

* Correlation is significant at $p > 0.05$.
** Correlation is significant at $p < 0.001$.

Table 5. Ordinary components for the principal component analysis (PCA) of selected soil properties measured for all the soil types grouped*.

Variable or factor	PC1	PC2	PC3	PC4
Thickness (cm)	−0.155	−0.303	0.023	−0.062
BD (g cm^{-3})	0.385	0.167	0.222	0.108
SOC (g kg^{-1})	0.380	0.255	−0.214	0.025
C/N	0.058	0.287	−0.054	−0.428
pH (H$_2$O)	0.277	0.026	0.215	0.061
Ca^{2+} (cmol kg^{-1})	0.302	−0.083	0.485	0.070
Na$^+$ (cmol kg^{-1})	0.211	−0.031	−0.108	0.365
Mg^{2+} (cmol kg^{-1})	0.153	−0.073	0.489	−0.234
K$^+$ (cmol kg^{-1})	0.419	0.103	−0.083	0.168
CEC (cmol kg^{-1})	0.237	−0.276	0.364	−0.260
BS (%)	0.023	0.286	0.252	0.350
HC (mm h^{-1})	0.170	0.181	0.083	0.380
Sand (%)	−0.260	0.414	0.241	0.034
Silt (%)	0.271	−0.271	−0.170	−0.287
Clay (%)	0.158	−0.407	−0.225	0.222
Eigenvalues	3.608	3.178	2.135	1.494
% variance	22.5	19.9	13.3	9.3
Cumulative explanation	22.5	42.4	55.8	65.1

SOC: soil organic carbon; exchangeable Ca^{2+}, Mg^{2+}, Na$^+$, K$^+$; CEC: cation exchange capacity; BS: base saturation (%); BD: bulk density; HC: hydraulic conductivity.
* Variables underlined with eigenvectors (coefficients) > 0.3 are considered significant.

3.2 Soil organic carbon, nitrogen and C : N ratio

SOC and N concentrations decreased with depth (Table 3). SOC and N content increased in the A horizon from OF to CT in all soil types except the Ah horizon in LPs for SOC (20.55 g kg^{-1} CT; 18.03 g kg^{-1} OF). Castro et al. (2008) observed a similar trend at different depths in soils with OGs. As for the surface horizons, SOC values were highly heterogeneous, ranging from 20.55 g kg^{-1} to 7.63 g kg^{-1} for LPs-CT and CMs-CT respectively. N had a similar trend, ranging from 1.70 g kg^{-1} to 0.76 g kg^{-1} (RGs-OF; CMs-CT). Normally, high SOC values involved high N values. The SOC in CT was < 10 g kg^{-1} with the LPs exception in Ap horizons (20.55 g kg^{-1}). Conversely, SOC was higher (> 10 g kg^{-1}) in OF for all soils. In this line, Bronick and Lal (2005), explain that CT (low OM inputs, ploughing and low vegetation cover) in OGs limits the incorporation of organic residues and enhances soil erosion risk. Also, the formation process between OM and mineral aggregates diminishes in the surface horizons in sandy soils (González and Candás, 2004). This justifies high levels of transformed OM and explains low OM concentrations at greater depths in the studied soils. In this sense, López-Garrido et al. (2011) observed similar results for different crops and management systems. Besides, Franzluebbers (2005) concludes that the topsoil is more vulnerable to changes in management practices, and the carbon sequestration occurs principally in the upper horizons. However, the surface horizon is not the layer with the higher SOC sequestration potential because SOC exists in stable forms in depth, which makes it highly recalcitrant to biodegradation processes (Lorenz and Lal, 2005). Nevertheless, CT promotes a fast SOM mineralisation in Mediterranean agricultural soils (Melero et al., 2009).

With regard to the management system, SOC and N showed significant differences ($p < 0.05$) in all horizons for CMs, LPs and RGs (Table 3). In the case of total SOC content, significant differences ($p < 0.05$) were only found in CMs. However, when total N was analysed, significant differences ($p < 0.05$) were found for management types in all cases. The C : N ratio at the surface compared to that in depth was generally higher in OF compared to CT. This coincides with Blanco-Canqui and Lal (2008) and could be explained by the higher contribution of residue input on the surface under OF compared to CT. Also, this may reflect less OM decomposed in the soil surface in OF. Additionally, the residue retention can increase SOC (Xu et al., 2011) with lower decomposition degree and higher C : N ratio (Yamashita et al., 2006). The C : N ratio tended to decrease with depth in CMs, LPs and RGs under OF. However, an opposite trend was observed (increase in the C : N ratio with depth) in CMs and LVs in CT. In this sense, Diekow et al. (2005) and Yamashita et al. (2006) explained that this decrease with depth could be associated with clay content (that increased with depth). High clay levels are associated with high decomposed OM and a lower C : N ratio (Diekow et al., 2005; Yamashita et al., 2006). On the other hand, an opposite trend was observed in soils under CT, which may be attributed to high C : N soluble organic compounds leaching into deeper layers (Diekow et al., 2005).

SOC had strong positive correlations (Table 4) ($p < 0.001$) with N ($r = 0.997$) and C/N ratios ($r = 0.624$). A similar trend was observed with respect to BD, but in this case the correlation was negative ($r = −0.455$); when SOC increased, BD decreased. However, when thickness horizon increased, SOC was scattered and was likely to have underwent a redistribution process. Other small correlations ($p < 0.001$) were observed for sand content ($r = 0.305$) and clay content ($r = −0.343$). Similarly, correlations between SOC and other soil parameters in the surface horizon were found. SOC was strongly correlated with N clay content and exchangeable Na$^+$ ($r = 0.999$, $p = 0.000$; $r = −0.350$, $p = 0.008$; $r = −0.342$, $p = 0.009$). In addition to this, SOC was moderately correlated with sand content and exchangeable K$^+$ and Ca^{2+} ($r = 0.263$, $p = 0.048$; $r = −0.334$, $p = 0.011$; $r = −0.279$, $p = 0.036$). This correlates with the results of Hontoria et al. (2004), which suggest that the variables that affect soil development in the driest areas of Spain the most are physical variables, rather than the management or climatic ones. SOC and N content was higher in the surface horizons for CMs and LVs in OF with

Table 6. General lineal model (GLM). ANOVA (soil type and management) for soil. Statistical analyses of soil were carried out for A horizon samples (r coefficient).

	One-way ANOVA					
	Soil-type management			Management horizon		
Parameter	S	M	S × M	M	H	M × H
Th	0.227	0.713	0.012*	0.713	0.000***	0.888
pH	0.028*	0.397	0.010**	0.397	0.002**	0.151
SOC	0.727	0.000***	0.003**	0.000***	0.000***	0.007**
TN	0.452	0.000***	0.001***	0.000***	0.000***	0.008**
C / N	0.001***	0.370	0.002**	0.370	0.070	0.043*
Ca^{2+}	0.239	0.083	0.149	0.083	0.214	0.881
Mg^{2+}	0.832	0.132	0.177	0.132	0.342	0.903
Na^+	0.216	0.151	0.970	0.151	0.652	0.656
K^+	0.014*	0.000***	0.089	0.000***	0.000***	0.837
CEC	0.507	0.718	0.084	0.718	0.581	0.811
BS	0.008**	0.005**	0.946	0.005**	0.359	0.503
BD	0.037*	0.811	0.105	0.811	0.000***	0.393
HC	0.026*	0.017*	0.199	0.017*	0.076	0.132
Sand	0.360	0.000***	0.030*	0.000***	0.119	0.073
Silt	0.012*	0.007**	0.001***	0.007**	0.964	0.441
Clay	0.002**	0.000***	0.077	0.000***	0.017*	0.091
PC1	0.458	0.000***	0.016*	0.000***	0.861	0.719
PC2	0.000***	0.020*	0.156	0.020*	0.000***	0.465
PC3	0.416	0.020*	0.281	0.020*	0.017*	0.272
PC4	0.521	0.154	0.322	0.154	0.095	0.124

Th: thickness (cm); pH: pH (H_2O); SOC: soil organic carbon (g kg^{-1}); TN: total nitrogen (g kg^{-1}); C / N: C : N ratio; exchangeable Ca^{2+}, Mg^{2+}, Na^+, K^+ (cmol kg^{-1}); CEC: cation exchange capacity (cmol kg^{-1}); BS: base saturation (%); BD: bulk density (Mg m^{-3}); HC: hydraulic conductivity (mm h^{-1}); sand, silt, clay (%); PC1, PC2, PC3, PC4: factors PCA.
* Correlation is significant at $P < 0.05$; ** Correlation is significant at $P < 0.01$; *** Correlation is significant at $P < 0.001$.

respect to CT. This was caused by the high OM concentration in the Ap horizons. SOC in CT in the surface horizons was < 10 g kg^{-1} in CMs and LVs. These low SOC concentrations are due to the high mineralisation of OM and the absence of harvest residues after periods of drought (Hernanz et al., 2009). In agricultural soils, low SOC levels have a negative impact on soil physical properties and nutrient cycling, mainly associated with soil degradation (Romanya and Rovira, 2011). However, SOC in OF in the surface horizons was > 10 g kg^{-1} in CMs, LPs and RGs in the Los Pedroches Valley during the whole 20-year period. Similar results were reported by Álvarez et al. (2007) in OGs in OF. According to Aranda et al. (2011), OF could affect carbon retention under stable forms and could enable SOCS increase, contributing to agro-environmental benefits such as increased soil fertility, erosion prevention, etc.

3.3 Management effect on soil organic carbon and nitrogen stocks

A critical issue was to analyse the influence of management on SOC stock (SOCS). The highest total SOCS was found in LVs-OF (95.4 Mg ha^{-1}) and the lowest in LPs-OF (42.77 Mg ha^{-1}) (Table 3). On average, the total SOCS for the main soil groups in Peninsular Spain (Rodríguez-Murillo, 2001) are 71.4 Mg ha^{-1}, 98.8 Mg ha^{-1}, 48.7 Mg ha^{-1} and 66 Mg ha^{-1} for CMs, LPs, RGs and LVs respectively and for soil used (OG) is 39.9 Mg ha^{-1}. These differences in total SOCS for soil groups in Peninsular Spain and the studied soils are likely to be caused by soil thickness, as we used complete soil profiles and Rodriguez-Murillo (2001) used descriptions of soil profiles deeper than 1 m.

Significant differences ($p < 0.05$) between SOCS under different management types were found in all soils. Therefore, the management system affected the total SOCS for the whole 20-year period. As for CMs and LVs, a higher total SOCS in OF than in CT was found. However, the trend was the opposite in LPs and RGs (more total SOCS in CT than OF). Similar results were obtained by Novara et al. (2012), attributing this phenomenon to the mixing of the upper soil layers during soil tillage. SOCS varies within the soil profile, with higher values in the topsoil in OF than in CT for CMs, LVs and RGs. Similar values were found in LPs for OF versus CT (31.65 Mg ha^{-1} CT; 30.0 Mg ha^{-1} OF). SOCS in the surface horizon varied between 34.95 Mg ha^{-1} for RGs-OF and 23.22 Mg ha^{-1} for LVs-CT. Nevertheless, high values of SOC were found in less sandy soils (RGs and LVs) (Tables 2 and 3). By contrast, SOCS was higher in the subsoil (Bt and C horizons) in LVs-OF and RGs-CT than in the surface horizon. Therefore, two trends can be observed in the studied soils. In the first case, soils with low SOC, which could be explained by their texture (sandy soils) and which are associated with vegetation losses and unsustainable soil management practices. This situation favours a continuous impoverishment of the SOM content causing low soil productivity and derived in unsuitable chemical properties. In the second case, soils with high SOC values. This was especially important in clayey soils (RGs and LVs), and is related to the clay stabilisation process in the soil, increasing the clay content with respect to CMs and LPs. Similar results were obtained by Leifeld et al. (2005). This is especially important in the Bt and C horizons in LVs-OF and RGs-CT. These have a higher SOCS than the surface horizon. In this line, Shrestha et al. (2004) explained that this increase could be due to the translocation of carbon in the form of dissolved organic carbon, soil fauna activity, and/or the effects of deep-rooting crops. Significant differences between horizons and soil types ($p < 0.05$) were found when the Ah horizon of LPs was not included in the analysis of management systems. Nevertheless, we found significant differences ($p < 0.05$) for total SOCS in all the studied soils.

As regards N stock (NS), the behaviour was similar to SOCS. The total NS showed significant differences ($p < 0.05$) in all soils with respect to management practices. When N was analysed, the results indicated a parallelism between N and SOC concentrations, which showed a positive C : N relation (Table 4). According to this, the clay content slowed the SOC oxidation and could have a positive relationship between clay and nitrogen. A similar result was obtained by Sakin et al. (2010). Other studies state that N mineralisation decreases when the clay content increases (Côté et al., 2000). This effect was particularly important in CT (N decreased when the clay content increased). In this line, McLauchlan (2006) explained that the aggregate formation increases and the potential N mineralisation decreases when the clay content increases in soil.

Melero et al. (2009) suggest that OF can increase SOC in longer experimental periods, which we can confirm because SOC and N stocks increased in OF during a long period (20 years).

3.4 Stratification ratio of soil organic carbon

In all soils, the SR of SOC increased with depth (Table 7). The SR of SOC of the surface with regard to depth [SR1] was higher in CT compared to OF, with the exception of RGs (SR1-RGs-CT 2.13; SR1-RGs-OF 3.33), in which the relation was the opposite due to the a low SOC concentration in Ah/C in CT. The [SR2] had a similar behaviour in

CMs and LVs. In LPs and RGs, this relation cannot be performed due to the lack of deeper horizons. Significant differences ($p < 0.05$) regarding the management system (by horizons) were found in CMs, LVs and RGs for SR of SOC. Many authors have shown SRs ranges from 1.1 to 1.9 for CT and between 2.1 and 4.1 for OF (Franzluebbers, 2002; Hernanz et al., 2009) for non-degraded soils. A good soil quality (SR > 2) was observed when the SR was applied for SOC in both management systems with the exception of SR1-LVs-OF (1.38). According to Franzluebbers (2005), the SR of SOC in depth under Mediterranean climatic conditions potentially affect the carbon incorporation in soils as the residue accumulation in the subsurface horizon has an effect on carbon incorporation. Also, the decomposition rates in deeper horizons are lower than in the upper soil horizons (Lorenz and Lal, 2005).

A critical issue was the reduction rates of SR of SOC in OF compared to CT in all soils except SR2-LVs and SR1-RGs, which increased. This scenario implied a reduction of soil quality when OF is applied for a long term (20 years) in CMs and LPs. The SR of N showed a similar trend to SR of SOC. The SR of the C : N ratio increased with depth with the exception of LVs-CT, which decreased with depth. Significant differences ($p < 0.05$) between management system (by horizons) were found in LVs and RGs for SR of N.

The SR of the C : N ratio was normally higher in OF than in CT. Significant differences ($p < 0.05$) between management systems (by horizons) were found in CMs, LVs and RGs for SR of the C : N ratio. This may be explained by a higher contribution of the residue relative to root inputs, leading to a higher soil C : N ratio (Puget and Lal, 2005). Under OF, the residue input could have been concentrated on the surface due to the straw soil surface coverage, leading to the stratification of the soil C : N ratio. This slight change in the C : N ratio suggests that the decomposition degree of SOC decreases toward the surface (Lou et al., 2012). This would involve little effect of the management system on the carbon accumulation in the soil. In this line, Balesdent and Balabane (1996) did not find any significant differences in SR, in a Geauga farm (Ohio).

In summary, the SR indexes in the studied soils showed three different results by soil type and management practices. In LPs, the management practices have little effect on carbon and N accumulation in the soil. In LVs and RGs, the management practices have an effect on carbon accumulation in OF: in LVs, the SR of SOC decreased in the topsoil but increased with depth. For RGs, land management changes increased the SR of SOC. Finally, in CMs a negative trend was observed, with the consequent SR decrease. Therefore, management practices affect SOC accumulation. The mean values of SR of SOC and TN were generally > 2, with the exception of LVs-OT (SR1-SOC 1.38; SR1-N 1.28) (Table 7).

Furthermore, the SR of SOC provided information about the SOC effects in the top soil layer, which could affect SOC accumulation in the soil profile. This is important in

Table 7. Stratification ratios of soil organic carbon concentration, total nitrogen concentration and C : N ratios in the Los Pedroches Valley (Mediterranean rangeland) in olive groves with conventional and organic farming. Data are mean $\pm$ SD**.

Soil	Tillage	Relations	SOC-SR	N-SR	C : N-SR
CM	CT (n = 32)	SR1 (Ap/Bw)	3.74 ± 0.55 a*	2.45 ± 0.37 a	1.52 ± 0.11 a*
		SR2 (Ap/C)	3.82 ± 0.15 a*	2.92 ± 0.52 b	1.31 ± 0.21 b*
	OF (n = 18)	SR1 (Ap/Bw)	2.92 ± 0.57 b	2.51 ± 0.57 a	1.16 ± 0.08 c
		SR2 (Ap/C)	3.40 ± 0.03 c	2.77 ± 0.61 b	1.22 ± 0.13 d
LV	CT (n = 12)	SR1 (Ap/Bt)	2.30 ± 0.35 a*	3.35 ± 0.50 a*	0.69 ± 0.05 a*
		SR2 (Ap/C)	2.41 ± 0.15 a*	2.03 ± 0.28 b*	1.19 ± 0.11 b*
	OF (n = 8)	SR1 (Ap/Bt)	1.38 ± 0.65 b	1.28 ± 0.64 c	1.07 ± 0.10 c
		SR 2 (Ap/C)	2.92 ± 0.46 c	2.85 ± 0.48 d	1.03 ± 0.21 c
LP	CT (n = 18)	SR1 (Ah/C)	3.84 ± 0.75 a	3.41 ± 0.53 a	1.12 ± 0.22 a
	OF (n = 10)	SR1 (Ah/C)	3.69 ± 0.80 a	3.33 ± 0.29 a	1.11 ± 0.17 a
RG	CT (n = 8)	SR1 (Ah/C)	2.13 ± 0.47 a*	2.05 ± 0.26 a*	1.04 ± 0.12 a*
	OF (n = 8)	SR1 (Ah/C)	3.33 ± 0.85 b	2.83 ± 0.38 b	1.16 ± 0.18 b

SOC-SR: Stratification ratio of soil organic carbon; N-SR: Stratification ratio of nitrogen; C : N-SR: Stratification ratio of the C : N ratio. CM: Cambisol; LP: Leptosol; RG: Regosol; LV: Luvisol. CT: Conventional tillage; OF: Organic farming.
n = Sample size.
** Standard deviation.
* Significant differences ($P < 0.05$) between CT and OF treatments (by horizons).
Numbers followed by different lower-case letters within the same column are significant differences ($P < 0.05$) between depth, considering the same soil type.

the studied soils (with high values of SR of SOC and N). According to Jobbágy and Jackson (2000), when reviewing over 2700 soil profiles worldwide, vegetation and climate are associated with the vertical distribution of SOC, but climate and clay content are more decisive in SOC storage.

4 Conclusions

The study concludes that the management system (CT and OF) in CMs, LVs, RGs and LPs for a long period in OGs in the Los Pedroches Valley (Mediterranean rangelands) affected SOC content, exchangeable macroelements (Ca^{2+} and K^+), texture (sand and clay) and N content especially. With regard to the top soil, there were significant differences ($p < 0.001$) between soil types and management systems in OGs, affecting SOC content, exchangeable K^+, BD, thickness, clay content and pH.

The main feature of the studied soils was the low OM concentrations in depth, conditioned by a high sand content and by the climate (semi-arid Mediterranean conditions). OF had a positive impact on CMs and LVs (increasing the total SOCS and total NS) with respect to CT. Conversely, there was a negative impact in RGs and LPs. This was caused by the mixing of the upper soil layers during soil tillage. Also, this SOC reduction can be explain by a process of soil degradation and by the reduced OM input, as well as by the reduced physical protection of soil from erosion and the increased decomposition rate as a consequence of tillage. As for N concentrations,

these were high in areas where the SOC was high, which showed a positive C : N ratio relation.

The SR indicated a good soil quality. In all cases, the SR of SOC and N increased with depth. The SR of SOC comparing the surface and values in depth was greater in CT than in OT. Our results indicate a preferential accumulation of SOC in the surface horizons, influenced by Mediterranean climate. In subsurface horizons, the carbon decomposition rates are lower than in the upper soil layers. The SR of C : N ratio increased with depth in some cases. Significant differences ($p < 0.001$) were found related to the management practices, which can be explained by a higher contribution of residue that implied a higher C : N ratio. Under OF the SOC is concentrated on the soil surface. Also, the soil C : N ratio is stratified. This slight change in the C : N ratio suggests a decomposition degree of SOC, decreasing towards the surface. This indicated little effect of the management system on the carbon accumulation in the topsoil.

This research corroborates the need to analyse entire soil profiles under different management systems, because large amounts of SOC can be transported to deeper soil horizons in temperate climates, contributing to subsoil carbon storage.

Acknowledgements. This study was supported by the Regional Government of Andalusia through projects OG-019/07, OG-018/06, OG-033/04 and OG-127/02.

Edited by: A. Jordán

References

Álvarez, S., Soriano, M. A., Landa, B. B., and Gómez, J. A.: Soil properties in organic olive groves compared with that in natural areas in a mountainous landscape in southern Spain, Soil Use Manage., 23, 404–416, 2007.

Aranda, V., Ayora-Cañada, M. J., Domínguez-Vidal, A., Martín-García, J. M., Calero, J., Delgado, R., Verdejo, T., and González-Vila, F. J.: Effect of soil type and management (organic vs. conventional) on soil organic matter quality in olive groves in a semi-arid environment in Sierra Mágina Natural Park (S Spain), Geoderma, 164, 54–63, 2011.

Balesdent, J. and Balabane, M.: Major contribution of roots to soil carbon storage inferred from maize cultivated soils, Soil Biol. Biochem., 28, 1261–1263, 1996.

Blanco-Canqui, H. and Lal, R.: No-tillage and soil-profile carbon sequestration: an on farm assessment, Soil Sci. Soc. Am. J., 72, 693–701, 2008.

Blake, G. R. and Hartge, K. H.: Particle density, in: Methods of soil analysis. Part I. Physical and mineralogical methods, edited by: Klute, A., Agronomy Monography no. 9, ASA, SSSA, Madison WI, USA, 377–382, 1986.

Bower, C. A., Reitemeier, R. F., and Fireman, M.: Exchangeable cation analysis of saline and alkali soil, Soil Sci., 73, 251–261, 1952.

Bradford, J. M. and Peterson, G. A.: Conservation tillage, in: Handbook of Soil Science, edited by: Sumner, M. E., CRC Press, Boca Raton, 247–269, 2000.

Bremner, J. M.: Total nitrogen, in: Methods of Soil Analysis: Chemical Methods, edited by: Sparks, D. L., Soil Science Society of America, Madison, WI, 1085–1086, 1996.

Bronick, C. J. and Lal, R.: Manuring and rotation effects on soil organic carbon concentration for different aggregate size fractions on two soils in northeastern Ohio, USA, Soil Till. Res., 81, 239–252, 2005.

Calatrava, J., Barberá, G. G., and Castillo, V. M.: Farming practices and policy measures for agricultural soil conservation in semiarid Mediterranean areas: the case of Guadalentín basin in Southeast Spain, Land Degrad. Develop., 22, 58–69, 2011.

Calatrava-Leyva, J., Franco-Martínez, J. A., and González-Roa, M. C.: Analysis of the adoption of soil conservation practices in olive groves: the case of mountainous areas in southern Spain, Span. J. Agr. Res., 5, 249–258, 2007.

Castro, J., Fernández-Ondoño, E., Rodríguez, C., Lallena, A. M., Sierra, M., and Aguilar, J.: Effects of different olive-grove management systems on the organic carbon and nitrogen content of the soil in Jaén (Spain), Soil Till. Res., 98, 56–67, 2008.

Cerdà, A. and Doerr, S. H.: Soil wettability, runoff and erodibility of major dry Mediterranean land use types on calcareous soils, Hydrol. Process., 21, 2325–2336, 2007.

Cerdà, A., Lavee, H., Romero-Díaz, A., Hooke, J., and Montanarella, L.: Preface, Land Degrad. Develop., 21, 71–74, 2010.

Colombo, S., Hanley, N., and Calatrava, J.: Designing policy for reducing the off farm effects of soil erosion using choice experiments, J. Agr. Econ., 56, 81–95, 2005.

Conant, R. T., Easter, M., Paustian, K., Swan, A., and Williams, S.: Impacts of periodic tillage on soil C stocks: A synthesis, Soil Till. Res., 95, 1–10, 2007.

Corral-Fernández, R., Parras-Alcántara, L., and Lozano-García, B.: Stratification ratio of soil organic C, N and C:N in Mediter-ranean evergreen oak woodland with conventional and organic tillage, Agr. Ecosyst. Environ., 164, 252–259, 2013.

Côté, L., Brown, S., Paré, D., Fyles, J., and Bauhus, J.: Dynamics of carbon and nitrogen mineralization in relation to stand type, stand age and soil texture in the boreal mixedwood, Soil Biol. Biochem., 32, 1079–1090, 2000.

Diekow, J., Mielniczuk, J., Knicker, H., Bayer, C., Dick, D. P., and Kögel-Knabner, I.: Soil C and N stocks as affected by cropping systems and nitrogen fertilisation in a southern Brazil Acrisol managed under no-tillage for 17 years, Soil Till. Res., 81, 87–95, 2005.

Duchaufour, P. H.: Manual de Edafología, Editorial Toray-Masson, Barcelona, 1975.

ESYRCE: Encuesta sobre superficies y rendimientos. Ministerio de agricultura, alimentación y medio ambiente, Madrid, 2012.

FAO: Guidelines for soil description. Food and Agriculture Organization of the United Nations, Rome, Italy, 2006.

Fernández-Escobar, R., Marina, L., Sánchez-Zamora, M. A., García-Novelo, J. M., Molina-Soria, C., and Parra, M. A.: Long-term effects of N fertilization on cropping and growth of olive trees and on N accumulation in soil profile, Eur. J. Agronom., 31, 223–232, 2009.

Franzluebbers, A. J.: Soil organic matter stratification ratio as an indicator of soil quality, Soil Till. Res., 66, 95–106, 2002.

Franzluebbers, A. J.: Soil organic carbon sequestration and agricultural greenhouse gas emissions in the south eastern USA, Soil Till. Res., 83, 120–147, 2005.

Gómez, J. A., Sobrinho, T. A., Giráldez, J. V., and Fereres, E.: Soil management effects on runoff, erosion and soil properties in an olive grove of Southern Spain, Soil Till. Res., 102, 5–13, 2009.

González, I., Grau, J. M., Fernández, A., Jiménez, R., and González, M. R.: Soil carbon stocks and soil solution chemistry in Quercus ilex stand in Mainland Spain, Eur. J. Forest. Res., 131, 1653–1667, doi:10.1007/s10342-012-0623-8, 2012.

González, J. and Candás, M.: Materia orgánica de suelos bajo encinas, Mineralización de carbono y nitrógeno, Invest. Agrar., 75–83, 2004.

Guitián, F. and Carballas, T.: Técnicas de Análisis de Suelos. Edit. Picro Sacro, Santiago de Compostela, España, 1976.

Hathaway-Jenkins, L., Sakrabani, R., Pearce, B., Whitmore, P., and Godwin, R.: A comparison of soil and water properties in organic and conventional farming systems in England, Soil Use Manage., 27, 133–142, 2011.

Hazelton, P. and Murphy, B.: Interpreting Soil Test Results: What Do All the Numbers Mean? CSIRO Publishing, Collingwood Victoria, Australia, 2007.

Hernanz, J. L., Sanchez-Giron, V., and Navarrete, L.: Soil carbon sequestration and stratification in a cereal/leguminous crop rotation with three tillage systems in semiarid conditions, Agr. Ecosyst. Environ., 133, 114–122, 2009.

Hontoria, C., Rodríguez-Murillo, J., and Saa, A.: Contenido de carbono orgánico en el suelo y factores de control en la España Peninsular, Edafología, 11, 149–155, 2004.

IOC, International Olive Council: The World Market in figures, OLIVAE, 117, 28–34, 2012.

IPCC, Intergovernmental Panel on Climate Change.: Good practice guidance for land use, land use change and forestry, in: IPCC/OECD/IEA/IGES, edited by: Penman, J., Gytarsky, M.,

Hiraishi, T., Krug, T., Kruger, D., Pipatti, R., Buendia, L., Miwa, K., Ngara, T., Tanabe, K., and Wagner, F., Hayama, Japan, 2003.

IUSS Working Group WRB: World Reference Base for Soil Resources 2006, World Soil Resources Reports No. 103, 2nd edition, FAO, Rome, Italy, 2006

Jobbágy, E. G. and Jackson, R. B.: The vertical distribution of soil organic carbon and its relation to climate and vegetation, Ecol. Appl., 104, 423–436, 2000.

Lal, R., Delgado, J. A., Groffman, P. M., Millar, N., Dell, C., and Rotz, A.: Management to mitigate and adapt to climate change, J. Soil Water Conserv., 66, 276–285, 2011.

Leifeld, J., Bassin, S., and Fuhrer, J.: Carbon Stocks in Swiss agricultural soils predicted by Land-use. Soil charactetistics and altitude, Agr. Ecosyst. Environ., 105, 255–266, 2005.

López-Garrido, R., Madejón, E., Murillo, J., and Moreno, F.: Short and long term distribution with depth of soil organic carbon and nutrients under traditional and conservation tillage in a Mediterranean environment (southwest Spain), Soil Use Manage., 27, 177–185, 2011.

Lorenz, K. and Lal, R.: The depth distribution of organic soil carbon in relation to land use and management and the potential of carbon sequestration in subsoil horizons, Adv. Agron., 88, 35–66, 2005.

Lou, Y., Xu, M., Chen, X., He, X., and Zhao, K.: Stratification of soil organic C, N and C : N ratio as affected by conservation tillage in two maize fields of China, Catena, 95, 124–130, 2012.

Louwagie, G., Gay, S. H., Sammeth, F., and Ratinger, T.: The potential of European Union policies to address soil degradation in agriculture, Land Degrad. Develop., 22, 5–17, 2011.

Lozano-García, B. and Parras-Alcántara, L.: Short-term effects of olive mill by products on soil organic carbon, total N, C : N ratio and stratification ratios in a Mediterranean olive grove, Agr. Ecosyst. Environ., 165, 68–73, 2013a.

Lozano-García, B. and Parras-Alcántara, L.: Land use and management effects on carbon and nitrogen in Mediterranean Cambisols, Agr. Ecosyst. Environ., 179, 208–214, 2013b.

Lozano-García, B. and Parras-Alcántara, L.: Variation in soil organic carbon and nitrogen stocks along a toposequence in a traditional mediterranean olive grove, Land Degrad. Develop., in press, available online: in Wiley Online Library (wileyonlinelibrary.com), doi:10.1002/ldr.2284, 2014.

Lozano-García, B., Parras-Alcántara, L., and del Toro, M.: Effects of oil mill by-products on surface soil properties, runoff and soil losses in traditional olive groves in southern Spain, Catena, 85, 187–193, 2011.

MAGRAMA: Avances de superficies y producciones de cultivos. Madrid: Ministerio de Agricultura, Alimentación y Medio Ambiente, Madrid, 2012.

Marañón, T.: Agro-sylvo-pastoral systems in the Iberian Peninsula: Dehesas and Montados, Rangelands, 10, 255–258, 1988.

McLauchlan, K. K.: Effect of soil texture on soil carbon and nitrogen dynamic after cessation of agriculture, Geoderma, 136, 289–299, 2006.

Melero, S., López-Garrido, R., Murillo, J. M., and Moreno, F.: Conservation tillage: Short- and long-term effects on soil carbon fractions and enzymatic activities under Mediterranean conditions, Soil Till. Res., 104, 292–298, 2009.

Metson, A. J.: Methods of chemical analysis for soil survey samples. Soil Bureau bulletin, 12. Department of Scientific and In-

dustrial Research and New Zealand Soil Bureau, Wellington, New Zealand, 1961.

Minitab.: Release 13. MINITAB Statistical Software, Minitab Inc., State College, PA, 2000.

Nelson, D. W. and Sommers, L. E.: Total carbon, organic carbon and organic matter. in: Methods of soil analysis, Part 2. Chemicaland microbiological properties, edited by: Page, A. L., Miller, R. H., and Keeney, D., Agronomy monograph, vol. 9, ASA and SSSA, Madison WI, 539–579, 1982.

Nerger, R., Rainer, M., Núñez, S., and Recio, J.: Presencia de carbonatos en suelos desarrollados sobre material granítico del Batolito de los Pedroches (Córdoba), in: Tendencias Actuales de la Ciencia del Suelo, edited by: Bellinfante, N. and Jordán, A., Sevilla, España, 2007.

Novara, A., La Mantia, T., Barbera, V., and Gristina, L.: Paired-site approach for studying soil organic carbon dynamics in a Mediterranean semiarid environment, Catena, 89, 1–7, 2012.

Parra-López, C., Groot, J. C. J., Carmona-Torres, C., and Rossing, W. A. H.: An integrated approach for ex-ante evaluation of public policies for sustainable agriculture at landscape level, Land Use Policy, 26, 1020–1030, 2009.

Parras-Alcántara, L., Díaz-Jaimes, L., and Lozano-García, B.: Organic farming affects C and N in soils under olive groves in Mediterranean areas, Land Degrad. Dev., in press, available online: in Wiley Online Library (wileyonlinelibrary.com), doi:10.1002/ldr.2231, 2013a.

Parras-Alcántara, L., Martín-Carrillo, M., and Lozano-García, B.: Impacts of land use change in soil carbon and nitrogen in a Mediterranean agricultural area (Southern Spain), Solid Earth, 4, 167–177, doi:10.5194/se-4-167-2013, 2013b.

Parras-Alcántara, L., Díaz-Jaimes, L., and Lozano-García, B.: Management effects on soil organic carbon stock in Mediterranean open rangelands-treeless grasslands, Land Degrad. Develop., in press, available online: in Wiley Online Library (wileyonlinelibrary.com), doi:10.1002/ldr.2269, 2014a.

Parras-Alcántara, L., Díaz-Jaimes, L., Lozano-García, B., Fernández, P., Moreno, F., and Carbonero, M. D.: Organic farming has little effect on carbon stock in a Mediterranean dehesa (southern Spain), Catena, 113, 9–17, 2014b.

Paustian, K., Six, J., Elliott, E. T., and Hunt, H. W.: Management options for reducing CO_2 emissions from agricultural soils, Biogeochemistry, 48, 147–163, 2000.

Puget, P. and Lal, R.: Soil organic carbon and nitrogen in a Mollisol in central Ohio as affected by tillage and land use, Soil Till. Res., 80, 201–213, 2005.

Pulido-Fernández, M., Schnabel, S., Lavado-Contador, J. F., Miralles, I., and Ortega, R.: Soil organic matter of Iberian open Woodland rangelands as influenced by vegetation cover and land management, Catena, 109, 13–24, 2013.

Recio, J. M., Corral, L., and Paneque, G.: Estudio de suelos en la Comarca de los Pedroches (Córdoba), Anal. Edaf. Agrob., 45, 989–1012, 1986.

Reynolds, W. D. and Elrick, D. E.: Determination of hydraulic conductivity using a tension infiltrometer, Soil Sci. Soc. Am. J., 55, 633–639, 1991.

Rodríguez-Entrena, M., Barreiro-Hurlé, J., Gómez-Limón, J. A., Espinosa-Goded, M., and Castro-Rodríguez, J.: Evaluating the demand for carbon sequestration in olive grove soils as a strategy toward mitigating climate change, J. Environ. Manage., 112, 368–376, 2012.

Rodríguez-Murillo, J. C.: Organic carbon content under different types of land use and soil in peninsular Spain, Biol. Fertil. Soils, 33, 53–61, 2001.

Romanya, J. and Rovira, P.: An appraisal of soil organic C content in Mediterranean agricultural soils, Soil Use Manage., 27, 321–332, 2011.

Ruiz, J. D., Pariente, S., Romero, A., and Martínez, J. F.: Variability of relationships between soil organic carbon and some soil properties in Mediterranean rangelands under different climatic conditions (South of Spain), Catena, 94, 17–25, 2012.

Sakin, E., Deliboran, A., Sakin, E. D., and Tutar, E.: Organic and inorganic carbon stock and balance of Adana city soils in Turkey, African J. Agr. Res., 5, 2737–2743, 2010.

Shrestha, B. M., Sitaula, B. K., Singh, B. R., and Bajracharya, R. M.: Soil organic carbon stocks in soil aggregates under different land use systems in Nepal, Nutr. Cycl. Agroecosys., 70, 201–213, 2004.

USDA: Soil survey laboratory methods manual, Soil survey investigation report No. 42, Version 4.0, USDA-NCRS, Lincoln, NE, 2004.

Wang, Y., Fu, B., Lü, Y., Song, C., and Luan, Y.: Local-scale spatial variability of soil organic carbon and its stock in the hilly area of the Loess Plateau, China, Quaternary Res., 73, 70–76, 2010.

Xu, M., Lou, Y., Sun, X., Wang, W., Baniyamuddin, M., and Zhao, K.: Soil organic carbon active fractions as early indicators for total carbon change under straw incorporation, Biol. Fertil. Soils, 47, 745–752, 2011.

Yamashita, T., Feiner, H., Bettina, J., Helfrich, M., and Ludwig, B.: Organic matter in density fractions of water-stable aggregates in silty soils: effect of land use, Soil Biol. Biochem., 38, 3222–3234, 2006.

Zdruli, P., Kapur, S., and Çelik, I.: Soils of the Mediterranean Region. Their characteristics, management and sustainable use, in: Sustainable land management learning from the past for the future, edited by: Kapur, S., Eswaran, H., and Blum, W. E. H., Springer-Verlag, Berlin, 2011.

7

Morphology and surface features of olivine in kimberlite: implications for ascent processes

T. J. Jones[1,2], J. K. Russell[1], L. A. Porritt[1,2], and R. J. Brown[3]

[1]Department of Earth, Ocean and Atmospheric Sciences, University of British Columbia, Vancouver, V6T 1 Z4, Canada
[2]School of Earth Sciences, University of Bristol, Wills Memorial Building, Bristol, BS8 1RJ, UK
[3]Department of Earth Sciences, Science Labs, Durham University, Durham, DH1 3LE, UK

Correspondence to: T. J. Jones (thomas.jones.2010@my.bristol.ac.uk)

Abstract. Most kimberlite rocks contain large proportions of ellipsoidal-shaped xenocrystic olivine grains that are derived mainly from disaggregation of peridotite. Here, we describe the shapes, sizes and surfaces of olivine grains recovered from kimberlite lavas erupted from the Quaternary Igwisi Hills volcano, Tanzania. The Igwisi Hills kimberlitic olivine grains are compared to phenocrystic olivine, liberated from picritic lavas, and mantle olivine, liberated from a fresh peridotite xenolith. Image analysis, scanning electron microscopy imagery and laser microscopy reveal significant differences in the morphologies and surface features of the three crystal populations. The kimberlitic olivine grains form smooth, rounded to ellipsoidal shapes and have rough flaky micro-surfaces that are populated by impact pits. Mantle olivine grains are characterised by flaked surfaces and indented shapes consistent with growth as a crystal aggregate. Phenocrystic olivine exhibit faceted, smooth-surfaced crystal faces. We suggest that the unique shape and surface properties of the Igwisi Hills kimberlitic olivine grains are products of the transport processes attending kimberlite ascent from mantle source to surface. We infer that the unique shapes and surfaces of kimberlitic olivine grains result from three distinct mechanical processes attending their rapid transport through the thick cratonic mantle lithosphere: (1) penetrative flaking from micro-tensile failure induced by rapid decompression; (2) sustained abrasion and attrition driven by particle–particle collisions between grains within a turbulent, volatile-rich flow regime; and (3) higher-energy particle–particle collisions producing impact cavities superimposed on decompression structures. The combination of these processes during the rapid ascent of kimberlite magmas is responsible for the distinctive ellipsoidal shape of olivine xenocrysts found in kimberlites worldwide.

1 Introduction

Kimberlite magmas, derived from low degrees of partial melting of the mantle, erode, carry and erupt significant amounts of crystalline lithospheric mantle as whole rock xenoliths and as single crystals or xenocrysts (e.g. Mitchell, 1986). They are consequently important geochemical and physical windows into the inner Earth. Abundant, subrounded to rounded, ovoid to elliptical grains of xenocrystic mantle olivine are characteristic of kimberlite intrusions, lavas and pyroclastic rocks (e.g. Brett et al., 2009; Clement and Skinner, 1979, 1985; Dawson and Hawthorne, 1973; Gernon et al., 2012; Kamenetsky et al., 2008; Mitchell, 1986, 2008; Moss and Russell, 2011).

However, the origin of their ellipsoidal morphologies remains somewhat enigmatic given that they represent disaggregated crystalline rocks. Explanations include rounding by magmatic corrosion and dissolution of the grains during ascent (Kamenetsky et al., 2008; Mitchell, 1986; Moore, 2012; Pilbeam et al., 2013) or mechanical milling (Arndt et al., 2006, 2010; Brett et al., 2009; Brett, 2009; Kamenetsky et al., 2008; Reid et al., 1975; Russell et al., 2012). Rapid CO_2 release following digestion of orthopyroxene was recently proposed as a mechanism for propelling kimberlite magmas rapidly to the surface over short timescales (Russell et al., 2012). Such a process is potentially a significant factor in increasing the erosive potential of the ascending magmas at depth. Intuitively, chemical corrosion and milling should leave different physical signatures on the exteriors of mantle-derived crystals, which, if not overprinted by late-stage crystal growth rims, or removed by alteration, should be discernible with scanning electron microscopy (SEM). Such features can provide insights into the nature of kimberlite magma transport from the mantle upwards: a topic still relatively poorly understood (e.g. Russell et al., 2012; Sparks, 2013; Sparks et al., 2006).

Here, we semi-quantitatively describe and analyse the shapes and surfaces of fresh xenocrystic olivine extracted from a Quaternary kimberlite lava in Tanzania. Our goals are two-fold: firstly, to understand the processes that create the highly distinctive shapes and surfaces of olivine crystals characteristic of kimberlite rocks and, secondly, to constrain ideas on the ascent of kimberlite magma.

2 Olivine in kimberlite

This study focusses on the physical and morphological features present on the surfaces of kimberlitic olivine. Our work builds on an extensive literature of petrographic and geochemical studies of kimberlitic olivine (Table 1) aimed at constraining the origins of kimberlite magmas (e.g. Arndt et al., 2006; Table 1; Arndt et al., 2010; Brett et al., 2009; Dawson, 1994; Kamenetsky et al., 2008; Mitchell, 1970, 1986, 1995, 2012; Pilbeam et al., 2013). The large number of detailed studies of olivine in kimberlite reflects the pervasively high abundance of olivine in these enigmatic rocks.

Mitchell (1986) separated olivine within kimberlite into phenocrystic and xenocrystic populations. Xenocrystic olivine, termed "macrocrysts", are rounded in shape and the most abundant volumetrically. "Phenocrystic" olivine was distinguished in the groundmass on the basis of a finer-grain size (< 0.5 mm) and its (sub-)euhedral character. However, in recent years, many workers have ascribed a xenocrystic origin to both olivine populations. Both populations are observed to have xenocrystic cores and magmatic rims (e.g. Arndt et al., 2010; Brett et al., 2009). Work by Arndt et al. (2010) has suggested mechanical abrasion as a process which operates during ascent. They observe rounded centimetre-scale aggregates of olivine, described as monomineralic dunitic nodules. The olivine is typically deformed, has variable composition, and has more Fe-rich rims and is therefore interpreted to be xenocrystic in origin. Arndt et al. (2010) suggested that kimberlite melts sampled and entrained fragments of a "defertilised", largely dunitic, mantle lithosphere. The mantle-derived dunitic xenoliths become disaggregated during ascent, and olivine crystallises as rims on the mantle olivine and as euhedral crystals during ascent. They noted that the late rims on xenocrystic olivines could be chemically heterogeneous and commonly absent on protruding portions of the rounded nodule. This is attributed to abrasion processes on ascent (Arndt et al., 2006, 2010).

Kamenetsky et al. (2008) argued for transport in an initially chloride–carbonate melt causing mechanical abrasion and chemical resorption of xenocrystic olivine and orthopyroxene before reaching olivine saturation and crystallisation. They ascribed a xenocrystic origin to both olivine populations (macrocrysts and phenocrysts), which they referred to as type 1 and 2, respectively. They also argued that, dur-

ing emplacement, the kimberlite reaches olivine saturation and crystallises rims on the xenocrystic olivine and minor groundmass phases (Kamenetsky et al., 2008).

Brett et al. (2009) used textural evidence combined with electron microprobe analysis of Ni contents in olivine to document the xenocrystic origins of kimberlitic olivine, regardless of grain size or shape. Their work corroborates the reported chemical trends observed in olivine from the Igwisi Hills kimberlite (Fig. 8; Dawson, 1994). Brett et al. (2009) demonstrate small volumes (≤ 5 %) of heterogeneous olivine crystallisation from the kimberlitic melt during ascent. This occurs on cores of pre-existing rounded xenocrystic olivines (at all sizes). The crystallisation is caused by orthopyroxene dissolution which drives the melt towards olivine saturation. The small volume of crystallisation only affects the overall shapes of the smallest xenocrysts to create the euhedral (apparent) "phenocryst" population of olivine grains.

Other studies have proposed a complex interplay between reabsorption and crystallisation during kimberlitic ascent (Moore, 2012). In Moore's model, rounding is caused by melt resorption, predicted to occur at multiple times and linked to superheating events during ascent. Pilbeam et al. (2013) also suggest resorption processes during ascent. Here, the digestion of a xenocrystic cargo coupled with the fractional crystallisation is thought to explain the marginal profiles on olivine. During reactive transport magmatic rims are crystallised on xenocrystic cores (Pilbeam et al., 2013).

Our study here aims to explore the evidence for mechanical processes which operate on olivine crystal cargo (i.e. xenocrysts) during kimberlite ascent. Our results, in part, answer some of the questions concerning the mechanisms of kimberlite ascent and reconcile some of the disparate observations and ideas found in the literature (Table 1).

3 Igwisi Hills volcanoes

The Quaternary Igwisi Hills monogenetic volcanoes (IHV), Tanzania, erupted a minimum of 3.4×10^7 m^3 of kimberlite magma. The eruptions produced three small volcanic edifices located along a ~ 2 km NE–SW-trending fissure (Fig. 1): the NE, SW and central volcano (Brown et al., 2012). The ~ 10 ka IHV are the youngest kimberlite volcanoes on Earth – postdating the next youngest kimberlite volcanoes by ~ 30 Ma (Brown et al., 2012). The Igwisi Hills volcanic deposits represent one of very few occurrences of fresh kimberlites showing remarkably low degrees of alteration/minimal serpentinisation (Brown et al., 2012).

Brown et al. (2012) proposed three eruptive phases for the IHV: an initial phreatomagmatic explosive phase of eruption, followed by weak explosive eruptions columns, and then effusion of lava. Brown et al. (2012) estimated that the lavas had high effective viscosities at emplacement ($> 10^2$ to 10^6 Pa s) relative to the range commonly ascribed to kimberlite magmas: ~ 1 Pa s (e.g. Sparks et al., 2006). The higher

Table 1. Results from published studies of kimberlite-hosted olivine. The observations and interpretations reported below are filtered for their relevance to the kimberlitic ascent processes elucidated in our study.

Source	Observations	Interpretations	Link to kimberlitic processes
Mitchell (1970, 1986, 1995)	– Macrocrysts: large sub-angular to sub-rounded grains; single crystals or polycrystalline aggregates; ± undulose extinction. – Non-macrocrysts: < 0.5 mm (sub-)euhedral unstrained, olivine grains with planar crystal faces.	– Larger-sized grains are xenocrystic. – Smaller-sized grains are "phenocrystic" and crystallised from kimberlite melt.	– Entrainment and break-up of mantle peridotite – Deep-seated crystallisation of olivine during transport.
Dawson (1994)	– Large xenocrystic olivines (~Fo$_{92}$) compositionally similar to olivine in peridotitic mantle xenoliths. – Large olivines are zoned and rims show CaO and MnO enrichment and NiO depletion. – Cores of "matrix" olivines are distinct from xenocrystic olivine (i.e. higher Fe, Ca, Ti). – Elongate matrix grains can show CaO zoning with CaO contents similar to neoblasts found on margins of xenoliths.	– Large xenocrystic olivines derive from a well-equilibrated phlogopite–calcite–garnet lherzolite protolith. – Some elongate matrix olivines might be rim fragments (xenocrysts) from the recrystallised margins of xenoliths. – Sits on the fence – it is also stated that some of the smaller olivines may have crystallised from the kimberlite magma.	– Entrainment and break-up of mantle peridotite producing xenocrystic olivine – Matrix olivine include neoblasts from peridotite and possible olivine crystallised from kimberlite.
Arndt et al. (2006)	– Rounded centimetre-scale aggregates of olivine (5–45 vol. %) described as nodules. – Fo compositions are constant within a nodule but vary greatly (Fo 81–93) between nodules. – Olivine in nodules and matrix record deformation. – Many nodules have thin rims of high Ca.	– Mechanical abrasion produced rounded nature of each nodule. – Both olivine in nodules and matrix are xenocrystic in origin. – Marginal rims represent crystallisation from the kimberlite melt.	– Mantle "defertilisation" removes all minerals but olivine from lithospheric mantle peridotite. – Defertilisation creates dunitic source and presages kimberlite magmatism. – Magma entrains and rounds dunitic mantle nodules.
Kamenetsky et al. (2008)	– Trace element abundances, oxygen isotopes ratios and Mg contents of olivine-I (macrocrysts) and cores of olivine-II ("groundmass olivine") are indistinguishable.	– Incorporation of peridotitic olivine (I). – Early crystallisation of olivine (II) at depth in small melt pockets. – Both olivine populations transported to surface within a chloride–carbonate melt – Additional low-pressure crystallisation of olivine rims.	– During transport mantle olivine and orthopyroxene are partially abraded, chemically reabsorbed, and recrystallised. – Olivine crystallises from kimberlite melt as rims on larger grains of olivine.
Brett et al. (2009)	– Disparities in Ni contents between rims and cores of two olivine size populations. Lower Ni within the rim. – Textural zonation of rims on rounded high-Ni cores defined by fluid/mineral inclusions.	– Macrocrysts and cores of smaller olivine grains are xenocrystic. – Rims of macrocrysts and smaller grains result from magmatic crystallisation of the kimberlitic melt. – The rims are thin; they do not change the shape of the previously rounded grains but trap mineral and fluid inclusions.	– Late, small-volume (≤ 5 %) heterogeneous crystallisation of phenocrystic olivine from kimberlitic melt during ascent. – Coupled orthopyroxene dissolution and olivine crystallisation during ascent. – For larger (~25 vol. %) amounts of orthopyroxene assimilation, kimberlite needs to be superliquidus (≥ 100 °C).
Arndt et al. (2010)	– Describes centimetre-scale monomineralic olivine nodules – Nodules have variable Fo compositions but lower-Fo rims. – Olivine rims have high CaO and MnO but low NiO. – Commonly chemical rims are absent on protruding portions of the nodule. – Nodule mineralogy and abundance suggest a non-garnet–peridotite source.	– Xenocrystic olivine is derived mainly from dunitic source. – Lower Fo-rich rims crystallised on ascent. – Euhedral grains re-crystallised on ascent. – Rounding of nodules by mechanical abrasion occurs during ascent, after liberation, and locally removes chemical rims.	– Mantle de-fertilisation (see above) creates a dunite lithospheric mantle source rock. – Kimberlite magmatism is associated with deformation of surrounding dunite and peridotite. – Fluid pressure increases in the pocket and fractures overlying mantle. Flow then entrains surrounding dunite and normal peridotite higher up.
Moore (2012)	– Cores and edges of most olivines cover a continuous compositional field. There is an overlap in chemical composition between the core and the rim in most Group 1 and Group 2 kimberlites. – The morphology of small "groundmass" olivine is comparable to those crystallised from a moderately under-cooled melt.	– Rounding result from chemical; resorption into melt. – Resorption of olivine linked to repeated superheating events during ascent. – Exsolution at shallow depths causes fluidisation and loss of CO$_2$. – Cooling due to CO$_2$ loss triggers late-stage (groundmass and rims) olivine crystallisation.	– All olivine results from polybaric crystallisation of kimberlite melt during ascent and involving interplay between supersaturation and dissolution.
Pilbeam et al. (2013)	– Olivine cores have variable composition and are determined to be xenocrystic. – Over 90 % of the olivine grains analysed show normal zoning. (high-Mg and high-Ni cores) – Measured core, margin, rim profiles are compared to model profiles for (1) equilibrium crystallisation, (2) fractional crystallisation, (3) digestion fractional crystallisation (DFC). – Observed profiles are best matched to DFC, modelling of minor diffusion post-DFC removes slight discrepancies.	– Fractional crystallisation coupled with digestion of xenocrysts (DFC), primarily orthopyroxene, explains observed chemical zoning of Mg, Ni, Ca, and Mn in olivine. – This is then followed by minor diffusive re-equilibration.	– Crystallisation of cognate phases occurs during ascent alongside the digestion of a xenocrystic cargo. – Melt evolves via orthopyroxene dissolution. – Magma emplaced as a slurry of xenocrystic olivine grains with magmatic margins. – Other xenocrysts and nodule types transported in a silica–calcium–carbonatite melt.

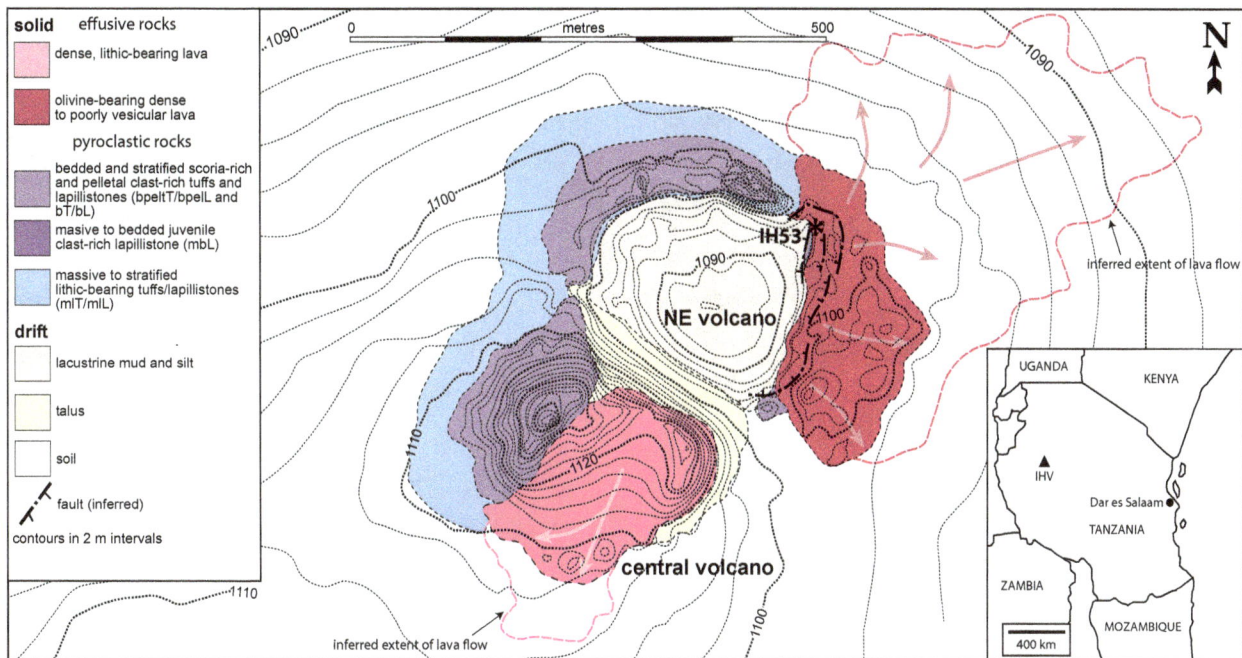

Figure 1. Map of Igwisi Hills volcanoes (IHV), showing the relative positions of the two northernmost volcanic edifices and the location of the lava (IH53) sampled in this study. The Igwisi Hills volcanoes are situated at latitude 4°51′ S, longitude 31°55′ E. Adapted from Brown et al. (2012).

effective viscosities are interpreted as resulting from the effects of shallow magma degassing and partial crystallisation of the groundmass. Brown et al. (2012) also provided a lower limit on magma viscosity by modelling olivine settling, which suggested a partially crystallised Bingham-like fluid with a yield strength and viscosities of 10^3 Pa s.

Dawson (1994) showed that the IHV showed strong affinities to low-alkali, calcite-rich kimberlites such as the Benfontein sills, South Africa. Reid et al. (1975) noted that the Igwisi material is characterised by ellipsoidal forsterite olivine hosted in a fine-grained carbonate–apatite–spinel–pervoskite–serpentine matrix. These xenocrysts and the polycrystalline xenoliths are ovoid and range up to 3 cm (i.e. "micro-xenoliths" in Dawson, 1994; "nodules" in Arndt et al., 2010). In a few cases olivine ellipsoids are rimmed or partially rimmed by pervoskite and Mg-Al spinel. Some olivine contains mineral inclusions, including chrome pyrope, Mg-Al chromite, low-Al enstatite, low-Al-Mg-Cr-diopside and high-Mg phlogopite (Reid et al., 1975). Mantle phases present in the micro-xenoliths provide estimates for pressure and temperature at formation. Under the assumption that the clinopyroxene coexisted with low-Ca pyroxene, thermobarometry places the source temperatures at $\sim 1000\,°\text{C}$ (Davis and Boyd, 1966). Additionally, assuming chemical equilibrium, the coexistence of low-Al_2O_3 enstatite with pyrope results in pressures of 50–60 kbar (Boyd, 1970).

4   Sample suite

In this study three sample sets were analysed. Fifteen individual olivine crystals from the Igwisi Hills (IH) lava are compared with olivine phenocrysts from an Icelandic lava and olivine grains disaggregated from mantle xenoliths. Phenocrysts from the Icelandic lava illustrate surface textures of crystals crystallising from magma stored in the crust, whilst olivine grains disaggregated from mantle peridotitic xenoliths serve as a proxy for the original material prior to kimberlite ascent. We used olivine from an IH lava rather than from the pyroclastic deposits to ensure that the surface textures represented magma transport processes and were not modified by explosive eruption processes. The secondary influence of sub-aerial eruption mechanisms, like secondary fragmentation, can therefore be ignored. Previous studies of the IH kimberlite deposits have also shown that the lavas record minimal alteration (Brown et al., 2012).

4.1   Phenocrystic olivine

The picritic lava flow from Iceland contains abundant ($> 20\,\%$) sub-hedral to euhedral forsteritic olivine phenocrysts with a grain size of approximately 2.5 mm. The sample was coarsely crushed to ~ 0.4–0.8 cm, and then olivine grains were hand-picked under a binocular microscope. This allowed the crystals to be selected individually to avoid grains with unnatural breakage surfaces.

Figure 2. Xenocrystic olivine grains within Igwisi Hills lava sample IH53. High-resolution scan of polished slab of lava sample showing abundance, distribution and highly ellipsoidal shapes of xenocrystic olivine.

4.2 Mantle olivine

These samples comprise pristine lithospheric mantle-derived peridotitic xenoliths collected from a basanitic dike from Mt Preston, western British Columbia (BC), Canada (Peterson, 2010). Detailed study of their mineralogy and textures shows that they record mantle equilibration conditions and have not been modified texturally or chemically during or post-emplacement (Peterson, 2010). As such, they provide a suitable reference material for the morphology of unadulterated mantle olivine. The peridotite xenolith chosen was friable enough to disaggregate by hand. Olivine grains, 2–4 mm in diameter, were handpicked under a binocular microscope.

4.3 Igwisi Hills volcanoes

Large (1–10 mm) spheroidal olivine crystals make up a large proportion (~ 26 vol.%) of the lower parts of a pāhoehoe-type lava flow from the NE volcano at Igwisi Hills (Fig. 2; Brown et al., 2012). Such a high concentration of olivine can be partially attributed to post-emplacement crystal settling and accumulation. The olivine grains are rounded to sub-rounded in shape and have aspect ratios of ~ 1.5. The olivine grains were carefully removed by cutting out a small volume of lava with the olivine enclosed by a thin layer of groundmass. The fine groundmass readily disaggregated on rinsing with water. The mineral surface was then checked for any residual attached groundmass using energy-dispersive X-ray spectroscopy (EDS) prior to textural analysis. This ensured any dissolution, mechanical or etching features observed were solely the result of the volcanic or magmatic processes rather than the experimental extraction method.

Dawson (1994) measured the compositions of olivine in the IHV lavas by electron microprobe and showed them to be $\sim Fo_{92}$, compositionally similar to olivine within peri-

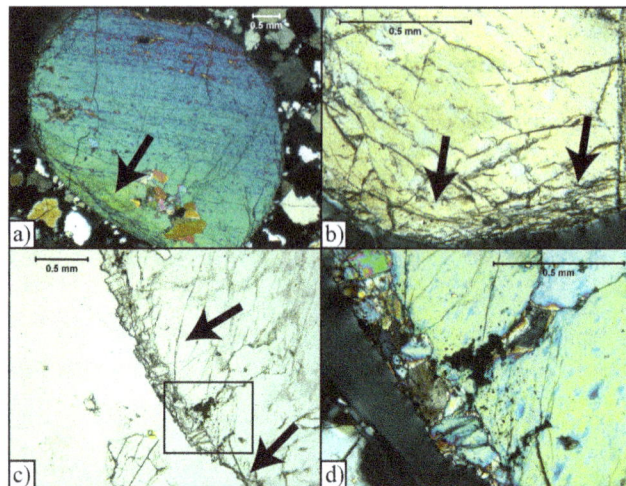

Figure 3. Photomicrographs of grains of olivine separated from sample of Igwisi Hills kimberlite. **(a)** Photomicrograph under cross-polarised light of a single xenocrystic olivine ellipsoid hosting smaller internal domains of recrystallised olivine. Olivine grain shows pronounced fractures parallel to the grain boundary. **(b)** High magnification view of olivine grain boundary recording brittle deformation manifest as arcuate fractures along the rim (< 0.1 mm) and at low angles to the grain boundary. **(c)** Grain boundary showing fine crystals grown along the rim of the larger olivine ellipsoid creating an irregular boundary. Fractures within the main crystal appear to bend towards the finer crystal rim and terminate within it. Again the fine-grained rim is not continuous (absent from the bottom of the photo). **(d)** Cross-polarised view of box in panel **(c)** highlighting the fine-grained crystal rim. Arrows in all figures point to the pronounced curved fractures.

dotitic mantle xenoliths. Reid et al. (1975) also measured olivine compositions in the IH volcanic rocks and, similarly, reported a narrow range (Fo_{90-92}). Several larger olivine grains contain small (< 1 mm) inclusions of spinel, clinopyroxene, garnet and/or apatite. These previous studies have also analysed the compositions of rims to the xenocrystic olivine grains from the IH lavas. Reid et al. (1975) and Dawson (1994) measure and describe the chemical zonation in IHV olivines as the rims being enriched in FeO, CaO and MnO and depleted in NiO, relative to their cores (Table 1).

Current ongoing work on the Igwisi Hills lavas sub-divides the olivine macrocrysts population into three groups: (1a) rounded, polycrystalline micro-xenoliths with inclusions of mantle minerals; (1b) Ni-rich, Ca-poor cores with overgrowth rims rich in CaO and MnO but poor in NiO; (2) unstrained $< 300 \mu$m phenocrysts which are chemically homogeneous; and (3) inclusions and intergrowths of olivine within a titanomagnetite host (Willcox et al., 2014). The olivine analysed in this study have been classified as type 1.

Thin section observations on olivine grains show intense fracturing localised on the rims (Fig. 3a and b). Transverse cracks across the individual olivine grains typically bend towards the heavily fractured/recrystallised rim and terminate within it, not at the edge of the ellipsoid (Fig. 3b). Fractures are commonly filled with opaque mineral inclusions resembling healed/sealed tension cracks (Brett, 2009). The margins of some of the larger rounded olivine grains comprise a randomly oriented aggregate of smaller, ~ 0.2 mm sub-grains. These grains (i.e. neoblasts) are the products of recrystallisation of the original strained olivine crystal. The neoblasts replace the original strained olivine grain but preserve the original overall rounded shape, indicating that recrystallisation post-dates or occurs concurrently to the rounding process. Replacement by these strain-free elongate neoblasts is commonly focussed at the grain boundary but is rarely continuous along the rim (Fig. 3c and d).

The groundmass of the host lava is very fine grained and, although difficult to resolve under a petrographic microscope, comprises carbonate (40 vol. %), small ~ 1 mm rounded olivine grains, spinel and perovskite.

5 Methodology

5.1 Image analysis

Olivine is the dominant phase in kimberlite, and therefore its abundance and grain size distributions are commonly used to characterise kimberlite units (e.g. Field et al., 2009; Jerram et al., 2009; Moss et al., 2010). The corresponding large olivine ellipsoid grain size distribution for the IHV lava is shown in Fig. 4b. The grain size and shape distributions of olivine are based on a high-resolution (1200 dpi) scan of a polished slab (Fig. 2). The resulting digital image was manually traced using Adobe Illustrator™ to provide a representation of the slab outline and each olivine grain. Olivine grains in the IHV rocks vary from 1 to 10 mm in diameter. Manual tracing captured all olivine grains with long axes < 1.6 mm. The digital representation of the olivine grains within the slab (Fig. 4a) was analysed using ImageJ software (http://rsbweb.nih.gov/ij/) for geometric parameters including circularity, axis length and area.

5.2 Scanning electron microscopy

Hand-picked olivine grains from all three sample sets were mounted, carbon-coated and studied under the Philips XL30 scanning electron microscope (SEM) at the University of British Columbia to document surface textures observed across all sample suites.

5.3 3-D Laser measuring microscopy

An Olympus LEXT 3-D Measuring Laser Microscope OLS4000 was used at the Advanced Materials and Process Engineering Laboratory (AMPEL), University of British Columbia, to collect data on the surface topography of the olivine grains. This device is calibrated in the same way as

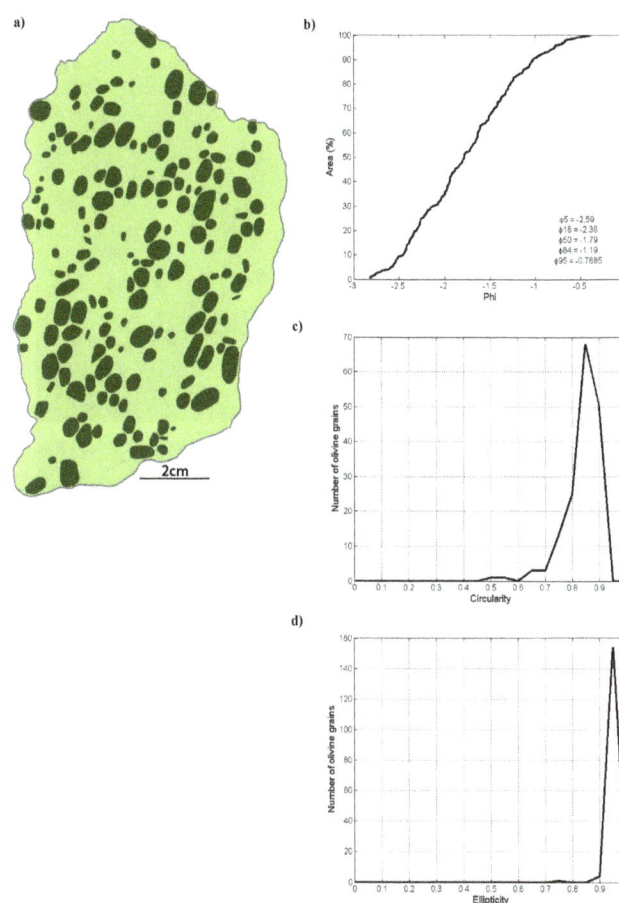

Figure 4. Grain size analysis of olivine grains in the IH53 lava. (**a**) False-colour digitised image of polished slab of IH53 lava; olivine are coloured dark green. Total slab area is 7138 mm^2 and comprises ~ 27 % xenocrystic olivine. (**b**) Histogram showing the olivine percentage area as a function of grain size. The grain size is expressed as $\varphi = -\log_2$ (**d**), where d is the olivine diameter. (**c**) Frequency distribution curve of circularity (C) of olivine grains defined as $C = (\text{Perimeter})^2 / (4\pi * \text{Area})$. $C = 1$ for a perfect circle, and values $0 < C < 1$ provide a relative quantification of roundness. (**d**) A distribution curve, showing the number of olivine grains having a specific ellipticity.

standard stylus instruments for surface measurements; however, it uses contactless measurement with an automatic line-stitching function to create a topographic map at a high resolution. These data sets were used to create models for the micron-scale topographic features of olivines from each of the three sample suites. The models were then used to make a quantitative comparison of the surface properties.

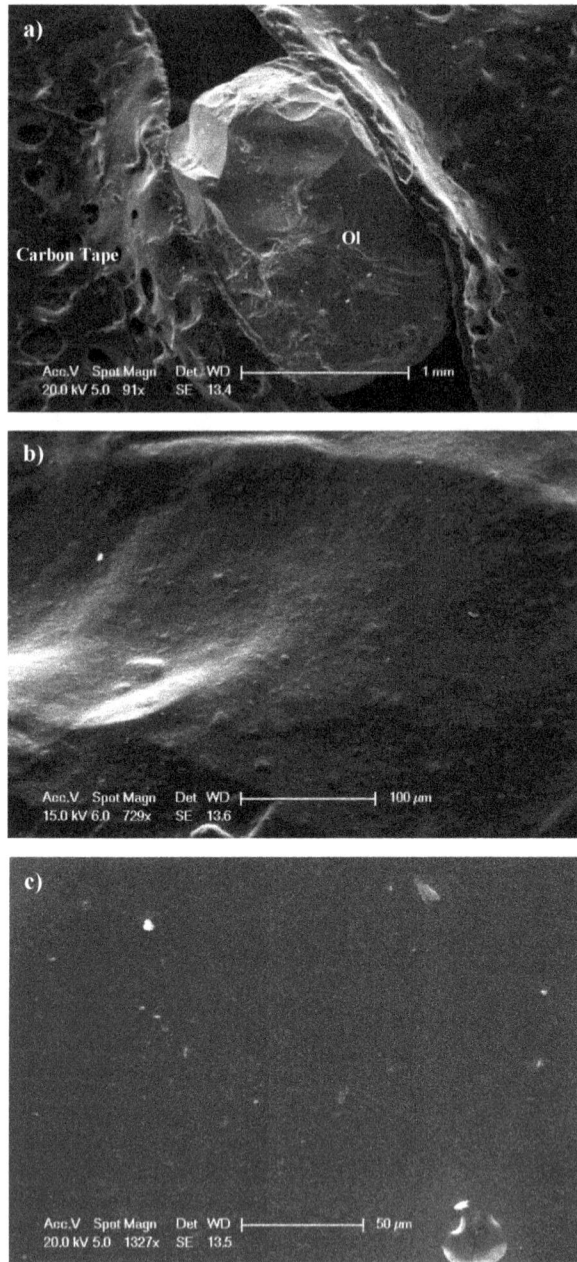

Figure 5. SEM imagery for an olivine phenocryst from an Icelandic picritic basalt lava. **(a)** Olivine grain bounded by mounting tape showing overall morphology of crystal grown from melt. **(b+c)** Detailed images of primary surfaces (unfractured) typical of phenocrystic olivine which shows minimal topography.

6   Results

6.1   Image analysis results

The cumulative frequency plot (Fig. 4b) shows the sphere-equivalent grain size plotted as a function of cumulative area percent. We have computed the Inman graphical standard de-

viation (σ_φ) for this 2-D analysis, as defined by

$$\sigma_\varphi = (\varphi_{84} - \varphi_{16})/2 \qquad (1)$$

to be 0.6 (Inman, 1952). On this basis, the olivine grains are very well sorted in accordance to pyroclastic classification and lie within the very well to moderately sorted descriptor for sedimentary rocks (Cas and Wright, 1987). We also calculated the circularity (C), a measure of roundness, for the olivine grains using

$$C = 4\pi(A/P_{\text{Trace}}^2), \qquad (2)$$

where A and P_{Trace} are the area and the perimeter of individual olivine grains recovered by image analysis via ImageJ. A circularity value of 1 indicates a perfect circle, and for all other shapes circularities are < 1. As the value tends to 0, an increasingly elongated polygon is formed. Figure 4c shows a sharp peak in olivine circularity at ca. 0.85; nearly all IHV olivine grains show circularity values within a narrow range of 0.7 to 0.95. This quantifies the highly rounded nature to be well rounded and close to a circle.

Lastly, we define a new parameter of ellipticity (E), by Eq. (3):

$$E = P_R/P_{\text{Trace}}, \qquad (3)$$

where P_R is the perimeter of a model ellipse which has the same major, a_{Trace}, and minor, b_{Trace}, axes as those measured on the olivine outline. The model ellipse is calculated by the Ramanujan approximation shown in Eq. (4) (Campbell, 2012; Ramanujan, 1962).

$$P_R = \pi \left(3\left(\frac{a_{\text{Trace}}}{2} + \frac{b_{\text{Trace}}}{2} \right) - \sqrt{\left(\frac{3a_{\text{Trace}}}{2} + \frac{b_{\text{Trace}}}{2} \right)\left(\frac{a_{\text{Trace}}}{2} + \frac{3b_{\text{Trace}}}{2} \right)} \right) \qquad (4)$$

Figure 4d shows the distribution of ellipticity values for the digitalised slab (Fig. 4a). For this parameter $E = 1$ is true for a perfect ellipse, and all other values < 1 represent other shapes which deviate from a perfect ellipse. The olivine grains have a mean ellipticity of 0.935, and ca. 97 % of the measured olivine grains have ellipticity values between 0.9 and 1.0. When comparing Fig. 4c and d it is clear that the olivine grains are better described in 2-D as an ellipse rather than a circle. Tables of values of circularity and ellipticity can be found in Appendix A.

6.2   Scanning electron microscopy

6.2.1   Phenocrystic olivine

Samples used for SEM imaging show little evidence of fractured surfaces resulting from the extraction of the crystals from the parent rock. They represent the primary surfaces of

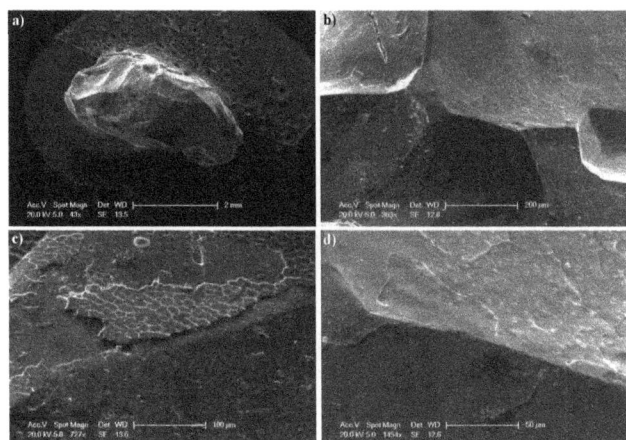

Figure 6. SEM imagery for an olivine grain from mantle-derived peridotite. (**a**) Olivine grain bounded by mounting tape (top) showing overall shape of grain. (**b**) Detailed image of primary surface (unfractured) of olivine typical of grains forming an interlocking mosaic within mantle peridotite. (**c**) Polygonal textured surface flaking. (**d**) Common small-scale surface flakes on the mantle olivine surface. Higher relief is observed at the edge of the flake, creating a stepped topography.

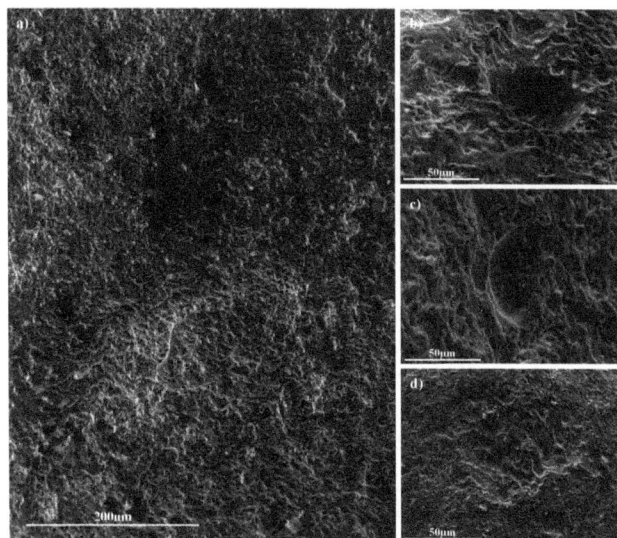

Figure 7. SEM imagery for Igwisi Hills lava olivine. (**a**) Large-scale image showing distinctive surface texture. (**b**) Hemispherical impact cavity featuring a smooth interior. (**c**) A common semi-hemispherical impact pit, again having a smooth interior. (**d**) Several semi-hemispherical excavations; the edge of this attrition feature demonstrates penetrative multiple layer flaking of the olivine surface.

the olivine crystals which crystallised as a phenocryst phase from the basaltic melt. The overall crystal morphology is shown in Fig. 5a. Figure 5b and 5c show a typical featureless surface at two different magnifications: it is very smooth, and flakes and pits are not present.

6.2.2 Mantle olivine

SEM imaging confirms that these samples represent primary olivine surfaces that have not experienced alteration or unnatural fracturing during extraction from the host rock (Fig. 6a). The overall morphology is governed by the interlocking of adjacent mantle olivine crystals within the peridotite forming the observed moulds (Fig. 6b). The mantle olivine is marked by little surface topographic relief despite common flake structures. In some places polygonal flakes are peeling away from the crystal (Fig. 6c and d). A clear topographic difference can be observed between the flaking layer and the smoother olivine surface.

6.2.3 Igwisi Hills lava olivine

The distinctive surfaces of the IHV olivine crystals are highly flaked, exfoliated, irregular and rough compared to the phenocrystic and mantle olivine. These irregular surfaces are characterised as a series of meandering ridges and arc-like steps, creating differences in relief. This surface feature is not confined to a small proportion of the sample; rather it occurs over the entire surface of each IHV olivine analysed.

The exterior surfaces of the IHV olivine grains also feature hemispherical impact cavities or pits (Fig. 7b). The interi-

ors of the pits are smooth, while the surface surrounding the pits is rough and flaky. The cavities are near-perfect hemispheres, with average diameters of $\sim 5\,\mu\text{m}$ superimposed on the flaked surfaces indicating a syn- to post-flaking series of events. Figure 7c shows a semi-hemispherical cavity, identical to Fig. 7b in terms of structure and overprinting relationships, but different in shape. In this image you are also able to observe that layering exists below the surface; the flaking is not just a superficial surface feature and is most likely related to the extensive fracturing observed at the grain edges in thin section (Fig. 3). Figure 7d shows a larger-scale feature, which is dominant on the Igwisi Hills lava olivine surface; it compromises several semi-hemispherical cavities similar to that shown in Fig. 7c. It creates a relatively smooth surface overprinting the irregular flakes and ridges.

6.2.4 3-D Laser measuring microscopy

The contour map (Fig. 8a and b) for the phenocrystic olivine reveals a surface with negligible ($\leq 3\,\mu\text{m}$) topography relief. Indeed, 98 % of the topography ranges within $\pm 2\,\mu\text{m}$ of the median surface. The contour map (Fig. 8c and 8d) for the mantle olivine shows a stepped appearance; all the measured mantle olivine surfaces show a decrease in surface elevation with increasing distance along the x axis. The steps are sub-parallel to the y axis and generate a decrease in topography of about $15\,\mu\text{m}$ across the field of view; it is believed that these steps represent surfaces parallel to crystal faces exposed by

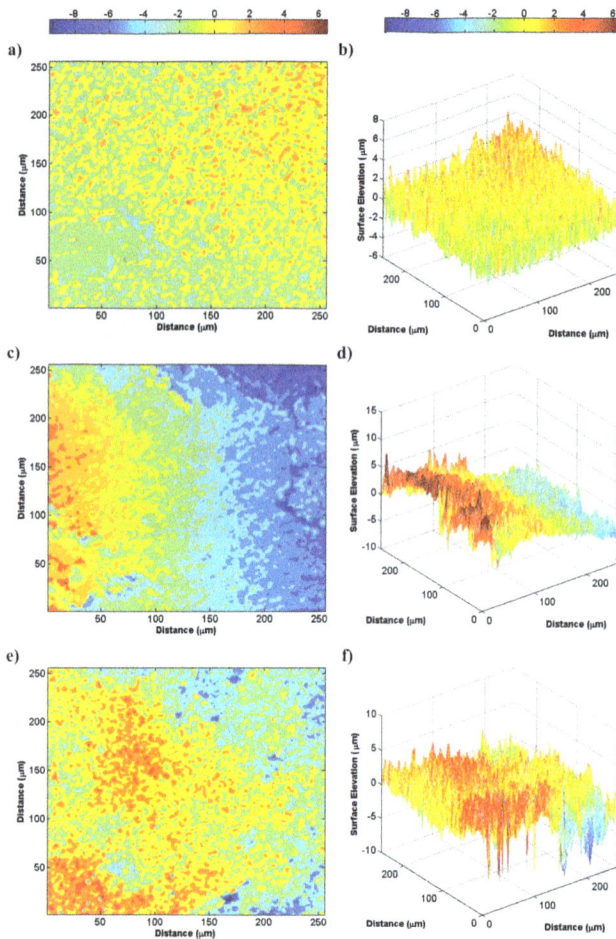

Figure 8. MATLAB-generated contour maps and topographic nets based on data from laser scanning of grain surfaces under microscope. Topographic variations (z axis) are illustrated by a common colour scale used for all images. Data sets are normalised to the mean topographic elevation; positive z axis values therefore correspond to relief greater than the mean. All maps represent a $256\,\mu m$ by $256\mu m$ area on the olivine surface. (**a + b**) Phenocrystic olivine, (**c + d**) mantle olivine and (**e + f**) Igwisi Hills lava olivine.

minor flaking of the olivine crystals exterior (cf. SEM images in Fig. 6).

The IHV samples have a greater surface roughness than both the phenocrystic and mantle olivine grains (Fig. 8f). Contour maps for the IHV olivine (Fig. 8e) also record circular depressions of variable diameter; these are interpreted as the craters/pits observed under the SEM (Fig. 7c). The features are approximately 8–$9\,\mu m$ in depth with diameters of ~ 10–$25\,\mu m$.

7 Discussion

7.1 Origin of IHV olivine surface features

Samples of phenocrystic olivine show smooth, nearly featureless surfaces with zero topography. These surfaces are attributed to free growth from a melt. They show no similarities to the Igwisi Hills samples. Mantle olivine samples commonly display flaked surfaces and a stepped topography (Fig. 6c). However, at a larger scale the mantle olivine grains have sharply faceted, sub-hedral to euhedal shapes, reflecting their textural equilibrium developed under the stable mantle high pressure-temperature conditions of formation. The micron-scale flaking is developed on, and post-dates, these sharply faceted crystal faces. These flaking structures are inferred to form during decompression when the crystal experiences lower pressures during ascent. EDS shows that flakes are identical in chemistry to that of the surrounding fresh olivine (Fig. 6c). Therefore, they are related to decompression rather than to chemical reaction between the crystal and the melt. For example, there is no evidence of melt infiltration and reaction during transport within the mantle nodules (Peterson, 2010).

The olivine grains in the IHV lava display shapes and surface morphologies which are significantly different from features characterising olivine phenocrysts and olivine in mantle peridotite. We observe a highly flaked, irregular and rough surface. The flaky texture parallels the exterior surface but is not just a superficial surface feature. Rather it is penetrative, in that it extends beneath the immediate exterior surface of the olivine grains (Figs. 7a and 3). The flaking structure has some similarities to the flaking observed in the mantle olivine but is much more intense and forms multiple layers. In the IH lava samples the flaking is pervasive, affecting every olivine we analysed.

We interpret the penetrative flaking found on the IH olivines to be related to rapid decompression of the olivine grains during ascent of the kimberlite magma. Olivine is incorporated into kimberlitic magmas at great depths as peridotitic xenoliths. Rapid ascent of 1–$20\,m\,s^{-1}$ (Sparks et al., 2006) of the kimberlite causes pronounced drops in pressure resulting in decompression at rates that can only be accommodated by brittle disaggregation of the xenoliths, liberating olivine grains (Brett, 2009). Tensile failure occurs when the change in stress, coupled to ascent rate, experienced by the olivine crystal is greater than it can viscously relax, defining the tensile strength for kimberlite olivine (Brett, 2009).

High rates of decompression due to rapid ascent rates can cause a rise in tensile stresses at rates faster than can be relaxed viscously. We suggest, therefore, that the flaky exfoliation surfaces on the olivine xenocrysts (Figs. 7 and 3b) are a result of the outer surfaces of the crystal experiencing a build-up of differential (tensile) stress at rates faster than the timescales needed for viscous relaxation. Rapid decompression of solids can cause differential expansion of the clast

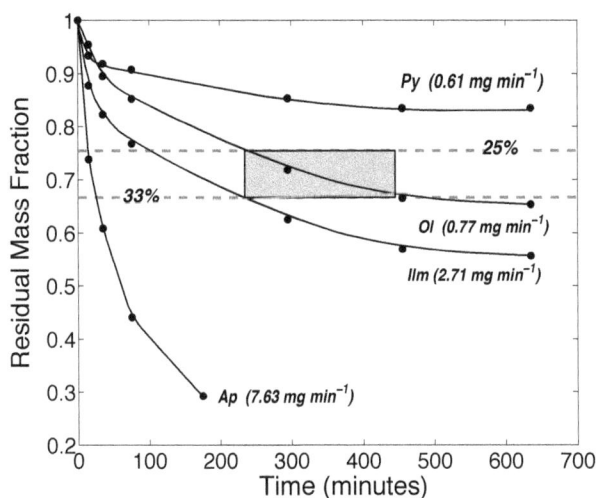

Figure 9. Experimentally measured mass loss rates for four common kimberlitic mineral phases (after Afanas'ev et al., 2008). Residual mass fractions of minerals pyrope (Py), picroilmenites (Ilm), olivine (Ol), and apatite (Ap) being abraded experimentally are plotted against duration (e.g. time). The slopes to the curves define the instantaneous mass loss rates; also shown are the average mass loss rates for each phase (mg min^{-1}). The grey box denotes the estimated mass loss for Igwisi Hills olivines based on their shape (e.g. smoothness, roundness) relative to the calibrated photographs of Afanas'ev et al. (2008).

rim versus its interior. This produces tangential compression within the exterior rim (Fig. 3b) and radial tension in the interior (Preston and White, 1934). We hypothesise that the stresses are partially released by exfoliation of the surface in a similar manner to thermal exfoliation (Preston and White, 1934; Thirumalai, 1969). These partial spalls would be easily removed via crystal–crystal collisions and abrasion during turbulent transport, thereby enhancing the overall rates of attrition (Campbell et al., 2013).

The recrystallisation of strained xenocrystic olivine is commonly localised at the exterior margins of the xenocrysts to create a finer-grained rim of neoblasts (Fig. 3c and d). This process may also facilitate mechanical milling and attrition of the olivine macrocrysts by plucking of the individual neoblasts. This idea was briefly mentioned by Dawson (1994), who suggested that these plucked recrystallised rims may contribute to the "groundmass" olivine population within the IH lavas.

When comparing IHV olivine surfaces to an experimental abrasion study on kimberlitic garnets (McCandless, 1990), great similarities in surface features are observed (Fig. 2d; McCandless, 1990). These have also been interpreted to form through an abrasion process. We also compare IHV olivine morphologies to another experimental study which compares the relative mass loss of kimberlitic minerals with increased abrasion time (Afanas'ev et al., 2008). When comparing observed morphologies to this experimental data set, we esti-

mate IHV olivines are comparable to those subject to 220–455 minutes of abrasion. Then, using a calculated mass loss rate of 0.77 mg min^{-1}, we suggest that the IHV olivine lost $\sim$ 25–33 % of its original mass during ascent (Fig. 9). These experiments are not thought to exactly replicate the process which we describe; they simply provide first-order estimates of mass loss.

Additionally the surfaces of the IHV olivine grains also feature discrete (semi-)hemispherical cavities or depressions (Fig. 7b and c) superimposed on the abrasion surface. We interpret these as impact pits produced by particle–particle collisions, involving higher energies than driving the more steady-state abrasion processes (Campbell et al., 2013; Dufek et al., 2012, 2009).

7.2 Mechanical versus chemical shaping

Olivine is well known to chemically react with or dissolve in kimberlitic melt during transport (e.g. Donaldson, 1990; Edwards and Russell, 1996, 1998). We therefore question to what extent the IHV olivine crystals have been modified by chemical processes rather than the mechanical processes described in this study. To investigate this, IHV olivine crystals were etched with 10 % hydrochloric acid on timescales varying from minutes to days. Their surfaces were then analysed under the SEM and compared to the natural samples.

Etch-pit formation by chemical dissolution is affected by the composition of the etchant, the crystal chemistry, crystallographic orientation, and the presence of impurities ("poisons") in the etchant. This poison may enhance the selectivity of the etching surface (Wegner and Christie, 1974). Thus, our experiments are not intended to replicate exactly magmatic dissolution processes; rather they serve as an analogue for comparative purposes.

Figure 10a shows an experimentally etched Igwisi Hills lava olivine; at this scale it resembles all other IH olivine in the sample suite displaying numerous characteristic features shown in Fig. 7. However at higher magnifications (Fig. 10b) previously unidentified features become apparent on the olivine surface. The etched olivine shows elongate structures strongly controlled by the crystallographic structure of the olivine (Fig. 10c); they do not resemble anything observed in the natural samples. Away from the main etch pit the original olivine surface has also been modified and is now more porous and exhibits a honeycomb-like texture; the natural surfaces of IHV olivine grains do not exhibit these textures.

We expect olivine to be subjected to both chemical and mechanical processes during transport. However the shapes and surfaces recorded within the erupted products studied are clearly dominated by mechanical processes; no similarities with the artificially etched samples are identified. We hypothesise that the chemical processes may dominate at greater depth (in the mantle lithosphere). Then, as the magma exsolves fluids and becomes more buoyant, its ascent velocity

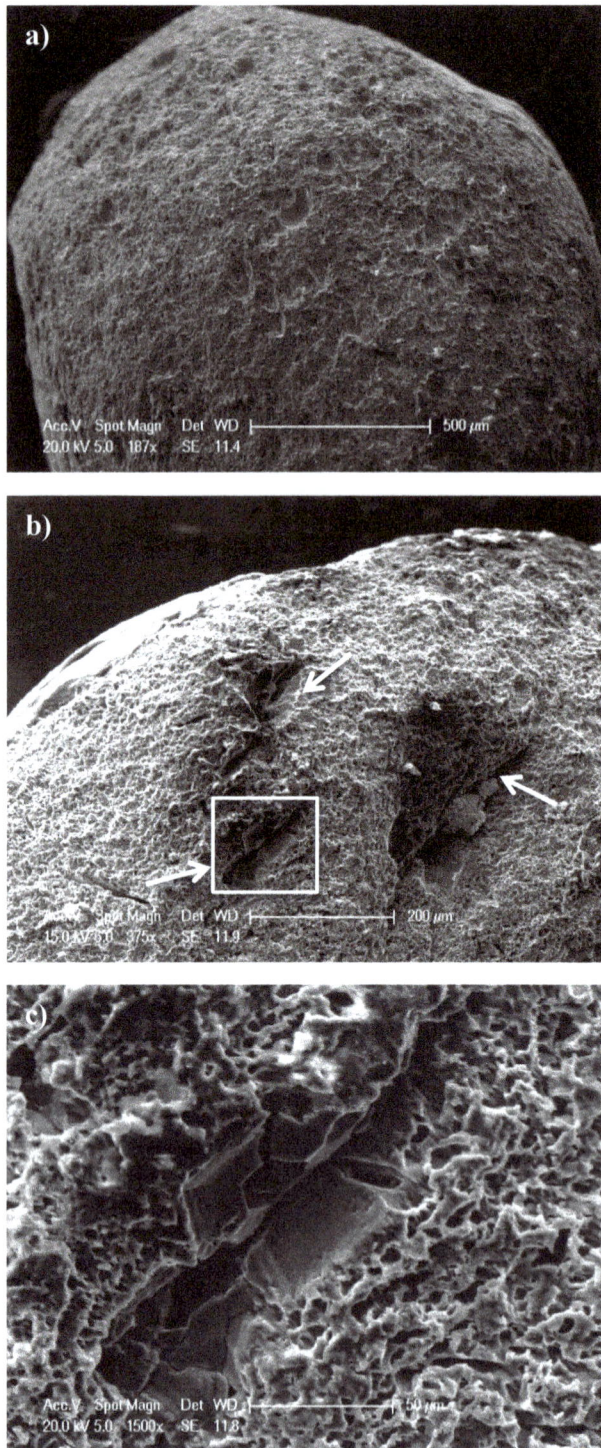

Figure 10. SEM imagery of olivine surfaces subjected to dissolution by weak acids (see text). (**a**) Overall morphology of etched grain. (**b**) Detailed image of etched olivine surface; white arrows highlight etch pits. The white box shows the area represented by part (**c**). (**c**) An etch pit developed on original surface of Igwisi Hills olivine; dissolution pit shape is strongly controlled by the crystallographic structure of the olivine.

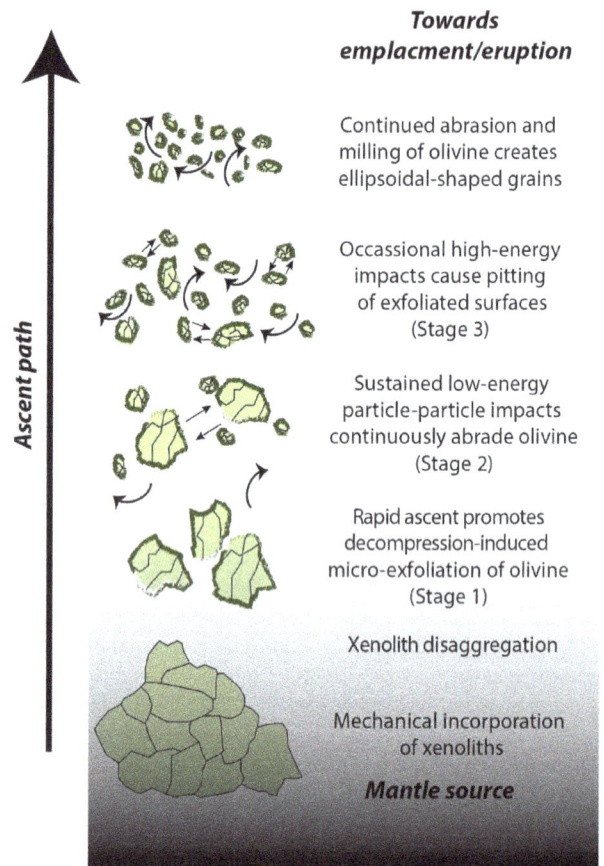

Figure 11. Summary model for the evolution of olivine during ascent of kimberlite. Kimberlite ascent processes cause brittle deformation of mantle peridotite, leading to production and entrainment of mantle xenoliths. Peridotitic xenoliths undergo mechanical disaggregation due to particle–particle collisions and decompression. The decompression is driven by rapid ascent and promotes tensile failure in liberated olivine crystals and is expressed as surface parallel micro-exfoliation. High-velocity (i.e. turbulent) transport of olivine in a solids-rich melt–gas mixture of kimberlite magma supports continual high-frequency low-energy impacts between particles, leading to abrasion and rounding of olivine grains. Periodic higher-energy impacts create impact pits on the olivine's exfoliated microsurfaces. The combination of these processes results in a high proportion of ellipsoidal-shaped, abraded and decompressed olivine grains within the IH lavas.

will rise. This results in a gas-rich mixture travelling at velocities that support turbulent flow and in which mechanical processes begin to dominate.

7.3 Implications for kimberlite ascent

Olivines from IHV lavas have elliptical morphologies and surfaces which preserve impact features that, to our knowledge, are unique to kimberlitic systems. Importantly, these features must be the result of magma transport processes, because these olivine samples derive from a pāhoehoe lava resulting from a quiescent effusion of kimberlite. We propose a three-part model to explain the distinctive attributes of the IHV olivine grains (Fig. 11). Firstly, olivine crystals become incorporated into a kimberlitic melt as whole rock mantle xenoliths. These xenoliths then undergo disaggregation facilitated by mineral expansion during ascent-driven continuous decompression. This imposes a build-up of internal stresses within the olivine crystal. The imposed timescales of ascent are faster than the olivine can viscously relax, and the stresses are relieved by surface parallel micro-exfoliation (Brett, 2009), resulting in the flaked surfaces (Fig. 7a) and intensely fractured olivine rims observed (Fig. 3b).

Secondly, intense decompression of the olivine exterior (Figures 7a and 3b) makes this xenocrystic cargo more susceptible for abrasion through particle–particle collisions sustained within a rapidly ascending, volatile-rich, turbulent mixture of solid, gas and melt driven by the assimilation of orthopyroxene (Russell et al., 2012). This gives rise to the final highly elliptical, near-spherical morphologies that have been documented. The hypothesis of abrasion producing rounding of angular clasts to spherical shapes is supported by experimental studies (Kuenen, 1960). Abrasion will exploit pre-existing weaknesses and preferentially erode weaker compounds (Suzuki and Takahashi, 1981).

Thirdly, higher-energy particle–particle impacts create the hemispherical and semi-hemispherical impact pits (Fig. 7b and c) that are superimposed on the rounded and exfoliated olivine. These events are of lower frequency and only occur when olivine grains collide at high velocities. However, there is an upper limit to the collisional energy – they must be lower than the threshold required for crystal breakage/disruptive fragmentation (Dufek et al., 2012).

The textures on the olivine grains at the Igwisi Hills volcanoes are unique to kimberlites; to our knowledge there is only one other study, Brett (2009), where similar textures and rounding have been briefly described. The features have not been observed on crystals from other mafic eruptions. This may result from the slower ascent rates and shallower melt sources (60–80 km) of other mafic volcanic systems (e.g. basalt, basanite, nephelinite), which lead to lower rates of decompression and the build-up of lower tensile stresses within crystals during transport. Additionally, these magmas in general have lower average solid-particle abundances, which reduces the frequency of particle interactions within a turbulent fluid. Lastly, the lower volatile component of non-kimberlitic magmas may result in more viscous magmas that dampen the energy of colliding solid particles.

Xenoliths and xenocrysts of minerals other than olivine are rare within the IHV rocks. This can be explained by the high abrasion stability of olivine (Fig. 9), which is third only to diamond and pyrope (in decreasing order: diamond–pyrope–olivine–picroilmenite–apatite–kimberlite; Afanas'ev et al., 2008). Therefore, during rapid turbulent ascent other mantle mineral phases are less likely to survive, relative to olivine.

The observations presented here pertain to the surface features of kimberlitic olivine after transport and eruption. Our model for the mechanical shaping of olivine xenocrysts during kimberlite ascent refines many of the ideas advanced previously in the literature (Table 1). Observations by Brett et al. (2009) suggest a milled xenocrystic core then subsequent crystallisation with or without further milling. Studies by Arndt et al. (2010) propose olivine rim crystallisation followed by abrasion with the removal of Fe-rich regions. We suggest either a continuous milling process occurring throughout xenocrystic transport and subsequent olivine rim crystallisation or repeated cycles of milling and olivine crystallisation during kimberlitic ascent. Both methods would leave rounded xenocrystic cores frequently observed in the literature and surface textures observed in this study.

8 Conclusions

The shapes and surface textures of xenocrystic olivine grains extracted from a young kimberlite lava have been described and interpreted. The xenocrystic olivine grains are ellipsoidal, exhibit rough flaky microsurfaces, covered with impact pits. These surface features differ significantly from both olivine phenocrysts and from olivine disaggregated from mantle nodules in the laboratory. A three-part model is proposed to explain these differences. (1) During rapid CO_2-driven magma ascent, decompression results in parallel micro-exfoliation on the surface of the xenocrystic olivine grains. (2) Turbulent suspension of solid, gas and melt phases during ascent promotes abrasion that results in the highly elliptical and sub-spherical grain morphology. (3) High-velocity particle–particle impacts result in impact pitting of the surface. Synthetic experiments rule out chemical dissolution and melt reabsorption as the causative process for rounding. The morphology and surface features of the xenocrystic olivine grains are consistent with mechanical processes operating during rapid CO_2-driven turbulent ascent of the magma from depth.

Acknowledgements. We thank AMPEL for support with the Laser Measuring Microscope. J. K. Russell is supported by the NSERC Discovery Grants programme. L. A. Porritt acknowledges support from a Marie Curie International Outgoing Postdoctoral Fellowship. Fieldwork on the Igwisi Hills Volcanoes was supported by a grant from National Geographic Committee for Research and Exploration awarded to R. J. Brown. We gratefully acknowledge critical reviews by Nicholas Arndt and an anonymous reviewer. We also thank Mike Heap for his editorial role.

Edited by: M. Heap

References

Afanas'ev, V. P., Nikolenko, E. I., Tychkov, N. S., Titov, A. T., Tolstov, A. V., Kornilova, V. P., and Sobolev, N. V.: Mechanical abrasion of kimberlite indicator minerals: experimental investigations, Russ. Geol. Geophys., 49, 91–97, 2008.

Arndt, N., Boullier, A.-M., Clement, J., Dubois, M., and Schissel, D.: What olivine, the neglected mineral, tells us about kimberlite petrogenesis, eEarth Discuss., 1, 37–50, 2006.

Arndt, N. T., Guitreau, M., Boullier, A.-M., Le Roex, A., Tommasi, A., Cordier, P., and Sobolev, A.: Olivine, and the Origin of Kimberlite, J. Petrol., 51, 573–602, 2010.

Boyd, F.: Garnet peridotites and the system CaSiO 3-MgSiO 3-Al₂O₃, Mineral. Soc. Amer. Spec. Pap, 3, 63–75, 1970.

Brett, R. C.: Kimberlitic olivine: Msc Thesis, University of British Columbia, 2009.

Brett, R., Russell, J., and Moss, S.: Origin of olivine in kimberlite: Phenocryst or impostor?, Lithos, 112, 201–212, 2009.

Brown, R. J., Manya, S., Buisman, I., Fontana, G., Field, M., Mac Niocaill, C., Sparks, R., and Stuart, F.: Eruption of kimberlite magmas: physical volcanology, geomorphology and age of the youngest kimberlitic volcanoes known on earth (the Upper Pleistocene/Holocene Igwisi Hills volcanoes, Tanzania), Bull. Volcan., 74, 1621–1643, 2012.

Campbell, M.: Thermomechanical milling of lithics in volcanic conduits: Msc Thesis, University of British Columbia, 2012.

Campbell, M. E., Russell, J. K., and Porritt, L. A.: Thermomechanical milling of accessory lithics in volcanic conduits, Earth Planet. Sci. Lett., 377/378, 276–286, 2013.

Cas, R. A. and Wright, J. V.: Volcanic successions, modern and ancient: A geological approach to processes, products, and successions, Allen & Unwin, 1987.

Clement, C. and Skinner, E.: A textural genetic classification of kimberlite rocks. Kimberlite Symposium II, Cambridge, in: Proceedings Extended Abstracts, 1979–1985.

Clement, C. and Skinner, E.: A textural-genetic classification of kimberlites, South Afr. J. Geol., 88, 403–409, 1985.

Davis, B., and Boyd, F.: The join Mg2Si2O6-CaMgSi2O6 at 30 kilobars pressure and its application to pyroxenes from kimberlites, J. Geophys. Res., 71, 3567–3576, 1966,

Dawson, J.: Quaternary kimberlitic volcanism on the Tanzania Craton, Contr. Mineral. Petrol., 116, 473–485, 1994.

Dawson, J. B. and Hawthorne, J. B.: Magmatic sedimentation and carbonatitic differentiation in kimberlite sills at Benfontein, South Africa, J. Geol. Soc., 129, 61–85, 1973.

Donaldson, C.: Forsterite dissolution in superheated basaltic, andesitic and rhyolitic melts, Mineral. Mag., 54, 67–74, 1990.

Dufek, J., Wexler, J., and Manga, M.: Transport capacity of pyroclastic density currents: Experiments and models of substrate-flow interaction, J. Geophys. Res., 114, B11203, doi:10.1029/2008JB006216, 2009.

Dufek, J., Manga, M., and Patel, A.: Granular disruption during explosive volcanic eruptions, Nature Geosci., 5, 561–564, 2012.

Edwards, B. and Russell, J.: A review and analysis of silicate mineral dissolution experiments in natural silicate melts, Chem. Geol., 130, 233–245, 1996.

Edwards, B. R. and Russell, J. K.: Time scales of magmatic processes: new insights from dynamic models for magmatic assimilation, Geology, 26, 1103–1106, 1998.

Field, M., Gernon, T. M., Mock, A., Walters, A., Sparks, R. S. J., and Jerram, D. A.: Variations of olivine abundance and grain size in the Snap Lake kimberlite intrusion, Northwest Territories, Canada: A possible proxy for diamonds, Lithos, 112, 23–35, 2009.

Gernon, T., Field, M., and Sparks, R.: Geology of the Snap Lake kimberlite intrusion, Northwest Territories, Canada: field observations and their interpretation, J. Geol. Soc., 169, 1–16, 2012.

Inman, D. L.: Measures for describing the size distribution of sediments,: J. Sediment. Res., 22, 125–145, 1952.

Jerram, D. A., Mock, A., Davis, G. R., Field, M., and Brown, R. J.: 3D crystal size distributions: A case study on quantifying olivine populations in kimberlites, Lithos, 112, 223–235, 2009.

Kamenetsky, V. S., Kamenetsky, M. B., Sobolev, A. V., Golovin, A. V., Demouchy, S., Faure, K., Sharygin, V. V., and Kuzmin, D. V.: Olivine in the Udachnaya-East kimberlite (Yakutia, Russia): types, compositions and origins, J. Petrol., 49, 823–839, 2008.

Kuenen, P. H.: Experimental Abrasion 4: Eolian Action: J. Geology, 68, 427–449, 1960.

McCandless, T. E.: Kimberlite xenocryst wear in high-energy fluvial systems: experimental studies, J. Geochem. Explor., 37, 323–331, 1990.

Mitchell, R.: Kimberlite and Related Rocks: A Critical Reappraisal, J. Geol., 78, 686–704, 1970.

Mitchell, R. H.: Kimberlites: mineralogy, geochemistry, and petrology, Plenum Press New York, 1986.

Mitchell, R. H.: Kimberlites, orangeites, and related rocks, Plenum Press New York, 1995.

Mitchell, R. H.: Petrology of hypabyssal kimberlites: relevance to primary magma compositions, J. Volcanol. Geothermal Res., 174, 1–8, 2008.

Moore, A. E.: The case for a cognate, polybaric origin for kimberlitic olivines, Lithos, 128, 1–10, 2012.

Moss, S. and Russell, J. K.: Fragmentation in kimberlite: products and intensity of explosive eruption, Bull. Volcanol., 73, 983–1003, 2011.

Moss, S., Russell, J. K., Smith, B. H. S., and Brett, R. C.: Olivine crystal size distributions in kimberlite, Am. Mineral., 95, 527–536, 2010.

Peterson, N. D.: Carbonated mantle lithosphere in the western Canadian Cordillera, 2010.

Pilbeam, L. H., Nielsen, T. F. D., and Waight, T. E.: Digestion Fractional Crystallization (DFC): an Important Process in the Genesis of Kimberlites. Evidence from Olivine in the Majuagaa Kimberlite, Southern West Greenland, J. Petrol., 54, 1399–1425, 2013.

Preston, F. W. and White, H. E.: OBSERVATIONS ON SPALLING*, J. Am. Ceram. Soc., 17, 137–144, 1934.

Ramanujan, S.: Ramanujan's collected works: New York: Chelsea, 1962.

Reid, A. M., Donaldson, C., Dawson, J., Brown, R., and Ridley, W.: The Igwisi Hills extrusive "kimberlites", Phys. Chem. Earth, 9, 199–218, 1975.

Russell, J. K., Porritt, L. A., Lavallée, Y., and Dingwell, D. B.: Kimberlite ascent by assimilation-fuelled buoyancy, Nature, 481, 352–356, 2012.

Sparks, R.: Kimberlite Volcanism, Ann. Rev. Earth Planet. Sci., 41, 497–528, 2013.

Sparks, R., Baker, L., Brown, R., Field, M., Schumacher, J., Stripp, G., and Walters, A.: Dynamical constraints on kimberlite volcanism, J. Volcanol. Geothermal Res., 155, 18–48, 2006.

Suzuki, T. and Takahashi, K. I.: An experimental study of wind abrasion, J. Geol., 89, 509–522, 1981.

Thirumalai, K.: Process Of Thermal Spalling Behavior In Rocks An Exploratory Study, in: Proceedings The 11th US Symposium on Rock Mechanics (USRMS), 1969.

Wegner, M. and Christie, J.: Preferential chemical etching of terrestrial and lunar olivines, Contribut. Mineral. Petrol., 43, 195–212, 1974.

Willcox, A., Buisman, I., Sparks, S., Brown, R., Manya, S., Schumacher, J., and Tuffen, H.: Petrology, geochemistry and low-temperature alteration of extrusive lavas and pyroclastic rocks of the Igwisi Hills kimberlites, Tanzania, Chem. Geol., in preparation, 2014.

8

Lithosphere and upper-mantle structure of the southern Baltic Sea estimated from modelling relative sea-level data with glacial isostatic adjustment

H. Steffen[1], G. Kaufmann[2], and R. Lampe[3]

[1]Lantmäteriet, Lantmäterigatan 2c, 80182 Gävle, Sweden
[2]Freie Universität Berlin, Institut für Geologische Wissenschaften, Fachrichtung Geophysik, Malteserstr. 74–100, Haus D, 12249 Berlin, Germany
[3]Ernst-Moritz-Arndt-Universität Greifswald, Institut für Geographie und Geologie, F.-L.-Jahn-Str. 16, 17487 Greifswald, Germany

Correspondence to: H. Steffen (holger-soren.steffen@lm.se)

Abstract. During the last glacial maximum, a large ice sheet covered Scandinavia, which depressed the earth's surface by several 100 m. In northern central Europe, mass redistribution in the upper mantle led to the development of a peripheral bulge. It has been subsiding since the begin of deglaciation due to the viscoelastic behaviour of the mantle.

We analyse relative sea-level (RSL) data of southern Sweden, Denmark, Germany, Poland and Lithuania to determine the lithospheric thickness and radial mantle viscosity structure for distinct regional RSL subsets. We load a 1-D Maxwell-viscoelastic earth model with a global ice-load history model of the last glaciation. We test two commonly used ice histories, RSES from the Australian National University and ICE-5G from the University of Toronto.

Our results indicate that the lithospheric thickness varies, depending on the ice model used, between 60 and 160 km. The lowest values are found in the Oslo Graben area and the western German Baltic Sea coast. In between, thickness increases by at least 30 km tracing the Ringkøbing-Fyn High. In Poland and Lithuania, lithospheric thickness reaches up to 160 km. However, the latter values are not well constrained as the confidence regions are large. Upper-mantle viscosity is found to bracket $[2–7] \times 10^{20}$ Pa s when using ICE-5G. Employing RSES much higher values of 2×10^{21} Pa s are obtained for the southern Baltic Sea. Further investigations should evaluate whether this ice-model version and/or the

RSL data need revision. We confirm that the lower-mantle viscosity in Fennoscandia can only be poorly resolved.

The lithospheric structure inferred from RSES partly supports structural features of regional and global lithosphere models based on thermal or seismological data. While there is agreement in eastern Europe and southwest Sweden, the structure in an area from south of Norway to northern Germany shows large discrepancies for two of the tested lithosphere models. The lithospheric thickness as determined with ICE-5G does not agree with the lithosphere models. Hence, more investigations have to be undertaken to sufficiently determine structures such as the Ringkøbing-Fyn High as seen with seismics with the help of glacial isostatic adjustment modelling.

1 Introduction

During the last colder climatic phase with average surface temperatures being about 10 °C lower than today (Petit et al., 1999), northern Europe – like other parts in the world – was covered by an extensive ice sheet. The mass of this so-called Fennoscandian ice sheet deformed the earth's crust into the mantle, leading to surface depressions of several hundreds of metres underneath the ice. Beyond the ice-covered area, a peripheral bulge developed around the ice sheet due to the bending of the elastic lithosphere outside the ice-covered area. This narrow band of 100–200 km width was uplifted up to a few tens of metres (Steffen and Wu, 2011). During and after the deglaciation phase, the mass redistribution is reversed, forcing uplift of the formerly glaciated areas and subsidence of the peripheral bulge. These changes are, due to the viscoelastic and thus time-delayed behaviour of the mantle, still observable today.

This dynamic response of the earth during glacial cycles is known as glacial isostatic adjustment (GIA). There are several observation methods for this process, and Fennoscandia has turned out to be the key area for GIA studies (e.g. Steffen and Wu, 2011, and references therein). Relative sea-level (RSL) data provide the longest observational data set from all observations, occasionally dating back several thousands of years. They document the movement of coastlines as a consequence of both the water redistribution between oceans and ice sheets and the deformation of the earth's surface that occurred in the past.

RSL data can be employed for the determination of the earth's internal structure, in particular the lithospheric thickness and mantle viscosities (e.g. Steffen and Wu, 2011, and references therein). Often, this is done in formerly glaciated areas, e.g. Fennoscandia, the Barents Sea or the British Isles. As an example, Steffen and Kaufmann (2005) subdivided the Fennoscandian RSL data set into RSL data located in the centre around the Baltic Sea and coastal data mainly along the Norwegian coast. They found clear differences in the earth's structure of the two regions. Vink et al. (2007) subdivided a RSL data set of the southern North Sea into three distinct regional subsets. A regional variation of the lithospheric thickness as well as regionally differing isostatic subsidence curves were determined.

The earth structure beneath northern Europe derived from GIA data can be summarized as follows: in Fennoscandia, the lithosphere is laterally varying with a thick root of more than 200 km in central-east Fennoscandia, becoming thinner towards the west (Steffen and Wu, 2011). Southwest Sweden is predicted to have a lithospheric thickness of about 100 km, and the German North Sea coast as well as the Norwegian Atlantic coast of about 80 km (Vink et al., 2007; Steffen and Wu, 2011). Note that we use the term lithosphere to refer to the strong outer shell of the earth composed of the crust and upper part of the mantle, which both have a purely elastic rheology on the GIA timescale.

Below the lithosphere, investigations have found upper-mantle viscosity to be between 10^{20} and 10^{21} Pa s (Steffen and Wu, 2011). The latest results calculated from different data are in the range $[3–8] \times 10^{20}$ Pa s. The viscosity increases towards the lower mantle (Steffen and Kaufmann, 2005). The lower-mantle viscosity is assumed to be around 1–2 orders of magnitude higher. Its determination, however, is complicated, as the resolving power of all data in Fennoscandia is too low to resolve more accurate values for the lower mantle (Steffen and Wu, 2011).

The values above have mainly been determined with spherically symmetric models using Maxwell rheology. However, other rheologies such as composite rheology (van der Wal et al., 2013) or models with laterally varying lithospheric thickness and/or mantle viscosities (Wu et al., 2005; Steffen et al., 2006; Wang et al., 2008; van der Wal et al., 2013; Wu et al., 2013) can also fit the observations in Fennoscandia reasonably well.

The lithosphere determined in GIA studies should be comparable to results from other studies, e.g. seismological studies. However, there are different geophysical definitions of the lithosphere depending on the method used for its determination. There are rheological, petrological, elastic, thermal, electrical and seismic definitions. It is beyond the scope of this paper to discuss individual definitions or their determination in detail, or the relation of one lithosphere definition to another. We therefore refer the interested reader to Tesauro et al. (2009), Eaton et al. (2009) and Artemieva (2009) for a detailed overview. But it has been noted that some of the definitions should coincide, such as the thermal definition and the seismological one (Tesauro et al., 2009). Eaton et al. (2009) define the lithosphere as "a rheological term referring to the strong outer shell of the earth composed of the crust and upper part of the mantle; also called a mechanical boundary layer". The seismological lithosphere is generally the high-velocity outer layer of the earth, approximately coincident with the lithosphere as a rheological term, which typically overlies a low-velocity zone (Eaton et al., 2009). The thermal lithosphere is defined by a depth to a constant isotherm or by the depth of the intersection of a continental geotherm either with a mantle adiabat or with a temperature close to mantle solidus (Artemieva, 2009). We will see that the lithospheric structure in northern Europe as derived with GIA modelling and outlined above, partly agrees with thermal and seismological studies on the lithosphere on a broad scale, but only in terms of lateral variation and not in an exact match of thicknesses.

The purpose of this study is to determine the earth's structure underneath the southern Baltic Sea with special attention given to the lateral variation of the lithosphere. We use RSL data that have emerged mainly in recent years. They are subdivided in regional subsets similar to the studies by Lambeck et al. (1998) and Vink et al. (2007) to derive radial profiles of the earth for five different regions of the southern Baltic Sea. The best-fitting models allow us to analyse the isostatic behaviour of each region, to highlight the lateral structure and to describe the peripheral bulge in northern Central Europe. We do not aim to investigate the presence of the asthenosphere in this area. Seismic tomographic imaging and a few GIA studies (e.g. Fjeldskaar, 1994) have indicated such an area of lower viscosity in western Fennoscandia (Steffen and Kaufmann, 2005). Unfortunately, the RSL data in the southern Baltic Sea cannot be used to accurately determine parameters for the asthenosphere as their time and depth range is small, see discussion in the next section. As an additional exercise, we compare the lithospheric thickness as derived in regional subsets to three lithospheric thickness models available to us.

In Sect. 2, we describe the RSL data used. This is followed by an overview of the modelling technique and the ice models implemented in this study (Sect. 3). Results are presented in Sect. 4 and discussed in Sect. 5. This includes a comparison to lithosphere models available to us. Finally, we summarize our main findings in Sect. 6.

2 Relative sea-level data

In the past decades mostly basal peat layers (sensu Lange and Menke, 1967) found in sediment cores were used to reconstruct the postglacial sea-level rise along the southern and western Baltic coast. However, these sea-level indicators, often scattered over larger areas, may have experienced different vertical movements due to isostasy and/or compaction and thus are compromised by large uncertainties in many respects. More recently, new sampling, positioning and dating techniques have allowed the detection of archaeological underwater finds such as settlement refuse, boats, fish weirs and fire places, or drowned in situ tree stumps (Tauber, 2007; Lübke et al., 2011). Such finds provide numerous samples for a distinct site and a specific elevation relative to modern sea level. Other approaches use a set of isolation basins or coastal mires to trace the sea-level variation over a longer period in a very limited area (Yu et al., 2004; Lampe et al., 2011). Such investigations allow the construction of sea-level curves owing to better resolution and minor altitude errors and thus higher precision. They provide an excellent base to test different ice-load history models and earth models as well.

For this study we use published data sets from Denmark (Great Belt and Halsskov Fjord: Christensen et al., 1997), northeastern Germany (Schleswig-Holstein: Winn et al., 1986; Jakobsen, 2004; Mecklenburg-Vorpommern: Lampe et al., 2007; Hoffmann et al., 2009; Poland: Uścinowicz, 2003 and a few data from Lithuania: Curonian Lagoon and adjacent areas: Bitinas et al., 2000, 2002). A common feature of the investigated regions is that the postglacial sea-level rise did not start until the transgressing ocean inundated the Danish Great Belt and invaded the Baltic Basin. Age determinations of the earliest marine influence in the southern Baltic therefore lie between 9.4 and 8.0 ka cal BP (Hofmann and Winn, 2000; Rößler et al., 2011; Bennike et al., 2004). Because the maximum depth of the Danish Great Belt amounts to 25 m below sea level, the rising ocean could not invade the Baltic Basin before it inundated this threshold and thus the sea-level change cannot be traced to greater depths. In coastal regions the Pleistocene relief further restricts the depth where the former sea level can be determined.

Therefore, the lowest sea-level indicators used in the study come from offshore areas in the Great Belt and Bay of Kiel, while all other indicators are from near-coastal on- and off-shore areas that are located at much lesser depths. Mostly, the data used belong to larger data sets compiled by archaeological, palaeoecological or geological investigations. From these sets data were chosen which are evaluated as reliably related to the former sea level, considering the kind of dated material and probability of relocation, sedimentary facies, accuracy of altitude determination and age–depth relations in the entire data set.

In addition to these new data for the southern Baltic Sea coast, we investigate RSL data in the southwestern part of Fennoscandia that were used by Steffen and Kaufmann (2005) and Schmitt et al. (2009). We group these data into five regional subsets according to dominant structures visible in the regional geology (Scheck-Wenderoth et al., 2005) and crust–mantle boundary (Dèzes and Ziegler, 2002), see our additional remarks on each subset below.

The first covers the Oslo Graben and the eastern part of the Norwegian–Danish Basin (Fig. 1). It contains 77 data from northern Denmark (Limfjord) and the Oslo Fjord. Lambeck et al. (1998) used a subset for the Oslo Fjord only while the Limfjord data were included in a Danish subset together with data from the Great Belt. We will see that both regions, Limfjord and Oslo Fjord, can be combined into one subset. The second subset includes 44 data from southwest (SW) Sweden that were used by Lambeck et al. (1998) in a subset for SW Sweden as well. In addition, 12 archaeological data from dated Hensbacka sites around the city of Gothenburg as described and used in Schmitt et al. (2009) are added resulting in a total of 56 data for this data set, which is located east of the Teisseyre–Tornquist Zone. The third subset, called Fyn, consists of 128 indicators from the Great Belt and northeastern Germany, but east of Rostock. These data are located within the Rinkøping-Fyn High and extend the area further east almost parallel to the former ice margin. The fourth subset contains 65 data of the bays of Kiel and Lübeck along the western coast of the German Baltic Sea. This area is part of the North German Basin. As there are RSL data which are at the border of the third and fourth subset, we test the influence of these data on the determined best-fitting earth model for each subset. These data are located at Rostock (yellow dots in Fig. 1), Körkwitz (light blue) and the Darss Peninsula (dark blue). As we test all three locations in each subset, this results in four different subset of "Fyn" and "bays of Kiel and Lübeck". The fifth subset encompasses 31 indicators from Poland and Lithuania. These data are found east of the Teisseyre–Tornquist Zone.

Figure 1 shows the spatial and temporal distribution of the data sets. One can clearly distinguish the characteristics of each data set. SW Sweden and the samples of the Oslo Fjord highlight land uplift over the last 15 000 years and thus are typical examples of near-field data. The Limfjord index points as well as the other data sets trace the sea-level rise in the last 12 000 years, here in conjunction with isostatic subsidence of the forebulge, and therefore illustrate the typical behaviour of far-field data. We also see that the vertical range

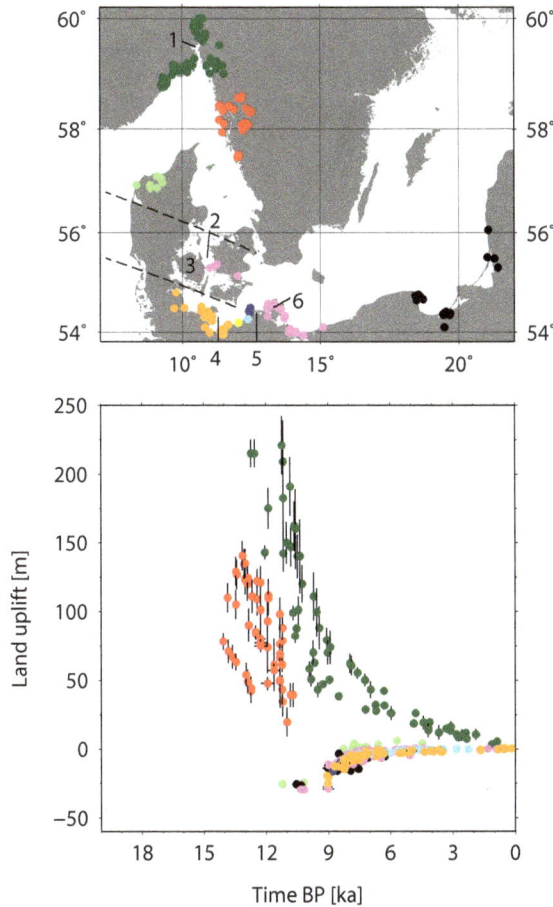

Figure 1. Spatial and temporal distribution of relative sea-level data used in this study. Colours indicate five regional subsets: (I) Southwest Sweden (red), (II) Oslo Graben (dark and light green for Oslo Fjord and Limfjord, respectively), (III) Fyn with Great Belt, Rügen, Usedom (violet), Darss Peninsula (dark blue) and Körkwitz (light blue), (IV) bays of Kiel and Lübeck (orange) and Rostock (yellow), (V) Poland and Lithuania (black). Data uncertainties are indicated by vertical error bars. Geographical information: 1 Oslo Fjord; 2 Great Belt; 3 Fyn; 4 Bay of Lübeck; 5 Darss Peninsula; 6 Rügen. Dashed lines mark the location of the Rinkøping-Fyn High.

of near-field data, here more than 200 m, is much larger than that of the far-field data, which has a range of less than 30 m. The main sea-level change visible in the latter data happens before 7 ka BP. After that, the change is in the metre range.

3 Modelling

3.1 Earth models

The modelling is undertaken with the software package ICEAGE (Kaufmann, 2004), which was successfully used in earlier GIA studies (e.g. Steffen and Kaufmann, 2005; Vink et al., 2007; Steffen et al., 2010). We briefly summarize the

main characteristics and methods only, and refer the reader to Steffen and Kaufmann (2005) for more information.

We employ a spherically symmetric (1-D), compressible, Maxwell-viscoelastic earth model having three layers to be varied; lithospheric thickness, upper- and lower-mantle viscosity. The depth of the boundary between upper and lower mantle is set to 670 km. An inviscid earth's core is set as lower boundary. The viscosity is kept constant within a layer. Elastic parameters are taken from the Preliminary Reference Earth Model (PREM Dziewonski and Anderson, 1981). Lithospheric thickness is varied between 60 and 160 km, upper-mantle viscosity between 10^{19} and 4×10^{21} Pa s, and lower-mantle viscosity between 10^{21} and 10^{23} Pa s. Based on former investigations (e.g. Steffen and Kaufmann, 2005; Vink et al., 2007) these values cover plausible values for three-layer models well.

We follow the pseudo-spectral approach described in Mitrovica et al. (1994) and Mitrovica and Milne (1998) for the calculation of relative sea levels with our models. It is an iterative procedure in the spectral domain with a spherical harmonic expansion up to degree 192, which solves the sea-level equation (Farrell and Clark, 1976) for a rotating earth. Relative sea levels are calculated for 1089 ($11 \times 11 \times 9$) different so-called three-layer earth models which are then compared to our regional RSL data sets based on a least-squares misfit

$$\chi = \sqrt{\frac{1}{n} \sum_{i=1}^{n} \left(\frac{o_i - p_i(a_j)}{\Delta o_i} \right)^2}, \quad (1)$$

with n the number of observations, o_i the observed RSL, $p_i(a_j)$ the predicted RSL for a specific earth model a_j, and Δo_i the error of the observed RSL. The lowest value of χ relates to the best-fitting earth model a_b out of the 1089 provided. In addition, we analyse the model confidence within the observational errors by calculating the confidence parameter

$$\psi = \sqrt{\frac{1}{n} \sum_{i=1}^{n} \left(\frac{p_i(a_b) - p_i(a_j)}{\Delta o_i} \right)^2} \quad (2)$$

of the predicted RSL for the best-fitting earth model $p_i(a_b)$ to all other earth models. We show the 1σ and 2σ uncertainty for models that obey $\psi \leq 1$ and $1 < \psi \leq 2$, respectively, of the best-fitting earth model.

3.2 Ice models

We apply two different global ice models as load on the earth models. First, as in Steffen and Kaufmann (2005) and Vink et al. (2007), we use the model RSES provided by Kurt Lambeck (Research School of Earth Sciences, Australian National University) (see e.g. Lambeck et al., 1998). It combines the extent and the melting history from different separate ice models around the world. It is an updated version of the one presented in Lambeck et al. (1998). The other global ice model is the commonly used ICE-5G ice history (Peltier, 2004). Both RSES and ICE-5G belong to the type of ice models which are constrained by solid-earth models. Hence, best-fitting models usually tend to converge to a radial profile of specific lithospheric thickness and several viscosity layers as used in the ice-model generation. This is especially the case when the same observational data are used in an investigation. In our case, we test a large set of RSL data that have not been used to generate the respective ice models. This may either imply modifications for the ice model if the best-fitting earth model is different, or may shed a light into lateral lithosphere and mantle viscosity variations if the ice model is assumed to be correct. RSES is associated with a 1-D earth model that has a lithospheric thickness of 65–85 km, an upper-mantle viscosity of $3–4 \times 10^{20}$ Pa s and a lower-mantle viscosity about one order of magnitude larger than the upper mantle. ICE-5G's underlying earth model, called VM2, has a lithospheric thickness of 90 km, and then several viscoelastic layers in the mantle. The average viscosities in the upper and lower mantle are about 6×10^{20} Pa s and 2×10^{21} Pa s, respectively.

We exemplarily show the extent of the Fennoscandian ice sheet at Last Glacial Maximum of the two models in Fig. 2. There are distinct differences in collapse history, ice height and extent of the models, such as the bridge between Fennoscandia and the British Isles. The ice-sheet maximum is located over the Gulf of Bothnia and central Sweden, with more ice in ICE-5G than RSES. Such differences between the ice models will consequently produce different patterns of rebound in the modelling.

4   Results

We start presentation of the results with a discussion of the best-fitting three-layer earth models (Table 1) for each ice model and regional RSL data set, which includes a brief presentation of results of the different groupings of sea-level data. We calculated the best-fitting earth model for two subsets of the Oslo Graben, the Oslo Fjord and Limfjord (see Table 1). We find almost the same best-fitting earth model for each RSL data subset, and thus combination of Oslo Fjord and Limfjord RSL data is possible. For the grouping of RSL data either in the Fyn or bays of Kiel and Lübeck subset we provide the results of four different combinations. For both ice models, we consider the combination with Rostock data in the bays of Kiel and Lübeck subset as best (Table 1). There is almost no change in the best-fitting earth model parameters for the Fyn subset using RSES, but the misfit gets worse the more data are moved to the other subset. The other subset (bays of Kiel and Lübeck) has the same best-fitting earth model parameters with and without the Rostock data set, but the misfit is better when including the Rostock data. Assigning more easterly located RSL data, of Körkwitz and

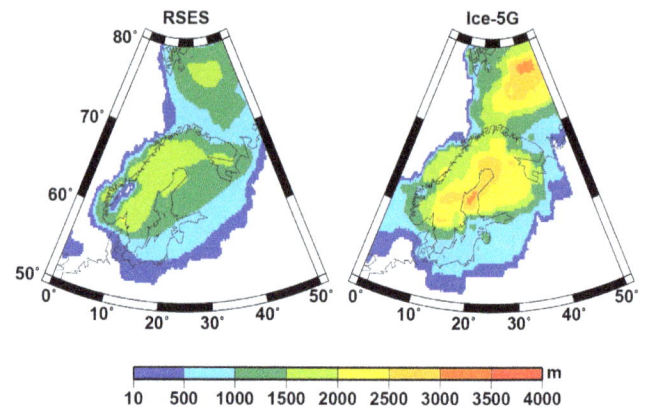

Figure 2. Ice extent at Last Glacial Maximum in Fennoscandia from global ice models (**a**) RSES (Lambeck et al., 1998) and (**b**) ICE-5G (Peltier, 2004).

Darss Peninsula, from the Fyn subset to this data set, the earth model parameters change abruptly and the misfit gets worse. Using ICE-5G, there is also a remarkable change in the earth model parameters if RSL data from Körkwitz and Darss Peninsula are moved from one subset to the other. As the misfit gets worse for Fyn when moving more data, and the misfit does not significantly change for the bays of Kiel and Lübeck subset, we use combination (2) in Table 1 in the discussion below.

Both ice models yield mainly similar earth structures for each region: a variation in lithospheric thickness from lower values along the Norwegian coast to higher values towards the Fennoscandian craton, and an increase in mantle viscosity from the upper to the lower mantle. However, distinct differences can be found, when comparing the results for the two ice models. While RSES shows a prominent increase in lithospheric thickness from west to east, thickness as determined with ICE-5G shows only a small increase with the highest value for Fyn. Both the Oslo Graben as well as the bays of Kiel and Lübeck are characterized by an at most 60 km thick lithosphere for both ice models. As 60 km is the lowermost tested value in our investigation, thicknesses lower than 60 km are also possible. In between, the Fyn subset yields a higher thickness of 90 (RSES) to 100 km (ICE-5G). SW Sweden reaches a higher thickness than the Oslo Graben, however, here the values of the two ice models diverge with 90 km for ICE-5G and 130 km for RSES. Towards Poland and Lithuania the thickness increases up to 160 km for RSES, but drops to 80 km for ICE-5G. Thus, thickness decreases from SW Sweden and Fyn to the southeastern Baltic Sea for ICE-5G, but increases for RSES. However, we note that the misfit for both ice models for the Polish data is much worse although the confidence areas are smaller than for other areas.

Table 1. Best-fitting three-layer 1-D earth models with RSES and ICE-5G ice-load history, respectively, as derived for each regional RSL data subset. Values in parentheses show the σ_1 range for each model parameter. If no parentheses appear, the σ_1 range encompasses the best-fitting model only. H_1 lithospheric thickness, η_{UM} upper-mantle viscosity, η_{LM} lower-mantle viscosity, χ misfit.

Region	H_1 in km	η_{UM} in 10^{20} Pa s	η_{LM} in 10^{22} Pa s	χ
RSES				
SW Sweden	130 (100–160)	4 (3–10)	0.1 (0.1–1)	1.18
Oslo Graben	60 (60–70)	2	4 (0.4–10)	1.58
Oslo Fjord	60 (60–90)	2	4 (0.4–10)	1.61
Limfjord	60 (60–70)	1 (0.5–2)	0.2 (0.2–10)	1.13
Fyn[1]	90 (70–150)	20 (7–20)	10 (0.7–10)	3.88
Fyn[2]	90 (70–150)	20 (7–20)	10 (0.7–10)	3.91
Fyn[3]	90 (70–140)	20 (7–20)	10 (0.7–10)	4.17
Fyn[4]	100 (80–160)	20 (7–20)	10 (1–10)	4.18
Bays of Kiel and Lübeck[1]	60 (60–150)	20	2 (2–3)	1.92
Bays of Kiel and Lübeck[2]	60 (60–150)	20	2 (2–3)	1.84
Bays of Kiel and Lübeck[3]	110 (60–150)	20 (7–20)	2 (0.3–3)	1.97
Bays of Kiel and Lübeck[4]	160 (120–160)	20	4 (3–7)	2.01
Polish Baltic Sea	160 (120–160)	20	10 (7–10)	5.70
ICE-5G				
SW Sweden	90 (60–140)	2 (0.6–2)	0.1 (0.1–10)	0.87
Oslo Graben	60 (60–70)	2	0.4 (0.3–0.9)	2.19
Oslo Fjord	70 (60–100)	2	1 (0.4–10)	1.44
Limfjord	60 (60–80)	0.7 (0.3–1)	0.1 (0.1–0.2)	1.82
Fyn[1]	100 (90–110)	2	0.1	3.19
Fyn[2]	100 (90–110)	2	0.1	3.25
Fyn[3]	80 (70–90)	4 (4–5)	7 (4–10)	3.47
Fyn[4]	80 (70–90)	4 (4–5)	7 (7–10)	3.48
Bays of Kiel and Lübeck[1]	60 (60–120)	7 (6–10)	0.7 (0.3–1)	1.95
Bays of Kiel and Lübeck[2]	60 (60–70)	4	4 (2–10)	1.95
Bays of Kiel and Lübeck[3]	100 (70–140)	2	0.1	1.90
Bays of Kiel and Lübeck[4]	100 (70–140)	2	0.1	1.91
Polish Baltic Sea	80	7 (6–7)	7 (6–9)	5.04

[1] Initial RSL data subsets of Fyn and Bays of Kiel and Lübeck.
[2] RSL data from Rostock (yellow dots in Fig. 1) are moved from Fyn[1] to Bays of Kiel and Lübeck[1].
[3] RSL data from Rostock and Körkwitz (yellow and light blue dots in Fig. 1) are moved from Fyn[1] to Bays of Kiel and Lübeck[1].
[4] RSL data from Rostock, Körkwitz and Darss Peninsula (yellow, light and dark blue dots in Fig. 1) are moved from Fyn[1] to Bays of Kiel and Lübeck[1].

Pronounced differences exist for the upper-mantle viscosity. While for ICE-5G only small variances between $[2–7] \times 10^{20}$ Pa s appear for the five investigated regions, the viscosity as determined with RSES varies by one order of magnitude with quite high upper-mantle viscosities of 2×10^{21} Pa s for southern Baltic Sea RSL data. In SW Sweden and Oslo Graben the viscosity values are comparable to those of ICE-5G. Lower-mantle viscosity also shows a wide range of values; however, it has already been often noted that lower-mantle viscosity cannot be well determined with Fennoscandian RSL data due to their low resolving power to such great depths. Lower-mantle viscosity is generally higher than the upper-mantle viscosity. For SW Sweden, this statement needs to be further evaluated as the lower-mantle viscosity is at the lower bound of our investigation area. A closer look at the 1σ range and the misfit maps (Fig. 3) shows that the lithospheric thickness and the upper-mantle viscosity in the Oslo Graben are quite well determined, while lower-mantle viscosity can be varied over a larger range, but would still give reasonable fits to the RSL data. In contrast, RSL data from SW Sweden highlight a larger variation of the three parameters. With the RSES ice model lithospheric thickness may range from 100 to 160 km and more and upper-mantle viscosity from $[3–10] \times 10^{20}$ Pa s. Using ICE-5G, this range is smaller, but lithospheric thickness can also reach higher values, providing an overlap to possible thicknesses as determined with RSES.

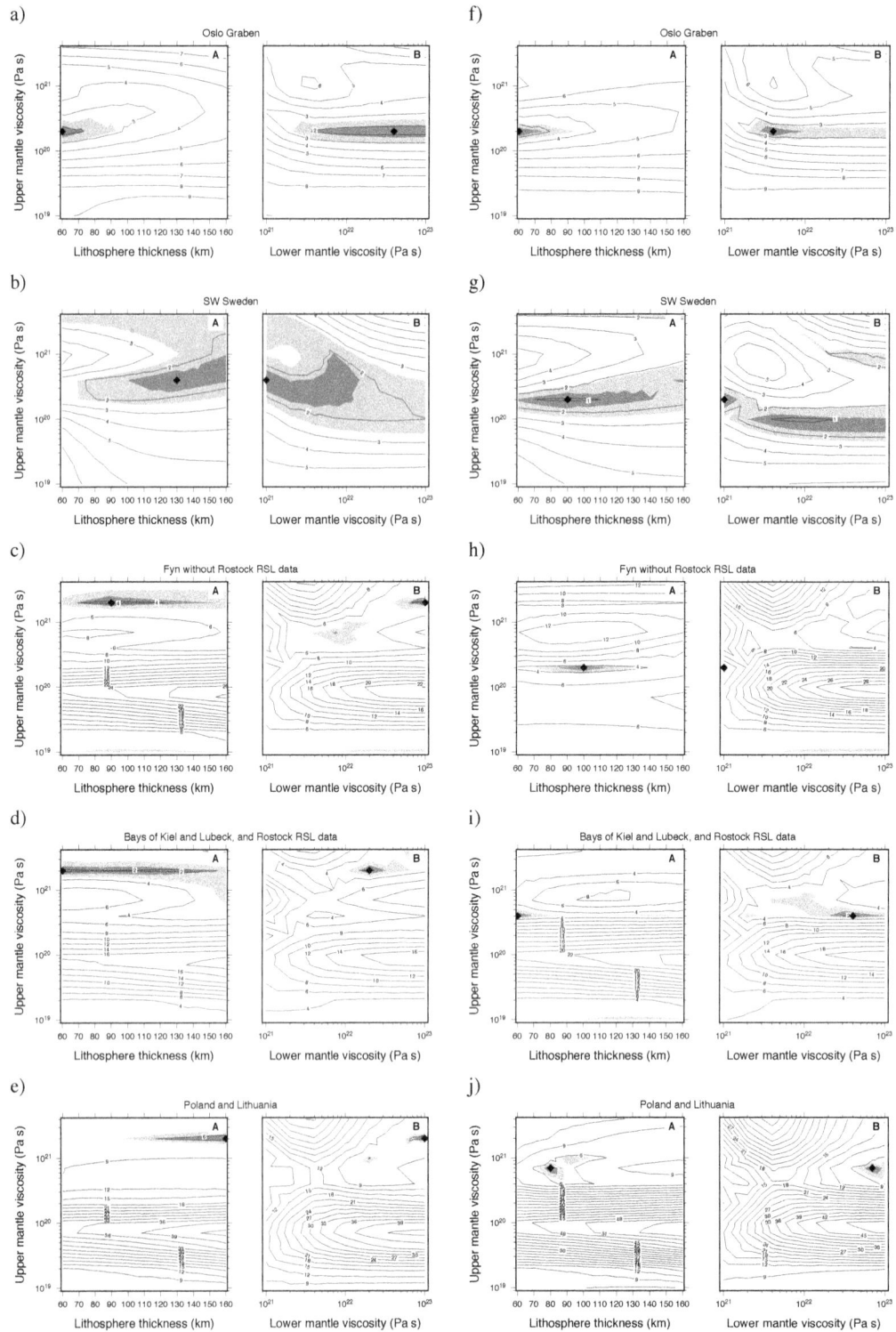

Figure 3. Misfit for ice models RSES (**a–e**) and ICE-5G (**f–j**), three-layer earth model and different data sets. Panel A: the misfit map as a function of lithospheric thickness and upper-mantle viscosity for a fixed lower-mantle viscosity according to the best-fitting earth model, see Table 1. Panel B: the misfit map as a function of upper- and lower-mantle viscosities according to the best-fitting earth model for a fixed lithospheric thickness, see Table 1. (**a, f**) Misfit map for Oslo Graben RSL data (light and dark green dots in Fig. 1). (**b, g**) Misfit map for SW Sweden RSL data (red dots in Fig. 1). (**c, h**) Misfit map for Fyn without Rostock RSL data (violet, dark and light blue dots in Fig. 1). (**d, i**) Misfit map for bays of Kiel and Lübeck and Rostock RSL data (orange and yellow dots in Fig. 1). (**e, j**) Misfit map for Polish and Lithuanian RSL data (black dots in Fig. 1). The best three-layer earth model is marked with a diamond, the light and dark shadings indicate the confidence regions $\psi \leq 1$ and $1 < \psi \leq 2$, respectively.

For Fyn as well as the bays of Kiel and Lübeck the 1σ ranges for the viscosities become much narrower than for SW Sweden. Only lithospheric thickness as determined with RSES may be varied over almost the whole tested parameter range. These two data sets as well as that of SW Sweden show the feature of bifurcation in the misfit maps of lithospheric thickness vs. upper-mantle viscosity. There are two regions of high misfits, one at about 10^{21} Pa s and thinner lithospheric thicknesses, and another one at about 10^{20} Pa s and lower covering the whole thickness range. This lower bound and the "island" at 10^{21} Pa s seem to force the best-fitting model to adopt upper-mantle viscosity values either of $[2–7] \times 10^{20}$ Pa s or of 2×10^{21} Pa s and larger. Lithospheric thickness is not strongly bounded for these two areas of low fits. While ICE-5G prefers the lower upper-mantle viscosity area, RSES tends to higher viscosities. Although the 1σ range for the RSES results does not cover the lower upper-mantle viscosity range, new deeper and older RSL data and an updated ice model may help shift the results to similar values as determined with ICE-5G.

Another interesting behaviour is that lower-mantle viscosity appears to be, except for SW Sweden, clearly determined. This also holds for the Polish and Lithuanian data. Instead, the island at 10^{21} Pa s for upper-mantle viscosity does not appear and lithospheric thickness is better determined (especially for ICE-5G) than for the other regions.

5 Discussion

In the previous section we derived bounds for lithospheric thickness and upper- and lower-mantle viscosity for the different regions. We now take closer look at the fitted RSL data. While the locations Oslo Graben and SW Sweden are mainly near-field data with a large time and height/depth range, the other three regional subsets contain far-field data of younger age and smaller depth ranges, i.e. there is only a window of about 4000 years where relative sea levels change by more than 30 m. Thus, it is challenging to identify the best-fitting modelled sea-level curve within the given error bars of the samples out of a large range of possible curves, despite the large number of samples within each subset. The determination of the best-fitting model can be much better achieved for Oslo Graben and SW Sweden. Here, we also note that the clear determination is much better for Oslo Graben as it contains a non-monotonic RSL change with rising and falling sea levels. We can only speculate for the reason of the poorer misfit to the Polish and Lithuanian data. It may be the RSL data themselves, which may be affected by unknown tectonic behaviour or subsidence, imperfections in the ice model, or a combination of both.

Further evaluation of our results is enabled by comparison of calculated sea-level curves from the best-fitting regional earth models to RSL data used. Figure 4 presents sea-level curves at eight selected locations. In the Oslo Fjord and in

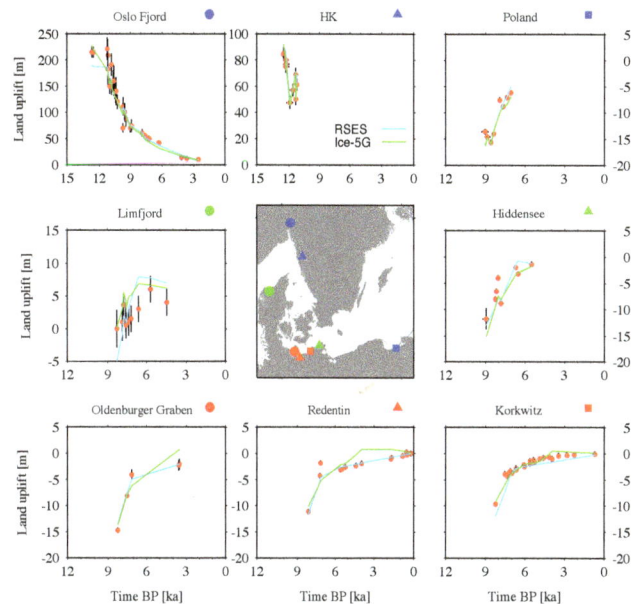

Figure 4. Comparison of RSL data (red dots) at selected locations to sea-level curves as calculated with the best earth model for a respective region and ice model RSES (Lambeck et al., 1998, blue) and ICE-5G (Peltier, 2004, green).

SW Sweden (HK, the archaeological data from Hensbacka culture sites), there is a very good fit between observations and the modelled curves. The RSL data from Limfjord in northern Denmark are not fitted well, but one has to note that there is only small variation of about 5 m in 5000 years in this data set, which is hard to trace for the model. Along the German Baltic Sea coast, this variation is much larger and thus better fits can be achieved. In Hiddensee both RSES and ICE-5G ice models result in a good match of the sea-level data, but partly outside the given error bars of the RSL data. In the Oldenburger Graben and Redentin, the RSES ice model traces the RSL data better than ICE-5G, while in Körkwitz the ICE-5G ice model performs better than RSES. In Poland both ice models predict the sea-level rise well. Our comparison shows that although good fits are achieved in some areas, each ice model cannot perfectly fit all data, and some sea-level curves as predicted by the models lie outside the error bars of the observations. Errors in the ice model affect the behaviour of calculated sea-level curves and may lead to a worse misfit, which eventually alters the confidence ranges in Fig. 3. This does not necessarily mean that another earth model would be preferred, but the RSL curve of this earth–ice model combination is disarranged.

We can compare our results to a former investigation by Lambeck et al. (1998), where the authors already used Fennoscandian RSL data divided into several subregions. However, data from the German, Polish and Lithuanian Baltic Sea coast were not used by Lambeck et al. (1998). In the southwest, RSL data were therefore grouped into these

Table 2. Overview of three-layer 1-D earth models derived for regional RSL data subsets in the southern Baltic Sea. H_l lithospheric thickness, η_{UM} upper-mantle viscosity, η_{LM} lower-mantle viscosity, χ misfit. * This regional result from Lambeck et al. (1998) contains additional RSL data that were considered to be less satisfactory by Lambeck et al. (1998).

Region	Reference	Ice model	H_l in km	η_{UM} in 10^{20} Pa s	η_{LM} in 10^{22} Pa s
SW Sweden	this study	RSES	130	4	0.1
	this study	ICE-5G	90	2	0.1
	Lambeck et al. (1998)	RSES	50	2.5	3
*	Lambeck et al. (1998)	RSES	80	1	1
Oslo Fjord	this study	RSES	60	2	4
	this study	ICE-5G	70	2	1
	Lambeck et al. (1998)	RSES	80	1.5	3
Fyn	this study	RSES	90	20	10
	this study	ICE-5G	100	2	0.1
(Denmark)	Lambeck et al. (1998)	RSES	150	4	3

three subsets with available data: Oslo Fjord, SW Sweden and Denmark. This choice is similar to our study, but the SW Sweden data set from Lambeck et al. (1998) did not contain RSL data from the Hensbacka sites and their Danish data set contained data from the Great Belt and the Limfjord. For Oslo Fjord, the authors found a 80 km thick lithosphere and an upper-mantle viscosity of 1.5×10^{20} Pa s with an older version of the RSES ice model (Table 2). In SW Sweden lithospheric thickness of 50 km thickness was 30 km thinner than that of Oslo Fjord. Upper-mantle viscosity is here slightly higher at 2.5×10^{20} Pa s. Higher values were found in Denmark. Lithospheric thickness was determined to be 150 km and upper-mantle viscosity with 4×10^{20} Pa s. While these results confirm the thicker lithosphere in Denmark/Rinkøping-Fyn High as well as the upper-mantle viscosities of our study, the differences between SW Sweden and the Oslo Fjord are large both in the lithospheric thickness estimate and also in the structural implications. These differences can be explained due to our slightly different grouping, the new data in the SW Sweden subset and the usage of an updated version of RSES that was available to us.

We note in this regard that Kaufmann and Wu (2002) showed that if the ice-load history is known, then it is only possible to accurately estimate lateral changes in lithospheric thickness with 1-D earth models and regional RSL data subsets if there is no lateral change in mantle viscosity below the lithosphere. Otherwise the inferred lateral variations in lithospheric thickness can only be estimated qualitatively. This condition is not met with our RSES ice-load history and thus these results have to be cautiously interpreted. ICE-5G shows smaller variations in upper-mantle viscosity for each region than RSES; therefore, these results are more reliable in view of the findings by Kaufmann and Wu (2002). However, the results from ICE-5G do not agree with seismological results,

which show large increase in lithospheric thickness towards the east.

To further evaluate this, we therefore turn to the lithosphere models derived from seismological data and compare them to our results. Gregersen et al. (2002) provided a NE–SW profile from southern Sweden to central Germany based on P-wave velocity perturbation. The generalized profile shows a 300 km thick lithosphere northeast of the Teisseyre–Tornquist Zone, but we note that the lower boundary cannot be clearly defined due to the relatively high velocities in the upper mantle. Therefore, the lithosphere might be thinner than 300 km. The lithospheric thickness then decreases to about 125 km between the Rinkøping-Fyn High and the Teisseyre–Tornquist Zone in Denmark, and about 80 km southwest of the Rinkøping-Fyn High in Germany.

Tesauro et al. (2009) showed a map of thermal lithospheric thickness in Europe south of 60° N latitude. The model is based on the inversion of a tomography model of Koulakov et al. (2009) and was provided to us in a 0.25×0.25 degree grid. In southern Sweden, they find a thickness exceeding 180 km (Fig. 5a, isolines). The thickness then decreases to about 120 km in northeastern Germany. In the southern North Sea, they find an average of about 135 km for Belgium and about 110 km for the Netherlands and northwest Germany. A comparison with receiver function data mirrored the lateral variation (Tesauro et al., 2009), and visual comparison with newer S-receiver function results (Geissler et al., 2010) supports the results as well. The British Isles have varying thicknesses between 100 and 180 km.

Hamza and Vieira (2012) developed a 2×2 degree global distribution map of the thermal lithospheric thickness based on global data sets for heat flow and crustal structure. In southern Sweden, lithospheric thickness is found to be between 170 and 210 km (Fig. 5b, isolines). Similar values arise for the German Baltic Sea coast and Denmark. The southern North Sea has a lithosphere of about 160 to 170 km thickness.

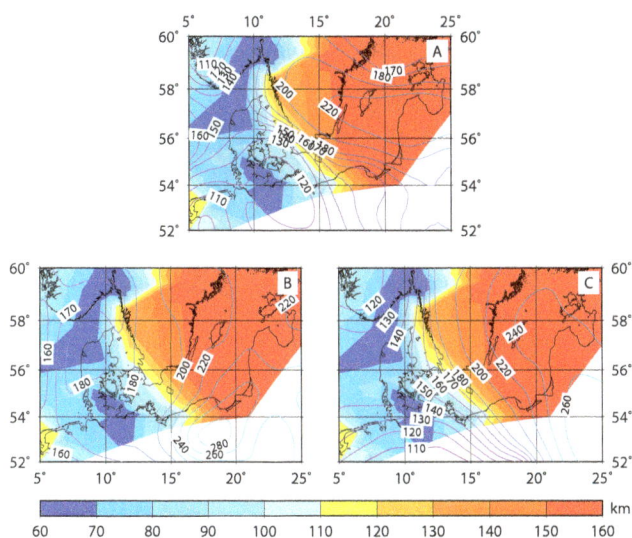

Figure 5. Comparison of calculated regional lithospheric thickness variations with the RSES ice model (filled contour maps) to seismically and thermally derived lithospheric thicknesses (solid lines) by (**A**) Tesauro et al. (2009), (**B**) Hamza and Vieira (2012) and (**C**) Priestley and McKenzie (2013). Contour maps are drawn with the GMT pscontour function.

Recently, Priestley and McKenzie (2013) introduced a 2×2 degree seismologically determined lithosphere model that also includes thermal information. They combined a surface wave tomography model with temperature (ocean and continents) and pressure (continents) estimates to generate shear-wave velocity estimates. These estimates and a description of their relaxation behaviour at high temperatures is then used to infer the lithospheric thickness. In the southern Baltic Sea area, there are two major structural features (Fig. 5c, isolines). First, lithospheric thickness decreases from 260 km in the east to 110–120 km in the west. The gradient is almost constant, but slightly steeper in SW Sweden. Second, from western central Denmark towards the North Sea, an area encompassing the Rinkøping-Fyn High, lithospheric thickness remains at an almost constant level of about 140 km. To the north and south it drops to about 110 km.

Figure 5 shows our results for the best lithospheric thickness estimates with the RSES ice model as coloured maps, with the additional estimates from Steffen and Kaufmann (2005) for Fennoscandia and from Vink et al. (2007) for the southern North Sea to give a more complete overview on GIA-inferred lithospheric thickness. We do not compare our ICE-5G results as (i) they do not show the pronounced thickness increase to the east and (ii) Steffen and Kaufmann (2005) and Vink et al. (2007) did not provide results for this ice model which would allow a comparison in the North Sea and central Fennoscandia. The GIA-inferred lithospheric thickness map is drawn using the GMT pscontour function (Wessel and Smith, 1998) by assigning the lithospheric thickness values of the best-fitting earth model for each region to the coordinates of each RSL data sample location. The results of Tesauro et al. (2009, A), Hamza and Vieira (2012, B) and Priestley and McKenzie (2013, C) are overlain with contour lines. In the south and east of the area shown no results exist for the GIA-inferred lithospheric thickness.

In general, the seismically and thermally inferred lithospheric thickness values do not show a good match to our GIA-model results. All these models show lithospheric thicknesses of at least 110 km in the area under investigation. Also, their maximum values exceed 200 km considerably. However, we note that these three lithosphere models also do not show a good fit to each other either, except for the general increase from west to east. The thicker lithospheres of the seismological and/or thermal models is due to the fact that a different timescale is addressed. Seismological results are related to observations and processes of seconds to minutes, while the GIA-inferred lithosphere is related to a process of 100 ka. Nonetheless, relative differences should agree.

The thickness according to Hamza and Vieira (2012) has a pronounced peak of 280 km in Poland and also shows decreasing values from east to west with no distinct change in the gradient except a kind of plateau with about 180 km in northwestern Denmark. Except the decrease in lithospheric thickness from east to west, there is no other similar feature when compared to our GIA-model results.

The lithospheric thickness by Tesauro et al. (2009) reaches its highest value of 220 km in a broad band from southeastern Sweden to Latvia. It also shows decreasing values from east to west; however, the gradient is much steeper at the southwestern Swedish coast. It thins to 150 km towards the northwest of Denmark, and then becoming thicker again. To the north and south of this area values drop to less than 110 km. There is a structural agreement in the form of the east–west decrease. The Rinkøping-Fyn High appears to lie further north in the thermal lithosphere. The thin GIA-lithosphere along the German Baltic Sea coast agrees with the plateau of 120 km in the thermal lithosphere. The structure of the Oslo Graben cannot be distinguished.

The best agreement of GIA-modelling-derived values is probably found in comparison to the new model by Priestley and McKenzie (2013). Both the EW-decrease trend and the location of the Rinkøping-Fyn High fit structurally well. Small differences are found in the northwest of our investigation area and in the German Bight. However, we also have to note that the spatial resolution of this model is two degrees and thus smaller features may not be clearly identified.

6 Conclusions

This is the first time that the regional earth structure in the southern Baltic Sea has been investigated with the help of regionally categorized RSL data and GIA modelling. Also, the lateral variation was visually compared to seismologically and/or thermally derived lithospheric thickness models for the first time. We employed the software ICEAGE and two different global ice models.

However, we made several assumptions and there are certain conditions to be kept in mind that may lead to different results in future investigations: we use ice models that are related to a certain earth model, and thus they are already biased by a certain lithospheric thickness and mantle viscosity. Our earth model is based on Maxwell rheology. Furthermore, it is possible that the ice models have imperfections that are absorbed by a wrong earth model, but anyhow lead to a good fit to the observations. We also note that variation in lithospheric thickness for regional subsets can only be clearly determined when mantle viscosity in each region is about constant (Kaufmann and Wu, 2002). This condition is not met for all regions. It is therefore possible that a 3-D earth model for the southern Baltic Sea with a different radial earth structure in each subregion than our determined 1-D earth models fits much better than a combination of all our 1-D models. All these items can increase the confidence regions of our study.

Within our calculated confidence levels, the following results were determined. The lithospheric thickness varies from 60 km in the Oslo Graben and the German Baltic Sea coast to up to 160 km in Poland. When only the best-fitting lithosphere is analysed, we see a trend to thicker lithosphere from west to east using the RSES ice model, but this is less pronounced with ICE-5G. The Rinkøping-Fyn High in between the Oslo Graben and Germany is at least 30 km thicker than the surrounding areas in the north and south. However, the confidence levels of the lithosphere are so large that an accurate determination is not possible. The variation in lithospheric thickness based on RSES agrees to a certain extent, when compared visually, to thickness models based on seismological and/or thermal investigation. A direct comparison of thicknesses is not possible due to the different definitions of lithosphere in seismological/thermal and GIA investigations.

Upper-mantle viscosity is about $[2–7] \times 10^{20}$ Pa s in the Oslo Graben and SW Sweden and thus confirms values found for Fennoscandia, the British Isles and the southern North Sea previously. In the southern Baltic Sea, similar values are obtained with ICE-5G, but we note quite high values of 2×10^{21} Pa s for this region when using the RSES ice history. Bifurcation indicates that lower values in the range of $[4–10] \times 10^{20}$ Pa s are likely. As expected, lower-mantle viscosity cannot be sufficiently determined.

Future investigations with hopefully more RSL data in the southern Baltic Sea and an updated ice model (both tested ice models have experienced major recent improvements, but these revised versions have not been published yet) may help to further confirm the results herein with smaller confidence regions than ours and also overcome the differences between the results from the two ice models in certain areas. However, it will not be possible to add RSL data in the southern Baltic Sea which are older and deeper than the ones used in our study as the Pleistocene relief with the threshold of 25 m in the Great Belt did not allow an earlier deposition.

Acknowledgements. We are grateful for the excellent reviews by Wouter van der Wal and Patrick Wu that helped improve the paper. We would like to thank Kurt Lambeck (Research School of Earth Sciences, Australian National University) and Magdala Tesauro (GFZ Potsdam) for kindly providing the RSES ice model and the thermal lithosphere model in central and southern Europe, respectively. Figures were prepared using GMT software (Wessel and Smith, 1998).

Special Issue: "Lithosphere-cryosphere interactions"
Edited by: M. Poutanen, B. Vermeersen, V. Klemann, and C. Pascal

References

Artemieva, I. M.: The continental lithosphere: Reconciling thermal, seismic, and petrologic data, Lithos, 109, 23–46, doi:10.1016/j.lithos.2008.09.015, 2009.

Bennike, O., Jensen, J. B., Lemke, W., Kuijpers, A., and Lomholt, S.: Late- and postglacial history of the Great Belt, Denmark, Boreas, 33, 18–33, doi:10.1111/j.1502-3885.2004.tb00993.x,2004.

Bitinas, A., Damušyte, A., Hütt, G., Martma, T., Ruplenaite, G., Stančikaite, M., Ūsaityte, D., and Vaikmäe, R.: Stratigraphic correlation of Late Weichselian and Holocene Deposits in the Lithuanian coastal region, P. Est. Acad. Sci.-Geol., 49, 200–217, 2000.

Bitinas, A., Damušyte, A., Stančikaite, M., and Aleksa P.: Geological development of the Nemunas River Delta and adjacent areas, West Lithuania, Geological Quarterly, 46, 375–389, 2002.

Christensen, Ch., Fischer, A., and Mathiassen, D. R.: The great sea rise in the Storebælt, in: The Danish Storebælt since the Ice Age – man, sea and forest, edited by: Pedersen, L., Fischer, A., and Aaby, B., A/S Storebælt Fixed Link, Copenhagen, 45–54, 1997.

Dèzes, P., and Ziegler, P. A.: Moho depth map of Western and Central Europe, avfailable at: https://comp1.geol.unibas.ch/, 2002.

Dziewonski, A. M. and Anderson D. L.: Preliminary reference Earth model, Phys. Earth Planet. Inter., 25, 297–356, doi:10.1016/0031-9201(81)90046-7, 1981.

Eaton, D. W., Darbyshire, F., Evans, R. L., Grütter, H., Jones, A. G., and Yuan, X.: The elusive lithosphere-asthenosphere boundary (LAB) beneath cratons, Lithos, 109, 1–22, doi:10.1016/j.lithos.2008.05.009, 2009.

Farrell, W. E. and Clark, J. A.: On postglacial sea level, Geophys. J. R. Astr. Soc., 46, 647–667, doi:10.1111/j.1365-246X.1976.tb01252.x, 1976.

Fjeldskaar, W.: Viscosity and thickness of the asthenosphere detected from the Fennoscandian uplift, Earth Planet. Sci. Lett., 126, 399–410, doi:10.1016/0012-821X(94)90120-1, 1994.

Geissler, W. H., Sodoudi, F., and Kind, R.: Thickness of the central and eastern european lithosphere as seen by S receiver functions, Geophys. J. Int., 181, 604–634, doi:10.1111/j.1365-246X.2010.04548.x, 2010.

Gregersen, S., Voss, P., and the TOR Working Group: Summary of project TOR: delineation of a stepwise, sharp, deep lithosphere transition across Germany – Denmark –

Sweden, Tectonophysics, 360, 61–73, doi:10.1016/S0040-1951(02)00347-5, 2002.

Hamza, V. M. and Vieira, F. P.: Global distribution of the lithosphere-asthenosphere boundary: a new look, Solid Earth, 3, 199–212, doi:10.5194/se-3-199-2012, 2012.

Hoffmann, G., Schmedemann, N., and Schafmeister, M.-Th.: Relative sea-level curve for SE Rügen and Usedom Island (SW Baltic Sea coast, Germany) using decompacted profiles, Z. dtsch. Ges. Geowiss., 160, 69–78, 2009.

Hofmann, W. and Winn, K.: The Littorina Transgression in the Western Baltic Sea as indicated by subfossil Chironomidae (Diptera) and Cladocera (Crustacea), Int. Rev. Hydrobiol., 85, 267–291, doi:10.1002/(SICI)1522-2632(200004)85:2/3<267::AID-IROH267>3.0.CO;2-Q, 2000.

Jakobsen, O.: Die Grube-Wesseker Niederung (Oldenburger Graben, Ostholstein): Quartärgeologische und geoarchäologische Untersuchungen zur Landschaftsgeschichte vor dem Hintergrund des anhaltenden postglazialen Meeresspiegelanstiegs, PhD.-Thesis, Univ. Kiel, 190 pp., 2004.

Kaufmann, G.: Program package ICEAGE, Version 2004, Manuscript, Institut für Geophysik der Universität Göttingen, 40 pp., 2004.

Kaufmann, G. and Wu, P.: Glacial isostatic adjustment in Fennoscandia with a three-dimensional viscosity structure as an inverse problem, Earth Planet. Sci. Lett., 197, 1–10, doi:10.1016/S0012-821X(02)00477-6, 2002.

Koulakov, I., Kaban, M. K., Tesauro, M., Cloetingh, S.: P and S velocity anomalies in the upper mantle beneath Europe from tomographic inversion of ISC data, Geophys. J. Int., 179, 345–366, doi:10.1111/j.1365-246X.2009.04279.x, 2009.

Lambeck K., Smither, C., and Johnston, P.: Sea-level change, glacial rebound and mantle viscosity for northern Europe, Geophys. J. Int., 134, 102–144, doi:10.1046/j.1365-246x.1998.00541.x, 1998.

Lampe, R., Meyer, H., Ziekur, R., Janke, W., and Endtmann, E.: Holocene evolution of an irregularly sinking coast and the interactions of sea-level rise, accumulation space and sediment supply, Bericht der Römisch-Germanischen Kommission, 88, 15–46, 2007.

Lampe, R., Endtmann, E., Janke, W., and Meyer, H.: Relative sea-level development and isostasy along the NE German Baltic Sea coast during the past 9 ka, Quaternary Sci. J., 59, 3–20, doi:10.3285/eg.59.1-2.01, 2011.

Lange, W. and Menke, B.: Beiträge zur frühpostglazialen erd- und vegetationsgeschichtlichen Entwicklung im Eidergebiet, insbesondere zur Flußgeschichte und zur Genese des sogenannten Basistorfes, Meyniana, 17, 29–44, 1967.

Lübke, H., Schmölcke, U., and Tauber, F.: Mesolithic Hunter-Fishers in a Changing World: a case study of submerged sites on the Jäckelberg, Wismar Bay, northeastern Germany, in: Submerged Prehistory, edited by: Benjamin, J., Bonsall, C., Pickard, C., and Fischer, A., 21–37, Oxbow Books, Oxford, 2011.

Mitrovica, J. X., and Milne, G. A.: Glaciation-induced perturbations in the Earth's rotation: a new appraisal, J. Geophys. Res., 103, 985–1005, doi:10.1029/97JB02121, 1998.

Mitrovica, J. X., Davis, J. L., and Shapiro, I. I.: A spectral formalism for computing three–dimensional deformations due to surface loads 1. Theory, J. Geophys. Res., 99, 7057–7073, doi:10.1029/93JB03128, 1994.

Peltier, W. R.: Global glacial isostasy and the surface of the Ice-Age Earth: the ICE-5G (VM2) model and GRACE, Annu. Rev. Earth Pl. Sc., 32, 111–149, doi:10.1146/annurev.earth.32.082503.144359, 2004.

Petit, J. R., Jouzel, J., Raynaud, D., Barkov, N. I., Barnola, J. M., Basile, I., Bender, M., Chappellaz, J., Davis, J., Delaygue, G., Delmotte, M., Kotlyakov, V. M., Legrand, M., Lipenkov, V., Lorius, C., Pépin, L., Ritz, C., Saltzman, E., and Stievenard, M.: Climate and Atmospheric History of the Past 420,000 years from the Vostok Ice Core, Antarctica, Nature, 399, 429–436, doi:10.1038/20859, 1999.

Priestley, K. and McKenzie, D.: The relationship between shear wave velocity, temperature, attenuation and viscosity in the shallow part of the mantle, Earth Planet. Sci. Lett., 381, 78–91, doi:10.1016/j.epsl.2013.08.022, 2013.

Rößler, D., Moros, M., and Lemke, W.: The Littorina transgression in the southwestern Baltic Sea: new insights based on proxy methods and radiocarbon dating of sediment cores,, Boreas, 40, 231–241, doi:10.1111/j.1502-3885.2010.00180.x, 2011.

Scheck-Wenderoth, M., and Lamarche, J.: Crustal memory and basin evolution in the Central European Basin System – new insights from a 3D structural model, Tectonophysics, 397, 143–165, doi:10.1016/j.tecto.2004.10.007, 2005.

Schmitt, L., Larsson, S., Burdukiewicz, J., Ziker, J., Svedhage, K., Zamon, J., Steffen, H.: Chronological insights, cultural change, and resource exploitation on the west coast of Sweden during the Late Paleolithic/early Mesolithic transition, Oxford J. Arch., 28, 1–27, doi:10.1111/j.1468-0092.2008.00317.x, 2009.

Steffen, H. and Kaufmann, G.: Glacial isostatic adjustment of Scandinavia and northwestern Europe and the radial viscosity structure of the Earth's mantle, Geophys. J. Int., 163, 801–812, doi:10.1111/j.1365-246X.2005.02740.x, 2005.

Steffen, H. and Wu, P.: Glacial isostatic adjustment in Fennoscandia – A review of data and modeling, J. Geodyn., 52, 169–204, doi:10.1016/j.jog.2011.03.002, 2011.

Steffen, H., Kaufmann, G., and Wu, P.: Three-dimensional finite-element modelling of the glacial isostatic adjustment in Fennoscandia, Earth Planet. Sci. Lett., 250, 358–375, doi:10.1016/j.epsl.2006.08.003, 2006.

Steffen, H., Wu, P., and Wang, H. S.: Determination of the Earth's structure in Fennoscandia from GRACE and implications on the optimal post-processing of GRACE data, Geophys. J. Int., 182, 1295–1310, doi:10.1111/j.1365-246X.2010.04718.x, 2010.

Tauber, F.: Seafloor exploration with sidescan sonar for geo-archaeological investigations, Berichte der Römisch-Germanischen Kommission, 88, 67–79, 2007.

Tesauro, M., Kaban, M. K., and Cloetingh, S. A. P. L.: A new thermal and rheological model of the European lithosphere, Tectonophysics, 476, 478–495, doi:10.1016/j.tecto.2009.07.022, 2009.

Uścinowicz, S.: Relative sea level changes, glacio-isostatic rebound and shoreline displacement in the Southern Baltic, Polish Geological Institute Special Papers, 10, 1–80, 2003.

van der Wal, W., Barnhoorn, A., Stocchi, P., Gradmann, S., Wu, P., Drury, M., and Vermeersen, L. L. A.: Glacial Isostatic Adjustment Model with Composite 3D Earth Rheology for Fennoscandia, Geophys. J. Int., 194, 61–77, doi:10.1093/gji/ggt099, 2013.

Vink, A., Steffen, H., Reinhardt L., and Kaufmann, G.: Holocene relative sea-level change, isostatic subsidence and the radial viscosity structure of the mantle of north-western Europe (Belgium, the Netherlands, Germany, southern North Sea), Quat. Sci. Rev., 26, 3249–3275, doi:10.1016/j.quascirev.2007.07.014, 2007.

Wang, H. S., Wu, P., and van der Wal, W.: Using postglacial sea level, crustal velocities and gravity-rate-of-change to constrain the influence of thermal effects on mantle lateral heterogeneities, J. Geodyn., 46, 104–117, doi:10.1016/j.jog.2008.03.003, 2008.

Wessel, P. and Smith, W. H. F.: New, improved version of generic mapping tools released, EOS Trans. AGU, 79, p. 579, doi:10.1029/98EO00426, 1998.

Winn, K., Averdieck, F.-R., Erlenkeuser, H. and Werner, F.: Holocene sea level rise in the western Baltic and the question of isostatic subsidence, Meyniana, 38, 61–80, 1986.

Wu, P., Wang, H. S., and Schotman, H.: Postglacial induced surface motions, sea levels and geoid rates on a spherical, self-gravitating laterally heterogeneous earth, J. Geodyn., 39, 127–142, doi:10.1016/j.jog.2004.08.006, 2005.

Wu, P., Wang, H., and Steffen, H.: The role of thermal effect on mantle seismic anomalies under Laurentia and Fennoscandia from observations of Glacial Isostatic Adjustment, Geophys. J. Int., 192, 7–17, doi:10.1093/gji/ggs009, 2013.

Yu, S.-Y., Berglund, B. E., Andrén, E., and Sandgren, P.: Mid-Holocene Baltic Sea transgression along the coast of Blekinge, SE Sweden – ancient lagoons correlated with beach ridges, GFF, 126, 257–272, doi:10.1080/11035890401263257, 2004.

Optimal locations of sea-level indicators in glacial isostatic adjustment investigations

H. Steffen[1], **P. Wu**[2,3], **and H. Wang**[4]

[1]Lantmäteriet, Lantmäterigatan 2c, 80182 Gävle, Sweden
[2]Department of Geoscience, University of Calgary, 2500 University Drive NW, Calgary, AB, T2N 1N4, Canada
[3]now at: Department of Earth Sciences, The University of Hong Kong, Pokfulam Road, Hong Kong
[4]State Key Laboratory of Geodesy and Earth's Dynamics, Institute of Geodesy and Geophysics, Chinese Academy of Sciences, Wuhan 430077, China

Correspondence to: H. Steffen (holger-soren.steffen@lm.se)

Abstract. Fréchet (sensitivity) kernels are an important tool in glacial isostatic adjustment (GIA) investigations to understand lithospheric thickness, mantle viscosity and ice-load model variations. These parameters influence the interpretation of geologic, geophysical and geodetic data, which contribute to our understanding of global change.

We discuss global sensitivities of relative sea-level (RSL) data of the last 18 000 years. This also includes indicative RSL-like data (e.g., lake levels) on the continents far off the coasts. We present detailed sensitivity maps for four parameters important in GIA investigations (ice-load history, lithospheric thickness, background viscosity, lateral viscosity variations) for up to nine dedicated times. Assuming an accuracy of 2 m of RSL data of all ages (based on analysis of currently available data), we highlight areas around the world where, if the environmental conditions allowed its deposition and survival until today, RSL data of at least this accuracy may help to quantify the GIA modeling parameters above.

The sensitivity to ice-load history variations is the dominating pattern covering almost the whole world before about 13 ka (calendar years before 1950). The other three parameters show distinct patterns, but are almost everywhere overlapped by the ice-load history pattern. The more recent the data are, the smaller the area of possible RSL locations that could provide enough information to a parameter. Such an area is mainly limited to the area of former glaciation, but we also note that when the accuracy of RSL data can be improved, e.g., from 2 m to 1 m, these areas become larger, allowing better inference of background viscosity and lateral heterogeneity. Although the patterns depend on the chosen models and error limit, our results are indicative enough to outline areas where one should look for helpful RSL data of a certain time period. Our results also indicate that as long as the ice-load history is not sufficiently known, the inference of lateral heterogeneities in mantle viscosity or lithospheric thickness will be interfered by the uncertainty of the ice model.

1 Introduction

Glacial isostatic adjustment (GIA) describes the response of the Earth to glacial loading and unloading processes. It includes changes in the Earth's deformation, gravity due to redistribution of mass, moment of inertia and state of stress. Hence, investigations of GIA address different fields giving among other things insight into ice-load dynamics and Earth rheology. For the latter, foci are mainly set with GIA models on lithospheric thickness and Newtonian mantle viscosities as well as their lateral variation in the Earth, respectively.

For an accurate determination of model parameters such as ice-load history, lithospheric thickness, radial and lateral variation of mantle viscosities, many geologic, geophysical and geodetic observations are used to constrain GIA models or identify the best-fitting one by comparing observations to model predictions (see, e.g., Steffen and Wu, 2011, for an overview). Nowadays, the most commonly used observations are GPS measurements, which provide a highly accurate current velocity/deformation field, and gravimetric observations based on terrestrial (absolute and relative gravimetry) and space techniques, which show the deviation from equilibrium and ongoing mass redistributions (Wu et al., 2013). It should be noted that both GPS land-uplift rate and gravity rate-of-change data only give the rate of change today, which is more than 8000 years after the end of deglaciation. On the other hand, relative sea-level (RSL) data record the deformation that occurred in the past (Wu et al., 2013), especially in the last 20 000 years or so since the Last Glacial Maximum. The determination of ice-load history, lithospheric thickness and mantle viscosity depends greatly on the quality and thus accuracy of the used data. Geodetic observations achieve sufficient accuracy for the detection of the GIA signal after a few years, i.e., about 5 years, of observation (Wu et al., 2010; Steffen et al., 2012). The longer the time span, the better the accuracy.

Wu et al. (2010) and Steffen et al. (2012) investigated the sensitivity of GPS and gravity observations, respectively, to four prominent GIA modeling parameters: ice-load history, lateral lithospheric thickness variation, background viscosity, and lateral viscosity variation. The major goal of the two studies was to identify optimal locations for these geodetic observations as economic, logistic and ecological reasons limit the capabilities to cover the (whole) Earth sufficiently with stations (Steffen et al., 2012). An optimal location is defined here by where sensitivity lies above the current detection accuracy of a selected geodetic observation (Wu et al., 2010).

Wu et al. (2010) studied the optimal locations for GPS measurements in North America and Fennoscandia, both areas with prominent GIA signals and already existing GPS networks. They clearly identified the region west of Hudson Bay until the Rocky Mountains as a major gap in the North American permanent GPS network. The network in northern Europe is almost adequate except in the northeast (Wu et al., 2010). Ice-load history appeared to be the best detectable parameter.

The study by Steffen et al. (2012) focused on optimal locations of terrestrial (absolute) gravity measurements in North America and northern Europe and also analyzed the sensitivity of the Gravity Recovery and Climate Experiment (GRACE) twin-satellite mission there to the four parameters. Both terrestrial measurements and GRACE observations sense the four parameters as their sensitivity is higher than the currently determined trend errors, with ice-load history being again the best detectable parameter (Steffen et al., 2012). The authors also suggested more absolute gravity stations in northwestern and Arctic Canada and a comprehensive data combination of all absolute gravity measurements in northern Europe.

This study now adds RSL data to the search for optimal locations of GIA observations to help constrain the four parameters above. RSL data have, since the beginning of GIA research, been an important data set in the understanding and modeling of the GIA process (Clark, 1980; Tushingham and Peltier, 1992, 1993; Steffen and Wu, 2011). Still, they help in constraining ancient ice history (Peltier, 2004; Horton et al., 2009; Engelhart et al., 2011), quantifying the timing of drainage of glacial lakes (Törnqvist & Hijma, 2012) or apparent uplift of the coast since the last interglacial 125 000 years ago (Pedoja et al., 2011).

The issue and analysis here is different to the former studies with geodetic data in at least three ways. First, GPS and gravity measurements represent recent measurements that determine the GIA signal today. The signal is small, i.e., about 1 cm a^{-1} vertical change and about 2 μGal a^{-1} gravity change, while RSL data may show a complete deformation curve over several thousands of years with occasionally several hundreds of meters. Thus, a geodetic signal can be considered as a snapshot of the time-delayed visco-elastic part of GIA, and the observations are "only" three-dimensional when compared to the four-dimensional (space and time) signal visible in RSL data. Hence, we compare something recent (GPS, gravity) with something from the past (RSL) (Wu et al., 2013).

Second, we cannot advise where to place instruments for adequate sea-level measurements as they have to be deposited under certain conditions in order to survive until today. While GPS and terrestrial gravity measurements are limited due to economic and/or logistic reasons, RSL data can potentially be found in all oceans and coastal areas. Also, RSL-like data such as lake levels can be found far off the coast, e.g., in Sweden (Lambeck et al., 1998a). However, there are different limitations depending on the sea-level indicator itself, the environment of its deposition, processes acting at the sample or in the area since its deposition, and many more. We thus can only indicate but not guarantee where RSL data with sufficient information could be found. In addition, we illustrate the sensitivity of RSL data on a global scale rather than the dedicated regions we had to use for geodetic observations.

Third, the sensitivity of RSL data varies with time. The same naturally holds for the sensitivity of geodetic observations as well; however, as aforementioned, geodetic measurements are only snapshots of today. Thus, we have to analyze different times when RSL data were likely deposited, but that also depends on the accuracy of current dating methods.

We will address the following questions in this paper:

– Where should RSL data be located to help constrain ice-load history models, lateral lithospheric thickness variations, background viscosity and lateral viscosity variations used in GIA modeling?

– At which times are RSL data at a certain location sensitive to one of the parameters?

– How accurate should they be?

– Where should new and helpful data be searched?

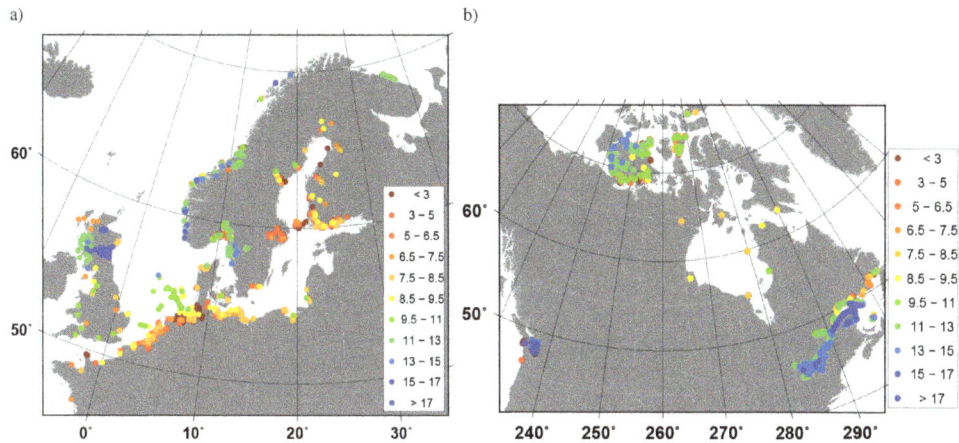

Figure 1. Exemplary overview of the location of relative sea-level data in **(a)** northern and central Europe and **(b)** North America. Colored dots highlight their age. Unit in ka (calendar years before 1950).

In the next section, we discuss RSL data, their errors and possible deposition times. This is followed by Sect. 3, which gives an introduction to the models used. Sections 4 and 5 present and discuss the results, respectively. Based on the discussion of RSL data in Sect. 2, we provide complete maps of RSL data sensitivities for nine different times in the past. Finally, the conclusion is given in Sect. 6.

2   Relative sea-level data

Relative sea levels or palaeo-strandlines document the crustal response of the Earth due to glaciation and subsequent water mass redistribution between the oceans and ice sheets. The sea level at a certain time and location can be dated by shells, corals, wood, whale bones or pollen (van de Plassche, 1986). Their great benefit is that they cover a long time period of deformation, occasionally dating back to several thousand years (Steffen and Wu, 2011). They are mostly dated by the ^{14}C method and thus need to be calibrated for use in GIA modeling (Fairbanks et al., 2005).

Sea-level indicators can be found in coastal and shelf areas all around the world. However, their quality and age vary from location to location as many processes such as changes in tidal range, storms, local tectonics, and compaction (see, e.g., Vink et al., 2007) influence their deposition and preservation. Also, the last ice sheets have destroyed evidence of previous shorelines, leading to a lack of data from before 20 ka (calendar years before 1950) in formerly glaciated areas (Steffen and Wu, 2011).

In northern Europe, for example, one can find about 4000 dated sea-level indicators, with most data going back to about 15 ka (Steffen and Wu, 2011). However, not all are publicly available (see Lambeck et al., 2010). All over the world, several thousand data have been collected so far (Klemann and Wolf, 2006), and new data are added occasionally.

Figure 1 shows the distribution of RSL data in our database in northern Europe and North America. We note that more data have been published for these regions, but those have not been added yet to our database. It can be seen that older data are found outside the former margin of glaciation. The closer the data are located to the last remnants of the ice sheets, the younger they are. The flooding of the southern North Sea is also mirrored in older data in the sea and younger data near the coast (Vink et al., 2007).

Now, each sample of a database has an associated error or uncertainty in height and time. This is different to GPS and gravity measurements, which are usually provided with an error in velocity or gravity rate of change, respectively. Thus, when investigating the observational error of RSL data one has to consider two errors. However, the time error of RSL data is often converted into an additional height error (Lambeck et al., 1998b) to ease a misfit calculation. The height error then includes the effect $|dh/dt|_t\sigma_t$ (Lambeck et al., 1998b), where $|dh/dt|_t$ is the rate of sea-level change at time t and σ_t the age error. The rate of sea-level change is usually taken from a rebound model, which is determined as part of an iterative solution in ice-model developments (Lambeck et al., 2010). Hence, the height error becomes larger, while the time error is set to zero. For further discussion of error sources in RSL data the reader is referred to Lambeck et al. (1998b).

As an example, we analyze our available data sets for North America and northern Europe (including the British Isles) for their errors. The aim of this exercise is to find a reliable average error that will be applied in this investigation. For the 11 time periods that we analyze in total (see Sect. 4), we group our data accordingly into subsets of 1000 or 2000 years in duration. Figure 2 shows the average and maximum RSL data errors in North America and northern Europe. About 3700 data samples were analyzed, which cover a large range in time and space. We thus consider our determined

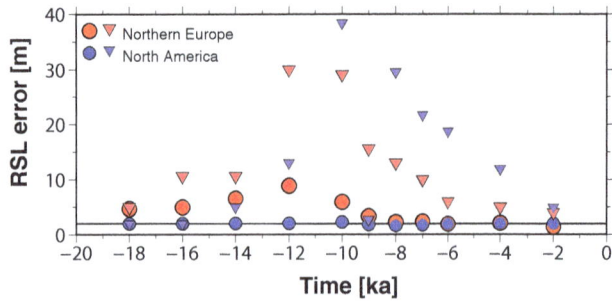

Figure 2. Average (circles) and maximum (inverted triangles) errors of relative sea-level data in North America (blue) and Europe (red) in time subsets of 1000 (between 10 and 6 ka) or 2000 years. Average calculated as arithmetic mean. In total about 3700 were analyzed for this example.

average value below to be representative of all possibly available RSL data.

Groups of younger samples contain many hundreds of samples, while groups with older samples, e.g., of 14 ka and older, envelop only a few. The maximum error becomes larger as the subset gets older, peaking at 10 (North America) and 12 ka (northern Europe), and then becoming much smaller (especially for North America). However, the number of older data is, as outlined above, much smaller than the number of younger data, therefore this error range is biased by the number of samples in each time span. One should also consider that the database partly contains samples analyzed a few decades ago when dating methods were not as sophisticated as today, thus such samples may have larger errors. These errors may increase the average error of a time span. It is beyond the aims of this study to evaluate each of the 3700 data samples to see how and when they were dated, so we shall use our database as a typical example and hope that this high number of samples allows us to perform a robust analysis.

North American data overall support an average error value of 2 m during all time subsets (thin solid black line in Fig. 2). Fennoscandian data show a higher average than 2 m for 10 ka and older. However, we hope as more newly determined data are added to these time subsets that the average will become lower. For example, new data for the southern North Sea show mainly errors of much less than a meter (Vink et al., 2007). Thus, we set 2 m as limit in this study, but we will also test in two examples how a better error of 1 m as well as an extreme value of 8 m (e.g., the average error of Fennoscandian data at 12 ka) affect our results.

3 Modeling

The models and approach used are taken from Wu et al. (2010) and Steffen et al. (2012). We use a reference model with 115 km lithospheric thickness as well as 6×10^{20} Pa s,

3×10^{21} Pa s and 6×10^{21} Pa s as background viscosity in the upper, shallower lower and deep lower mantle, respectively. The ice-load history is taken from model ICE-4G (Peltier, 1994). It is employed as surface load on a 2-degree grid of a non-rotating, spherical, self-gravitating, Maxwell visco-elastic finite-element earth model that includes material compressibility and self-gravitating oceans. We systematically vary, one at a time, the four previously mentioned parameters in the model to test its sensitivity in the global RSL predictions. The reference model and all the varied parameters can be found in Table 1.

For the sensitivity to the ice model, we compare the response between ICE-4G and ICE-5G (Peltier, 2004) globally (which differ not only in the Northern Hemisphere but also in Antarctica). For the other three parameters we apply the same changes as in Steffen et al. (2012). The model of lateral heterogeneous lithospheric thickness in Wu et al. (2005) is used instead of a 115 km uniformly thick lithosphere. The background viscosity is changed to 7×10^{20} Pa s in the upper mantle and 10^{22} Pa s throughout the lower mantle. Thus, we modify a VM2-like model (Wu et al., 2013) with a slight gradual viscosity change from the upper to the lower mantle to one with a higher viscosity contrast with depth. The lateral heterogeneous mantle viscosity is implemented from model RF3S20 by Wang et al. (2008).

As in former studies, we caution that the model parameters used represent typical cases only. We do not provide definitive sensitivity results as we apply selected models for ice-load history, lateral lithospheric thickness and viscosity, and there exists a broad variety of models and opinions for each parameter. There is, for example, still no consensus about how viscosity increases with depth in the mantle (Steffen and Wu, 2011; Wu et al., 2013). Hence, it is rather our goal to give a feel of what sensitivity one may expect in general, and also where we can expect or look for RSL data that may help solve problems still under debate.

4 Results

We plotted the sensitivity kernels at 11 different times between 18 ka and 2 ka. Time steps are 2000 years, but we also included the sensitivity for 9 and 7 ka, as the large continental ice sheets vanished rapidly from 10 ka until 6 ka. For this paper, we only show two distinct examples out of the large number of 44 figures or subplots. The first is an overview of 6 sensitivity patterns for a changed ice-load history at 18, 16, 14, 12, 10 and 8 ka in Fig. 3 to show the temporal pattern change of a parameter. The other example is the sensitivity of each parameter at 7 ka to compare four patterns at a dedicated time. As deposition of sea-level indicators or similar samples is not possible in glaciated areas on land, we mark these areas in the figures by drawing the extent of the ice at that time from model ICE-5G.

Table 1. Model parameters for the reference model and other models for sensitivity tests. LT: lithospheric thickness; UM: upper-mantle viscosity (above 670 km depth); LM1: shallow lower-mantle viscosity (670–1171 km depth); LM2: deep lower-mantle viscosity (1171 km to core-mantle boundary).

Effect of	Ice model	LT [km]	UM [Pa s]	LM1 [Pa s]	LM2 [Pa s]
Reference model	ICE-4G	115	6×10^{20}	3×10^{21}	6×10^{21}
Ice model	ICE-5G	115	6×10^{20}	3×10^{21}	6×10^{21}
Lat. heterogeneous lithosphere	ICE-4G	Lat. het. lith (Wu et al., 2005)	6×10^{20}	3×10^{21}	6×10^{21}
Background viscosity	ICE-4G	115	7×10^{20}	10^{22}	10^{22}
Lat. heterogeneous viscosity	ICE-4G	115	Lat. het. mantle RF3S20 with $\beta = 0.4$ (Wang et al., 2008)		

Figure 3 clearly shows the areas of highest sensitivity to changes in ice-load history, e.g., more than 600 m are located under the ice in North America at 18 ka. As it is unlikely to find samples under ice coverage, we focus on ice-free areas. At 18 ka, significant sensitivities are found in northern Russia, which is related to differences in the ice models. We therefore draw the ice extent according to model ICE-4G with a green line to allow a rigorous analysis. The extent of the Barents and Kara seas ice sheet in ICE-4G at 18 ka is much farther to the east, resulting in a notable sensitivity signal. Another area is found farther east in the Chukchi Sea, where ICE-4G contains a glaciation. Both areas show sensitivities of more than 200 m, while it is much less than 100 m in all other areas (e.g., in Antarctica). This behavior continues through time as long as the ice sheets remain significantly on land. At about 12 ka (Fig. 3d) we find a prominent retreat east of the Rocky Mountains uncovering high sensitivities of up to 400 m due to significant differences in ice thickness west of Hudson Bay between the two ice models used. Sensitivities of 100 m and more still exist at 7 ka (Fig. 4a). In Scandinavia, sensitivities are not that large, but can also reach 50 m at 10 ka (Fig. 3e). Similar features are found around Antarctica. In all other areas sensitivities are much lower.

Compared to the solid Earth parameters (see Fig. 4), ice-load history has significantly larger sensitivity. RSL data are mainly sensitive to lithospheric thickness variations in formerly glaciated areas and also around still glaciated ones. Values of about 12 m are reached. Sensitivity to background viscosity is constrained to the Hudson Bay area and the Antarctic coast. Areas of lower sensitivity can be found around the Arctic and in British Columbia. For sensitivity to lateral viscosity variations, RSL data should be checked in North America, Fennoscandia and the Barents Sea.

Next, we show the places where the sensitivity of the RSL data exceeds 2 m. Figures 5–7 show the superposition of the sensitivity pattern (above 2 m error) of all four parameters at eight selected times. As it may be possible one day to determine heights above sea level in past times far inland and to allow a better comparison of the pattern change over time, contours on-land are also shown.

As mentioned earlier, the dominant parameter in these figures is ice-load history. Samples dated to 18 ka are sensitive to it almost everywhere in the world (Fig. 5a, red lines), with the exception of the southern Indian Ocean. As we shall see in Fig. 8, the highlighted area will change if the error of the RSL data is different from 2 m. At later times (Fig. 5b and c), RSL data from all over the world are sensitive to ice-load history. At 12 ka (Fig. 5c), the pattern shows low sensitivities in the circum-antarctic oceans. This white space is shifted 2000 years later to north of the Equator, with a low-sensitivity region around some parts of the Mediterranean and the Black Sea (e.g., Fig. 6). Thereafter, the whole white space expands until 2 ka (Fig. 7b), pushing back areas of higher sensitivity to the (formerly) glaciated regions and leaving local sensitivity areas above 2 m error at certain times. The latter can be found, for example, at 7 ka in South America, southern Africa and Australia (Fig. 6c). Most coastal areas far away from the former glaciation are insensitive. This has held, for example, since 10 ka for a major part of the Mediterranean and some parts of the Caribbean. In comparison to areas sensitive to ice-load history, areas sensitive to lithospheric thickness variations are much smaller. They are found near the ice sheets or formerly glaciated areas (Fig. 5a, green lines), and the behavior of the pattern remains throughout all times. At 2 ka (Fig. 7b), sensitive areas remain at the Antarctic Peninsula, the northern Gulf of Bothnia and Baffin Bay.

Sensitivity to background viscosity covers larger areas than sensitivity to lithospheric thickness variations. Almost all areas north of 45° N, South America, parts of Africa, East Asia, Australia and Antarctica show a sensitivity above 2 m at 18 ka (Fig. 5a, blue dots). This pattern does not change significantly until 12 ka (Fig. 5c). Thereafter, the behavior is similar to lithospheric thickness variations, although they cover larger areas. At 2 ka (Fig. 7b), only a few spots (southern James Bay, the northern Gulf of Bothnia and the Barents Sea) are left in the Northern Hemisphere. Lateral variations

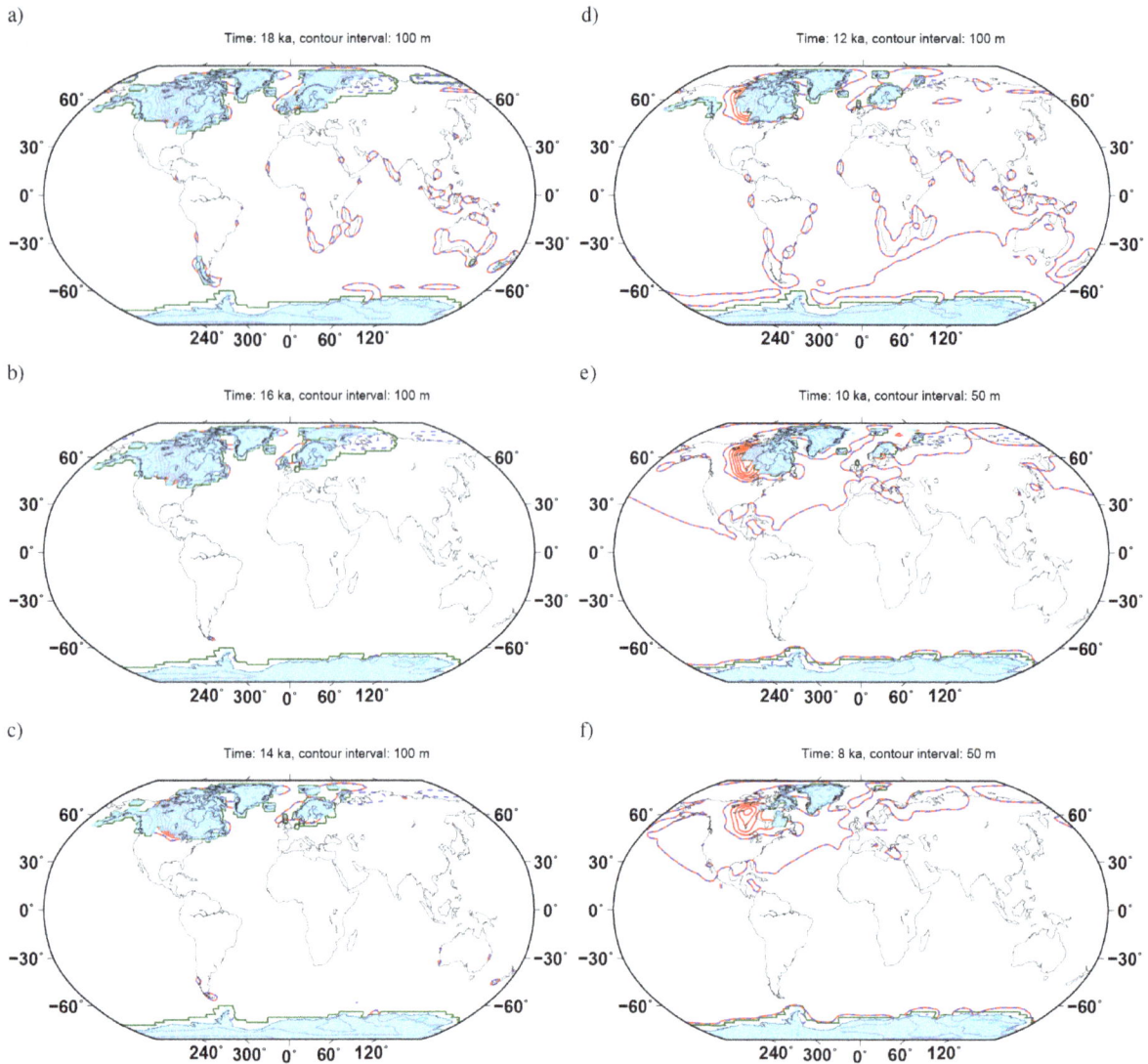

Figure 3. Sensitivity of relative sea-level data around the world to changes in ice model at **(a)** 18, **(b)** 16, **(c)** 14, **(d)** 12, **(e)** 10 and **(c)** 8 ka. Light blue areas mark the extent of ice sheets at the time, taken from the ICE-5G model (Peltier, 2004). Green solid line marks the ice extent from the ICE-4G model (Peltier, 1994). Red and blue-dashed lines are contours with positive and negative sensitivity, respectively. The red-blue-dashed line marks zero sensitivity. Contour intervals indicated on top. Unit in m. To read the sensitivity of a certain line, count the number of lines from the zero-sensitivity line and multiply by the contour interval.

in viscosity show the most diverse sensitivity patterns. From 18 ka (Fig. 5a, purple lines) until 14 ka (Fig. 5b), many sensitivity areas are determined next to the immediate surrounding of the ice sheets, e.g., the western and northeastern coasts of South America or the northwestern coasts of Africa and Australia. In the following millennia the areas are more constrained to the near surrounding of the (formerly) glaciated areas. At 2 ka (Fig. 7b), there are only a few very small areas on land left in North America and the Lofoten in Fennoscandia.

In the following we analyze how the pattern at a specific time changes if a different error is assumed. Figure 8 shows the effect of error size (1 m for (a), 2 m for (b) and 8 m for

(c)) on the pattern for 10 ka. The latter represents a rather extreme case, while an error of 1 m is a likely improvement for more recently discovered and dated samples. Any pattern at a specific time will not change significantly if the error value is changed moderately, e.g., by a few decimeters. If the value is changed significantly to higher or lower values, the pattern of a parameter will decrease or increase its sensitivity area accordingly. To understand why the area increases when the error value decreases, note that the plotted areas have sensitivity values (e.g., Figs. 3 and 4) above the error value. Thus a smaller error value means more area can be sensitive to that parameter variation. When the error changes from 2 to 1 m, the global sensitivity pattern of

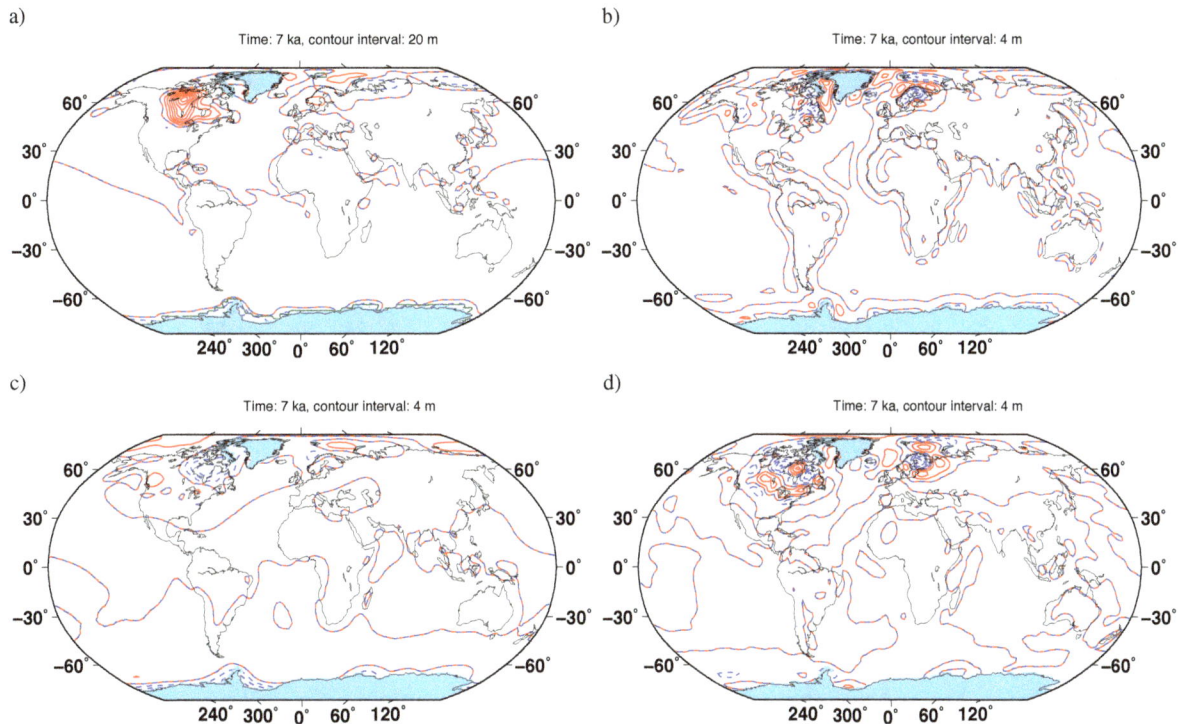

Figure 4. Sensitivity of relative sea-level data around the world to changes in ice-load history model (**a**), lithospheric thickness variations (**b**), background viscosity (**c**), and lateral viscosity variations (**d**) at 7 ka. Light blue areas mark the extent of ice sheets at the time, taken from the ICE-5G model (Peltier, 2004). Red and blue-dashed lines are contours with positive and negative sensitivity, respectively. The red-blue-dashed line marks zero sensitivity. Contour intervals indicated on top. Unit in m. To read the sensitivity of a certain line, count the number of lines from the zero-sensitivity line and multiply by the contour interval.

ice-load history shows mainly the same signature as for an error of 2 m, but the area becomes larger, reducing the insensitive areas in the Caribbean and the Mediterranean. For the solid Earth parameters the patterns increase more drastically around the Equator. When raising the error to 8 m, the area for all parameters is reduced significantly. Sensitivity to the solid Earth parameters is now mainly found near glaciated areas, whereas background viscosity sensitivity areas are quite small and restricted.

5 Discussion

The high sensitivity of RSL data to ice-load history changes over all millennia and almost independent of the chosen error confirms that RSL data play an outstandingly important role in the development of ice models, especially on a global scale. The reason is due to the relationship between the sea-level changes and ice coverage via the sea-level equation (Farrell and Clark, 1976): the higher the amount of ocean water bound in ice sheets at a certain time, the larger the sensitivity areas. Well-known sea-level fingerprints from the ice sheets (e.g., Mitrovica et al., 2001) appear in the sensitivity pattern of the RSL data, which confirms a link of selected, but not all RSL data to a certain ice sheet (Peltier, 2004; Horton

et al., 2009). Areas of interest for improving ice-load history are the eastern coast of the United States, the southern coasts of South America, Africa and Australia as well as the coast of Antarctica. Southern Hemisphere RSL data of 7 ka and older probably help in constraining the Antarctic Ice Sheet history. Data from the US east coast (from 18 ka until 6 ka), the Canadian coast and shelves (from 10 ka until 4 ka) and the Hudson Bay (from about 8 ka on) should help in constraining the Laurentide Ice Sheet, which confirms Horton et al. (2009) and Simon et al. (2011). We also note a corridor between the Rocky Mountains and Hudson Bay from about 12 ka on, where lake-level data of former and still existing lakes may be found. In Fennoscandia both the North and Baltic seas highlight sufficient sensitivities from 14 ka on. RSL data that are sensitive to lithospheric thickness can only help in quantifying variations near the ice sheets if the ice-load history is accurately known. This is due to the overlap between the ice-load history sensitivity pattern and that due to lithospheric variations. Sites far away from any ice sheets (e.g., Africa) will not provide insight into the underlying lithosphere structure. As background viscosity controls the amount of lithospheric depression due to the ice load and thus influences vertical movements and ocean geometry, the pattern at glacial maximum is clearly characterized by a

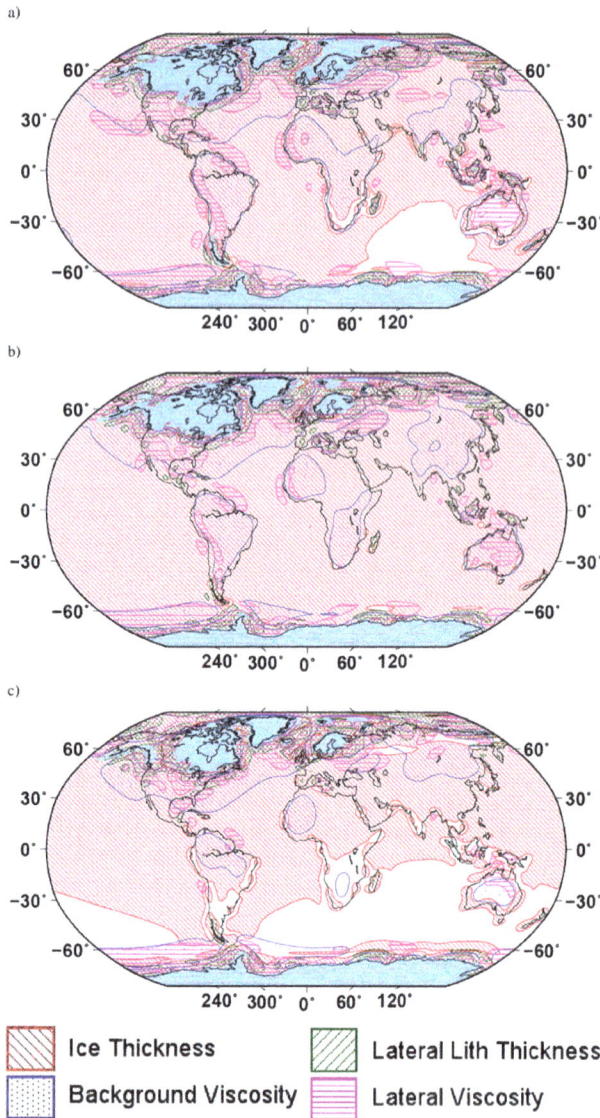

Figure 5. Sensitivity of relative sea-level data around the world above an assumed error of 2 m to changes in ice-load history model (see text, red area, lines from top left to bottom right), lithospheric thickness variations (green, lines from top right to bottom left), background viscosity (blue dots), and lateral viscosity variations (pink, horizontal lines) at **(a)** 18, **(b)** 14 and **(c)** 12 ka. If a color does not appear, then the sensitivity of this parameter lies below the error. Light blue areas mark the extent of ice sheets at the time, taken from the ICE-5G model (Peltier, 2004).

mixture of high sensitivities in and around the glaciated areas as well as in other high sensitivities. Thus, older far field RSL data may help determine background viscosity if the ice thickness is known satisfactorily. This statement may be altered if the error of RSL data decreases to 1 m or smaller. This can be seen in Fig. 8a in an area in the northern Pacific, where the patterns of background viscosity and ice-load history do not overlap. Such a non-overlapping area also exists

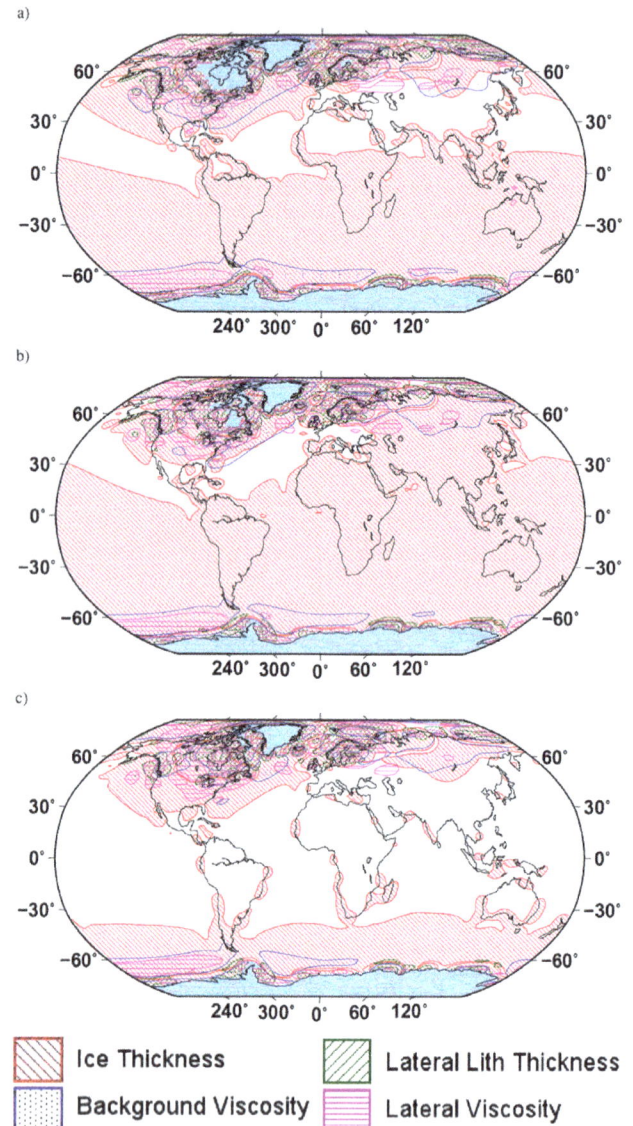

Figure 6. Same as Fig. 5, but for **(a)** 9, **(b)** 8 and **(c)** 7 ka.

for other times if the RSL error is 1 m or smaller. The area for lateral variations in mantle viscosity also overlaps with that for ice-load history, so lateral viscosity variations can only be determined if the ice-load history is known accurately. However, one should caution that the sensitivity pattern of lateral variations in mantle viscosity is affected by the model of lateral variations. The 2 m error needs to be compared to the deformation and/or sea-level change at a certain time in an area of interest. Sensitivity exceeds 2 m during glaciation (18–7 ka) almost everywhere including where RSL data can be expected. After glaciation (7 ka until the present day) the sensitivity area becomes smaller, as the (calculated) deformation or sea-level change can be less than 2 m. However, more recent RSL data often have errors smaller than 2 m, which enlarges the sensitivity pattern for each parameter shown in

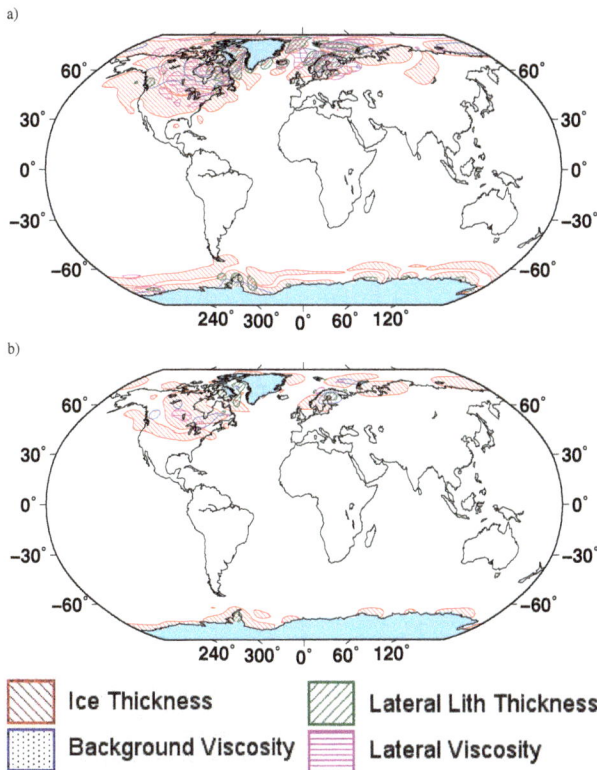

Figure 7. Same as Fig. 5, but for (**a**) 4, and (**b**) 2 ka.

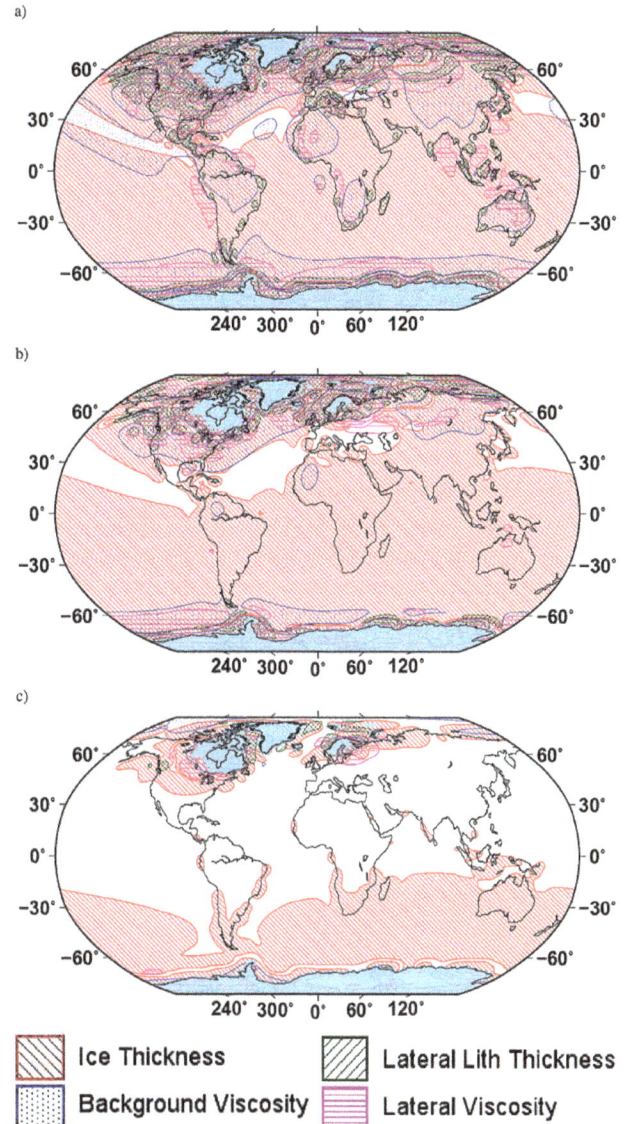

Figure 8. Same as Fig. 5, but for 10 ka and different RSL data errors of (**a**) 1, (**b**) 2 and (**c**) 8 m.

Figs. 5–7. Thus, samples from other areas may be used in case their error is smaller than the new limit. In our example using a 1 m error limit (Fig. 8), the pattern for ice-load history shows the smallest variation as the sensitivity in sea level at a specific time between the two tested ice models reaches several hundreds of meters, see Fig. 3. In comparison to that, the other three parameters have smaller sensitivities (Fig. 4), and thus a small change in the error limit can lead to significant changes in each pattern. The general findings of our study will thus not be affected if a moderately different error (e.g., a difference of a few decimeters) than 2 m would be chosen. The difference can be larger though when investigating ice-load history, as an increase by a factor of 4 (from 2 to 8 m) still highlights its typical pattern, but is reduced in the equatorial area. The other three parameters need accurate RSL data and a precise ice-load model for their determination. The dominant sensitivity signal of ice-load history supports the findings of Wu et al. (2010) to GPS measurements and Steffen et al. (2012) to gravity observations. As RSL data illustrate vertical deformation, the pattern shape of all sensitivities in Fennoscandia and North America has strong similarities to the sensitivity pattern of the vertical component from GPS and gravity measurements. This holds especially for 4 and 2 ka, the times closest to GPS and gravity measurements today.

6 Conclusions

We provide global sensitivity pattern maps of RSL data from the time of the Last Glacial Maximum until 2 ka for four parameters that are important in GIA modeling: ice-load history, lateral lithospheric thickness variations, background viscosity, and lateral mantle viscosity variations. Our maps do not exclude the deep sea and the continents as we hope that future methods will give similar information as near-coastal RSL data today.

Ice-load history dominates the sensitivity maps and generally overlaps with the patterns of the other three parameters. This has implications for studies of the other three parameters: as long as the ice-load history is not sufficiently known, lateral heterogeneities in mantle viscosity or lithospheric thickness (also background viscosity, but to a lesser degree) can only be poorly determined, as their influence is rather low when compared to the effect of ice-load history, which is dominant if the difference between ICE-4G and ICE-5G is representative of the ice thickness uncertainty. Recent studies (e.g., Argus and Peltier, 2010) indicate that it is likely an over-estimate. Also, it should be noted that the level of interference depends on the magnitude of the uncertainty in ice thickness. The level of interference decreases rapidly as the difference between the time of the ice-thickness uncertainty and the time of the RSL data increases. In addition, it should be evaluated whether rheologic changes in the oceans, e.g., due to subduction zones (Austermann et al., 2013), influence our assumptions.

In view of the dominant ice-load history sensitivity pattern, we speculate that for investigations of glacial cycles older than the last Pleistocene one, it may not be necessary to include lateral heterogeneities as the ice history of these glacial cycles is less well constrained than the late Pleistocene. However, further research is recommended. The three solid-earth parameters are mainly constrained to areas of former glaciation. The area of all patterns decreases with time. These distinct patterns depend on the background models and the chosen error limit. The latter can be changed within a few decimeters to give similar results, which especially holds for sensitivity to ice-load history, but a larger change in the error limit alters the pattern significantly.

In view of improvements in the data error, e.g., when reducing the error from 2 m to 1 m, more locations, even outside the near field of GIA, can be used to infer parameters such as background viscosity and lateral heterogeneity. In particular, studies of background viscosity can be in a better situation if the error for RSL is reduced to 1 m or less.

Due to the dominant overlapping signal of ice-load history, one has to distinguish between regions sensitive to one, two, three or all four parameters. Assuming that ice-load history is thoroughly investigated and well determined in the future, RSL data sensitive to only one of the other three may help to constrain that particular parameter. The results will improve GIA modeling significantly and may also help in initiatives such as PALSEA (Siddall and Milne, 2012), i.e., may guide coastal geomorphologists and ocean scientists to check locations of potential RSL data helpful in GIA studies and thus may foster and trigger new mutually beneficial cooperation between the GIA modeling community and the deep-sea drilling community.

Our sensitivity study suggests the value of collecting and interpreting RSL data in coastal areas that are surrounded by deeper ocean and that non-marine fresh water lakes also provide valuable new information to constrain models.

At least 14 000 RSL data samples have been determined in the last decades around the world (see, e.g., Klemann and Wolf, 2006; Lambeck et al., 2010). However, not all are easily accessible for everyone, thus we cannot clearly evaluate if this database is sufficient and present a definite recommen-

dation for new data to be looked for. Of course, more data are always better, needed and greatly appreciated! However, one has to investigate thoroughly if new data improve our understanding of the GIA and the Earth's interior. Wu et al. (2013), for example, noted that sensitivity of RSL data to lower-mantle viscosity is constrained to lie in formerly glaciated areas. Our results indicate now that this argument is valid for RSL data from about 6 ka until today, but is not the case for much older RSL data. In any case, adding hundreds of newly determined far-field data that are dated to about 6 ka and younger may introduce error to such an investigation.

As RSL data cover both spatial and temporal effects of GIA and therefore provide additional information to geodetic measurements performed on land, a combined solution from many different GIA observations is recommended in GIA investigations as long as their measurement errors allow such an analysis.

Acknowledgements. We are grateful for the constructive comments by two anonymous reviewers. The finite-element calculation was performed with the ABAQUS package from Simulia. This research is supported by an operating grant from NSERC of Canada to Patrick Wu and Hansheng Wang is supported by the National Natural Science Foundation of China (grant nos. 40825012, 41021003, 41174016, 41274026 and 41204013), the National Key Basic Research Program of China (973 Program, grant no. 2012CB957703) and the CAS/SAFEA International Partnership Program for Creative Research Teams (KZZD-EW-TZ-05). The figures in this paper are drawn using the GMT graphics package (Wessel & Smith, 1998).

Edited by: B. Vermeersen

References

Argus, D. F. and Peltier, W. R.: Constraining models of postglacial rebound using space geodesy: a detailed assessment of model ICE-5G (VM2) and its relatives. Geophys. J. Int., 181, 697–723, doi:10.1111/j.1365-246X.2010.04562.x, 2010.

Austermann, J., Mitrovica, J. X., Latychev, K., and Milne, G. A.: Barbados-based estimate of ice volume at Last Glacial Maximum affected by subducted plate, Nat. Geosci., 6, 553–557, doi:10.1038/ngeo1859, 2013.

Clark, J. A.: The reconstruction of the Laurentide Ice Sheet of North America from sea level data: Method and preliminary results, J. Geophys. Res., 85, 4307–4323, doi:10.1029/JB085iB08p04307, 1980.

Engelhart, S. E., Peltier, W. R., and Horton, B. P.: Holocene relative sea-level changes and glacial isostatic adjustment of the U.S. Atlantic coast, Geology, 39, 751–754, doi:10.1130/G31857.1, 2011.

Fairbanks, R. G., Mortlock, R. A., Chiu, T. C., Cao, L., Kaplan, A., Guilderson, T. P., Fairbanks, T. W., Bloom, A. L., Grootes, P. M., and Nadeau, M. J.: Radiocarbon calibration curve spanning 0 to 50,000 years BP based on paired ^{230}Th/^{234}U/^{238}U and ^{14}C dates on pristine corals. Quat. Sci. Rev., 24, 1781–1796, doi:10.1016/j.quascirev.2005.04.007, 2005.

Farrell, W. E., and Clark, J. A.: On postglacial sea level, Geophys. J. R. Astr. Soc., 46, 647–667, doi:10.1111/j.1365-246X.1976.tb01252.x, 1976.

Horton, B. P., Peltier, W. R., Culver, S. J., Drummond, R., Engelhart, S. E., Kemp, A. C., Mallinson, D., Thieler, E. R., Riggs, S. R., Ames, D. V., and Thomson, K. H.: Holocene sea-level changes along the North Carolina Coastline and their implications for glacial isostatic adjustment models, Quat. Sci. Rev., 28, 1725–1736, doi:10.1016/j.quascirev.2009.02.002, 2009.

Klemann, V. and Wolf, D.: A global data base for late-glacial and Holocene sea-level indicators. Poster presented at WCRP Workshop "Understanding Sea-Level Rise and Variability", Paris, 6–9 June, 2006.

Lambeck, K., Smither, C., and Ekman, M.: Tests of glacial rebound models for Fennoscandia based on instrumented sea- and lake-level records. Geophys. J. Int., 135, 375–387, doi:10.1046/j.1365-246X.1998.00643.x, 1998a.

Lambeck, K., Smither, C., and Johnston, P.: Sea-level change, glacial rebound and mantle viscosity for northern Europe, Geophys. J. Int., 134, 102–144, doi:10.1046/j.1365-246x.1998.00541.x, 1998b.

Lambeck, K., Purcell, A., Zhao, J., and Svensson, N.-O.: The Scandinavian Ice Sheet: from MIS 4 to the end of the Last Glacial Maximum, Boreas, 39, 410–435, doi:10.1111/j.1502-3885.2010.00140.x, 2010.

Lambeck, K., Purcell, A., and Dutton, A.: The anatomy of interglacial sea level: The relationship between sea levels and ice volumes during the Last Interglacial, Earth Planet. Sci. Lett., 4, 315–316, doi:10.1016/j.epsl.2011.08.026, 2012.

Lidberg, M., Johansson, J. M., Scherneck, H.-G., and Milne, G. A.: Recent results based on continuous GPS observations of the GIA process in Fennoscandia from BIFROST, J. Geodyn., 50, 8–18, doi:10.1016/j.jog.2009.11.010, 2010.

Mitrovica, J. X., Tamisiea, M. E., Davis, J. L., and Milne, G. A.: Recent mass balance of polar ice sheets inferred from patterns of global sea-level change, Nature, 409, 1026–1029, doi:10.1038/35059054, 2001.

Pedoja, K., Husson, L., Regard, V., Cobbold, P. R., Ostanciaux, E., Johnson, M. E., Kershaw, S., Saillard, M., Martinod, J., Furgerot, L., Weill, P., and Delcaillau, B.: Relative sea-level fall since the last interglacial stage: Are coasts uplifting worldwide? Earth Sci. Rev., 108, 1–15, doi:10.1016/j.earscirev.2011.05.002, 2011.

Peltier, W. R.: Ice age paleotopography, Science, 265, 195–201, doi:10.1126/science.265.5169.195, 1994.

Peltier, W. R.: Global glacial isostasy and the surface of the ice-age Earth: the ICE-5G(VM2) model and GRACE. Annu. Rev. Earth Planet. Sci., 32, 111–149, doi:10.1146/annurev.earth.32.082503.144359, 2004.

Siddall, M. and Milne, G. A.: Understanding sea-level change is impossible without both insights from paleo studies and working across disciplines, Earth Planet. Sci. Lett., 315–316, 2–3, doi:10.1016/j.epsl.2011.10.023, 2012.

Simon, K. M., James, T. S., Dyke, A., and Forbes, D. L. Refining Glacial Isostatic Adjustment Models in Northern Canada: Implications for Ice Sheet History, Sea-Level Change, and Land Emergence Along the West Coast of Hudson Bay, AGU Fall Meeting Abstracts, G23B-04, 2011.

Steffen, H. and Kaufmann, G.: Glacial isostatic adjustment of Scandinavia and northwestern Europe and the radial viscosity structure of the Earth's mantle, Geophys. J. Int., 163, 801–812, doi:10.1111/j.1365-246X.2005.02740.x, 2005.

Steffen, H. and Wu, P.: Glacial isostatic adjustment in Fennoscandia – A review of data and modeling, J. Geodyn., 52, 169–204, doi:10.1016/j.jog.2011.03.002, 2011.

Steffen, H., Wu, P., and Wang, H.: Optimal locations for absolute gravity measurements and sensitivity of GRACE observations for constraining glacial isostatic adjustment on the northern hemisphere, Geophys. J. Int., 190, 1483–1494, doi:10.1111/j.1365-246X.2012.05563.x, 2012.

Törnqvist, T. E. and Hijma, M. P.: Links between early Holocene ice-sheet decay, sea-level rise and abrupt climate change. Nat. Geosci., 5, 601–606, doi:10.1038/ngeo1536, 2012.

Tushingham, A. M. and Peltier, W. R.: Validation of the ICE-3G model of Würm-Wisconsin deglaciation using a global data base of relative sea level histories, J. Geophys. Res., 97, 3285–3304, doi:10.1029/91JB02176, 1992.

Tushingham, A. M. and Peltier, W. R.: Implications of the Radiocarbon Timescale for Ice-Sheet Chronology and Sea-Level Change, Quat. Res., 39, 125–129, doi:10.1006/qres.1993.1015, 1993.

van de Plassche, O.: Sea-Level Research: A Manual for the Collection and Evaluation of Data, GeoBooks, Norwich, 1986.

Vink, A., Steffen, H., Reinhardt L., and Kaufmann, G.: Holocene relative sea-level change, isostatic subsidence and the radial viscosity structure of the mantle of north-western Europe (Belgium, the Netherlands, Germany, southern North Sea), Quat. Sci. Rev., 26, 3249–3275, doi:10.1016/j.quascirev.2007.07.014, 2007.

Wang, H. S., Wu, P., and van der Wal, W.: Using postglacial sea level, crustal velocities and gravity-rate-of-change to constrain the influence of thermal effects on mantle lateral heterogeneities, J. Geodyn., 46, 104–117, doi:10.1016/j.jog.2008.03.003, 2008.

Wessel, P. and Smith, W. H. F.: New, improved version of generic mapping tools released, EOS, 79, 579, doi:10.1029/98EO00426, 1998.

Wu, P., Wang, H. S., and Schotman, H.: Postglacial induced surface motions, sea levels and geoid rates on a spherical, self-gravitating laterally heterogeneous earth, J. Geodyn., 39, 127–142, doi:10.1016/j.jog.2004.08.006, 2005.

Wu, P., Steffen, H., and Wang, H. S.: Optimal locations for GPS measurements in North America and northern Europe for constraining Glacial Isostatic Adjustment, Geophys. J. Int., 181, 653–664, doi:10.1111/j.1365-246X.2010.04545.x, 2010.

Wu, P., Wang, H., and Steffen, H.: The role of thermal effect on mantle seismic anomalies under Laurentia and Fennoscandia from observations of Glacial Isostatic Adjustment, Geophys. J. Int., 192, 7–17, doi:10.1093/gji/ggs009, 2013.

10

A new model of the upper mantle structure beneath the western rim of the East European Craton

M. Dec[1], **M. Malinowski**[1], and **E. Perchuc**[2]

[1]Institute of Geophysics, Polish Academy of Sciences, Ks. Janusza 64, 01-452 Warsaw, Poland
[2]Bernardynska 21/57, 02-904 Warsaw, Poland

Correspondence to: M. Dec (monikadec@igf.edu.pl) and E. Perchuc (perchuc.ed.geo@gmail.com)

Abstract. We present a new 1-D P wave seismic velocity model (called MP1-SUW) of the upper mantle structure beneath the western rim of the East European Craton (EEC) based on the analysis of the earthquakes recorded at the Suwałki (SUW) seismic station located in NE Poland which belongs to the Polish Seismological Network (PLSN). Motivation for this study arises from the observation of a group of reflected waves after expected $P_{410}P$ at epicentral distances 2300–2800 km from the SUW station. Although the existing global models represent the first-arrival traveltimes, they do not represent the full wavefield with all reflected waves because they do not take into account the structural features occurring regionally such as 300 km discontinuity. We perform P wave traveltime analysis using 1-D and 2-D forward ray-tracing modelling for the distances of up to 3000 km. We analysed 249 natural seismic events from four azimuthal spans with epicentres in the western Mediterranean Sea region (WMSR), the Greece and Turkey region (GTR), the Caucasus region (CR) and the part of the northern Mid-Atlantic Ridge near the Jan Mayen Island (JMR). For all chosen regions, except the JMR group for which 2-D modelling was performed, we estimate a 1-D average velocity model which will characterize the main seismic discontinuities. It appears that a single 1-D model (MP1-SUW model) explains well the observed traveltimes for the analysed groups of events. Differences resulting from the different azimuth range of earthquakes are close to the assumed picking uncertainty. The MP1-SUW model documents the bottom of the asthenospheric low-velocity zone (LVZ) at the depth of 220 km, 335 km discontinuity and the zone with the reduction of P wave velocity atop 410 km discontinuity which is depressed to 440 km depth. The nature of the regionally oc-

curring 300 km boundary is explained here by tracing the ancient subduction regime related to the closure of the Iapetus Ocean, the Rheic Ocean and the Tornquist Sea.

1 Introduction

One-dimensional reference models are employed almost in every seismological method aimed at imaging of Earth's interior (tomography, receiver functions, underside reflections). However, results of those methods can be biased by the choice of the background velocity model (e.g. Bastow, 2012). Therefore, in the regional studies, it might be more appropriate to use modified reference models taking into account, e.g. the tectonic regime of the area (e.g. Perchuc and Thybo, 1996). Following this strategy, we attempt to derive a one-dimensional upper mantle P wave velocity model for the areas surrounding the East European Craton (EEC) to the west and to the south. Toward this end, we use the data recorded at the Suwałki (SUW) station belonging to the Polish Seismological Network. The data recorded at the SUW station were used in few seismological studies (e.g. Bock et al., 1997; Świeczak et al., 2004; Wilde-Piórko, 2005), but none of these were focusing on the detailed interpretation of the recorded traveltimes in the far-regional mode.

Due to the advancement in instrumentation and access to infrastructure, array seismology is developing rapidly (e.g. the Earthscope USArray project; Levander, 2003). However, the cost of experiments remain very high. Here we would like to explore the concept of using single-station data from the existing seismological network to study the upper mantle structure based on the traveltimes and amplitudes of the

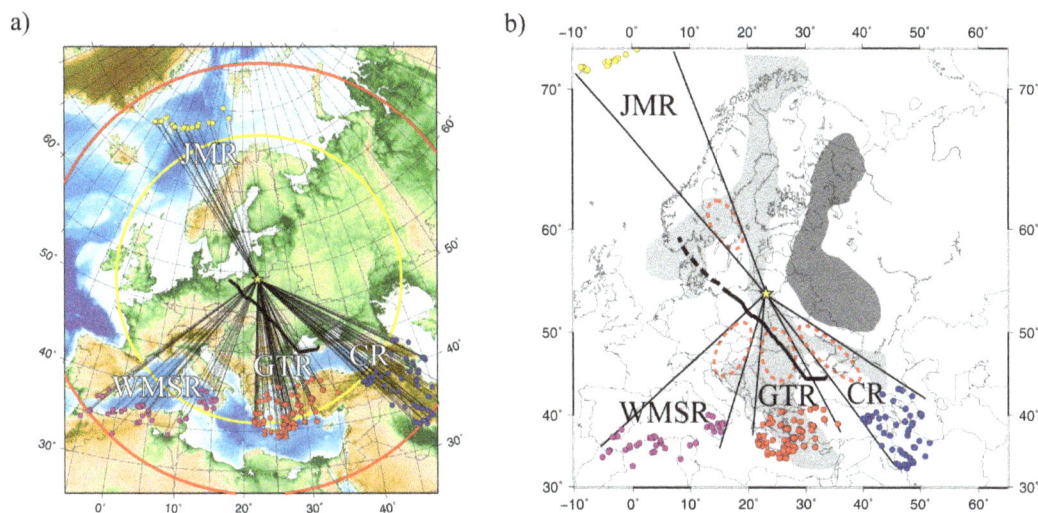

Figure 1. (**a**) Four groups of earthquakes recorded at the SUW station (yellow asterisk): JMR (yellow points), WMSR (violet points), GTR (red points) and CR (blue points). Circles centred at SUW mark the distance of 2000 (yellow) and 3000 km (red). Thick black line represents the TTZ. (**b**) Map of the study area with indication of the regions to which our upper mantle model pertains (red dashed ellipses). Areas shaded in grey are characterized by higher S wave velocities at 250 km depth according to the model of Shapiro and Ritzwoller (2002). Light grey represents velocities about 4.65 km s^{-1} and dark grey about 4.80 km s^{-1}.

body waves. The necessary prerequisite for performing such a study is the selection of the station that is located optimally for imaging particular mantle structure by providing proper azimuthal span and epicentral distances of the recorded earthquakes (see e.g. Nita et al., 2012).

The choice of the SUW station for this analysis is twofold. First of all, it is characterized by a good signal-to-noise ratio and relatively simple wavefield, as it is located on the stable part of the EEC with a thin sedimentary cover. Secondly, the earthquakes recorded from the four azimuthal sectors (northern Mid-Atlantic Ridge, Western Mediterranean Sea, Greece–Turkey, and Caucasus) are located in the 1500–3000 km epicentral distance range, providing proper illumination of the upper mantle structure between the base of the asthenosphere and the 410 km discontinuity. The bottoming points of the refracted waves and midpoints of the reflected waves fall in to the south-western rim of the East European Craton as indicated in Fig. 1. Therefore, we can study the influence of the Teisseyre–Tornquist zone (TTZ) on the recorded wavefield and mantle structure.

First, we present the tectonic setting and a review of the studies pertaining to the investigated area, which mainly focused on Earth's crust. Subsequently, we describe data (seismic waveforms from natural seismic events) and methods used in this analysis (1-D and 2-D ray-tracing modelling). In the results section we present all analysed seismic sections together with the traveltimes calculated for our preferred upper mantle model. It is followed by error analysis and traveltime residuals statistics. Finally, we discuss our results in the context of the geodynamics of the circum-EEC region.

2 Tectonic setting and previous geophysical investigations

In order to understand the structure and nature of the upper mantle below the Baltica, it is necessary to understand its history as a palaeocontinent which was moving across Iapetus Ocean with other palaeocontinents and microplates which are nowadays neighbouring Baltica. The EEC is the oldest part of eastern Europe. It formed about 1.8–1.7 Ga (years ago) as a result of a collision between three separate plates: Fenoscandia, Sarmatia and Volgo-Uralia (Bogdanova et al., 2001, Bogdanova et al., 2005). The history of the palaeoplate movements from Vendian to Permian was described by Cocks and Torsvik (2006). There were two major palaeotectonical events. First, about 450 Ma in the Ordovician, Avalonia separated from Gondwana and docked with Baltica and afterwards they both joined Laurentia. The last event was connected with the closure of the Iapetus Ocean which ended about 420 Ma and the closure of the Tornquist Sea – a branch of Iapetus. A different situation had place in the SW part of the cratonic rim. In this area there were Rheic Ocean structures formed from Devonian to Permian and then the Palaeotethyan, which were evolving up to the TTZ (Ziegler et al., 2006). Caledonian structures: Northwest Highlands (Gee, 1975) and Polish Caledonides (Znosko, 1986) were imposed on the NW and SW margins of the EEC.

Geophysical studies are mainly seismic. There were numerous deep seismic sounding profiles recorded both west and east of the TTZ (see recent summary in Guterch et al., 2010). Starting with the LT profiles located in Poland (Guterch et al., 1986), through the EUGENO-S project in Denmark and southern Sweden (EUGENO-S Working Group et al., 1988), Polish projects POLONAISE'97 (Guterch et al., 1999) and CELEBRATION 2000 (Guterch et al., 2001), the FENNOLORA transect (Guggisberg and Berthelsen, 1987; Abramovitz et al., 2002), the BABEL project (BABEL Working Group, 1991), the MONA LISA (Abramovitz et al., 1998) or EUROBRIDGE projects (EUROBRIDGE Seismic Working Group, 1999) we have a good control on the crustal structure of the EEC and surrounding areas.

All of these projects provided gross crustal structure down to the Moho and only in few cases were deeper boundaries interpreted (e.g. Grad et al., 2002). During the acquisition of the deep seismic sounding profiles, some upper mantle phases (both reflections and refractions) were traced in the offset range beyond 400 km (e.g. in project POLONAISE'97, Grad et al., 2002; EUGENO-S Working Group, 1988; BABEL Working Group, 1993; MONA LISA, Abramovitz et al., 1999). Apart from the controlled-source seismology, some passive-source experiments were conducted with the TOR project being the most significant (Gregersen et al., 2010; Plomerova et al., 2002). The TOR project spanning about 1000 km from northern Germany, through Denmark up to southern Sweden, crossed principal tectonic zones inside the rim of the EEC such as the Variscan Front, the Caledonian Deformation Front and the Sorgenfrei–Tornquist zone – the northern extension of the TTZ. This allowed us to image seismic structure of the crust and upper mantle (e.g. Somali et al., 2006, Gregersen et al., 2010). The recently conducted PASSEQ project (Wilde-Piórko et al., 2008) provided new data for studying mantle structure across the TTZ, e.g. using teleseismic P wave receiver functions (Knapmeyer-Endrun et al., 2013).

3   Data and method

The SUW seismic station was chosen for this study because of the good recording conditions in the remote part of NE Poland and the thin sediment cover resulting in high-quality seismic signal registrations. The station is equipped with the STS-2 sensor and the data are recorded at 20 Hz. We used broadband data recorded between 1997 and 2010 for earthquakes with magnitude greater than 4.5. The hypocentre locations were taken from the ISC bulletin (International Seismological Centre; http://www.isc.ac.uk/iscbulletin). The retrieved waveforms were converted from the native MSEED to SEG-Y format in order to make seismic record sections suitable for interpretation and modelling using standard tools used in controlled-source seismology (Zelt, 1994).

The 249 events analysed (see Supplement and Fig. 1) are located in the following areas:

Table 1. Mean signal-to-noise ratio for the analysed groups of events.

Region	Number of events	Signal-to-noise ratio		
		Raw data	After filtering (0.5–2.0 Hz)	Improvement (%)
JMR	34	4.70	8.17	73.83
CR 020	34	2.57	7.50	191.83
CR 2050	23	3.52	6.66	89.20
CR 50+	12	5.54	12.66	128.52
GTR 020	53	6.59	4.85	−26.40
GTR 2050	36	3.61	5.26	45.71
GTR 50+	14	13.84	18.02	30.20
WMSR	43	6.30	11.54	83.17
Average		5.38	8.16	67.76

1. rift zone around the Jan Mayen region in the northern Atlantic (JMR)

2. Western Mediterranean Sea region (WMSR)

3. Greece and Turkey region (GTR)

4. Caucasus region (CR).

All earthquakes generated in the JMR (part of the Mid-Atlantic Ridge) (Fig. 2a) were shallow crustal ones (depth down to 10 km). The earthquakes in the WMSR were also shallow, with focal depths down to 20 km (Fig. 2b). However, the data from the GTR (Fig. 2c) and CR (Fig. 2d), due to a large variability of the source depths, were separated into three groups with the focal depth ranges: 0–20 km, 20–50 km and > 50 km. We divided earthquakes into these focal depth ranges after evaluating differences in traveltimes calculated for different focal depths in the AK135 model (Kennett et al., 1995). Modelling was performed for each group separately. It allowed us to limit static shifts in the record sections to 1 s. We did not apply focal depth corrections to the data belonging to each epicentral group.

Seismic record sections sorted according to the increasing epicentral distance consist of the following number of seismograms: 34 (JMR), 43 (WMR), 103 (GTR), 69 (CR). We tested a range of bandpass filter frequencies for displaying data and finally we concluded that the best results were obtained for a 0.5–2.0 Hz bandwidth. This bandpass filter is commonly used in analysing data recorded in the far-regional mode (e.g. Chu et al., 2012; Świeczak et al., 2004). As a data quality check we calculated the signal-to-noise (SNR) ratio, estimated using energy of the signal in the 2 s wide window before and after first arrivals. Calculations were made using the ZPLOT software (Zelt, 1994). Table 1 shows mean SNRs calculated for the individual group of events. Group of the deepest earthquakes from the GTR are characterized by the best SNR for raw data. The best improvement after filtering was observed for the CR group for events with focal depths between 0 and 20 km. The average SNR for all raw data is

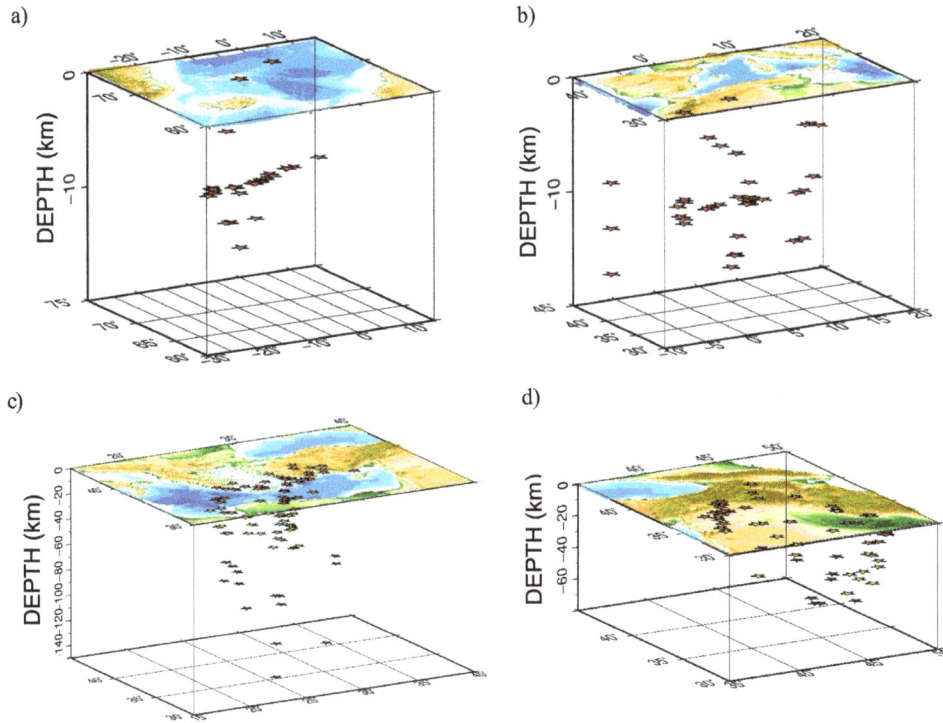

Figure 2. Distribution of the focal depths for the analysed groups of earthquakes: (**a**) JMR, (**b**) WMSR, (**c**) GTR, (**d**) CR.

Table 2. List of seismic events shown in Fig. 3. Numbers from column 1 correspond to numbers of seismic records in Fig. 3. (UTC – coordinated universal time)

No.	Offset (km)	Date	Origin time (UTC)	Lat	Long	Depth (km)	Mag (mb)	Seismic region
1	2259.18	24 Jun 2002	01:20:37.60	35.92	9.88	10.0	4.7	Tunisia
2	2267.73	25 Feb 2007	21:53:13.80	73.18	6.77	10.0	5.1	Greenland Sea
3	2275.16	07 Jun 1999	16:10:33.66	73.02	5.35	10.0	5.2	Greenland Sea
4	2307.27	30 Aug 2005	20:53:48.17	71.91	−1.09	10.0	4.8	Jan Mayen Island region
5	2322.30	08 Sep 2008	21:17:15.10	72.63	0.96	10.0	5.0	Norwegian Sea
6	2323.20	26 Aug 1999	05:03:05.04	71.70	−2.43	10.0	5.1	Jan Mayen Island region
7	2334.51	02 Dec 1997	00:02:03.51	71.65	−3.03	10.0	5.0	Jan Mayen Island region
8	2340.29	13 Aug 2006	19:03:08.50	71.39	−4.00	13.1	4.8	Jan Mayen Island region
9	2360.94	23 Mar 1998	20:19:27.74	71.50	−4.47	10.0	5.1	Jan Mayen Island region
10	2392.71	22 May 2003	13:57:21.27	37.12	3.84	10.0	4.8	Western Mediterranean Sea
11	2395.07	16 Nov 2000	11:33:08.87	36.63	4.79	8.7	4.8	Northern Algeria
12	2402.14	22 May 2003	03:14:04.85	37.16	3.57	15.0	5.2	Western Mediterranean Sea
13	2412.09	02 Sep 2008	20:00:50.82	38.72	45.79	3.0	4.9	Iran–Armenia–Azerbaijan border
14	2421.17	19 Jun 2003	12:59:23.14	71.08	−7.64	0.5	5.6	Jan Mayen Island region
15	2422.18	14 Apr 2004	23:07:37.81	71.05	−7.74	10.5	5.7	Jan Mayen Island region
16	2424.21	27 May 2003	17:11:28.35	36.94	3.54	6.1	5.5	Northern Algeria
17	2430.11	18 Aug 2000	18:15:06.54	36.19	4.96	10.0	4.9	Northern Algeria

5.38 and after bandpass filtering we achieve improvement of ∼ 68 %. Therefore, all presented seismic sections are displayed with a 0.5–2.0 Hz bandpass filter.

In this work we analysed earthquakes recorded in the far-regional mode, i.e. with epicentral distances of 1500–3000 km. As shown, for example, by Thybo and Perchuc (1997), this offset range allows one to study upper mantle structure between the Lehman discontinuity at ca. 200 km depth and the 410 km discontinuity. Phase identification was supported by aligning the observed reflectivity based on the maximum amplitude (Fig. 3) for earthquakes from different azimuthal spans (Table 2). Good alignment was obtained for the phase emerging at the time characteristic of the 410 km discontinuity (Fig. 3a). There are also some other

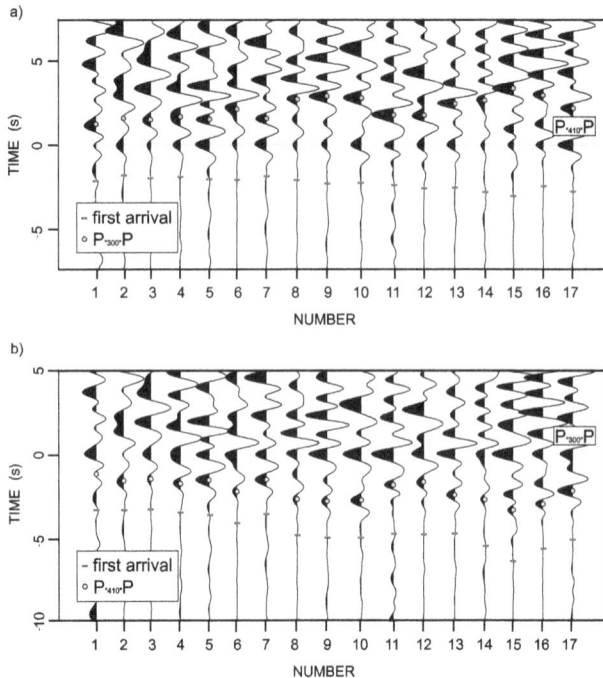

Figure 3. Seismograms aligned by the maximum amplitudes of (a) $P_{410}P$ (green line) and (b) $P_{300}P$ (blue line). Seismograms are ordered by the event's number (see Table 2). Red dashes represent first arrivals. Blue dots in (a) correspond to blue line in (b). Green dots in (b) correspond to green line in (a).

strong phases that can be aligned in the epicentral distance of 2260–2430 km (Fig. 3b), later interpreted as being the reflection phases from around 300 km depth. The discontinuity at ca. 300 km was introduced based on earlier results of Nita et al. (2012) and the presence of the reflected arrivals above the $P_{410}P$ reflected phase. The following reflected and refracted seismic phases were interpreted in the record sections (Figs. 4–7) during subsequent modelling: P_{220} – refracted wave from the bottom of the asthenosphere (LVZ), $P_{335}P$ – reflected wave from the regionally occurring discontinuity located at the depth of about 335 km, and P_{440} and $P_{440}P$ – refracted and reflected waves from depressed 410 km discontinuity, modelled at ca. 440 km depth.

We used forward trial-and-error one-dimensional (Kamiński and Muller, 1979) or two-dimensional ray-tracing modelling (Gorman, 2002) in order to fit the observed traveltimes of the refracted and reflected P wave mantle phases and derive upper mantle P wave velocity models beneath the areas indicated in Fig. 1b. We use spherical earth conversion in 1-D modelling and spherical earth 2-D ray-tracing in case of 2-D modelling. We picked first arrivals (P_{220} and P_{440}) and reflected waves from discontinuities at the depths of 220, 335 and 440 km ($P_{220}P$, $P_{335}P$ and $P_{440}P$ respectively). Finally, models were verified by calculating synthetic seismograms using the reflectivity method (Fuchs and Muller, 1971; Muller, 1985).

4 Results

We begin our modelling with the analysis of the MP-1 model derived for the seismically active part of the United States (Perchuc et al., 2008), west of the North American Craton. The North American Craton was created in the same time as the EEC (Hoffman, 1988) and is characterized by similar features, which justifies such an analysis. For all chosen regions in this analysis we try to estimate a 1-D average model which will characterize the main seismic discontinuities. It came out during the modelling that a single 1-D model (MP1-SUW model) explains well the observed traveltimes for the four groups of events analysed. Differences resulting from the different azimuth range of earthquakes are close to the assumed picking uncertainty (0.1 s).

Figure 4a shows the obtained velocity distribution beneath the western rim of the EEC (MP1-SUW model, Table 3) resulting from the forward ray-tracing modelling and fitting of the calculated traveltimes to the observed mantle arrivals (Figs. 4–7). Because of the far-regional offset range used in this study, the velocities in the crust and down to 8° discontinuity (ca. 100 km depth; Thybo and Perchuc, 1997) and within the asthenosphere were introduced a priori and fixed during modelling. The crustal part of the models was based on the published deep seismic sounding results (e.g. Guterch et al., 1996; Grad et al., 2009). In case of the JMR group, the waves leaving the source propagated within the oceanic crust first and then went through the continental-type lithosphere. Therefore, 2-D modelling was necessary to account for the variability between the oceanic and continental lithosphere encountered along this transect (Fig. 7b). The continental part has a similar vertical distribution of velocity to 1-D models for other azimuths.

Figure 4 shows the seismic sections for the GTR area. Most of the events in this region occur down to 20 km depth and the epicentral distances for those events are from 1500 to 2300 km (Fig. 4b). There is a good fit between the data and the traveltimes refracted at the base of the mantle LVZ (red dashed line) at the depth of 220 km. It is observed in the first arrivals up to 2100 km. The $P_{335}P$ (blue line) occurs as secondary arrivals. At an epicentral distance of about 2100 km, there is an intersection of the traveltime branches of P_{220} and P_{440} (green dashed line) phases. After 2100 km, the offset branch of P_{440} (green dashed line) is observed in first arrivals.

Figure 4c shows the earthquakes from the focal depth range of 20–50 km. These events were recorded at epicentral distances ranging from 1500 to 2300 km. As in the previous group, the P_{220} phase is observed in first arrivals at distances up to about 2100 km. After 2100 km the P_{440} phase arrives first. The $P_{335}P$ wave is also visible as a second group. $P_{440}P$ is also recorded in the form of high-amplitude signals observed at distances from 1950 to 2300 km. The closer the distance of 2250 km, the more difficult it is to separate $P_{335}P$

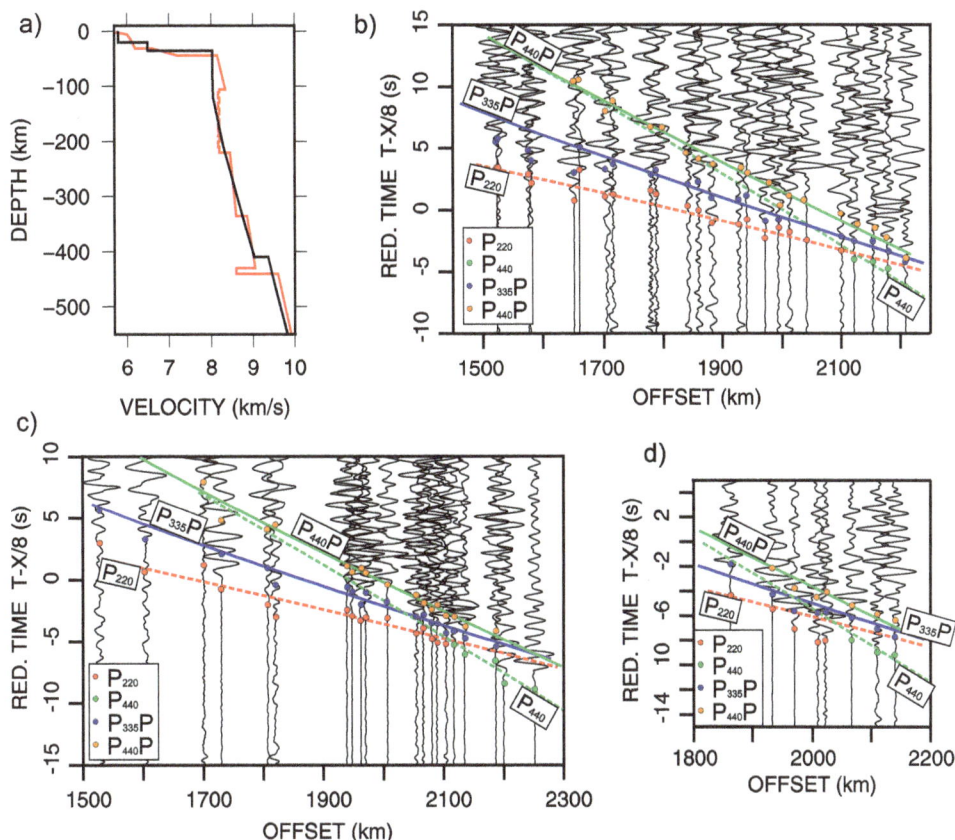

Figure 4. (a) The new MP1-SUW model (red line) compared with the AK 135 model (black line). **(b–d)** Seismic sections with traveltimes calculated for the MP1-SUW model for three focal depth ranges **(b)** 0–20 km, **(c)** 20–50 km, and **(d)** >50 km. Red dots represent P_{200}, green – P_{440}, blue – $P_{335}P$, and orange – $P_{440}P$ phases.

and $P_{440}P$ because these two branches intersect each other at this point.

Figure 4d shows the deepest group of earthquakes (focal depth >50 km) recorded at the epicentral distances of 1860 to 2140 km. P_{220} can be seen only on two seismograms. Although, its presence at this depth is documented in the previous two sections. On this section the $P_{335}P$ and $P_{440}P$ reflected phases are also observed.

Figure 5 shows the seismic sections for the CR. Records from these events start at about 2000 and end at about 3000 km epicentral distance. Here we also divided data into three focal depth ranges: 0–20 km depth (Fig. 5a), 20–50 km depth (Fig. 5b) and >50 km (Fig. 5c). In this case, we can only observe first arrivals of P_{220} in the first two sections (red dashed line). However, analysis of the wavefield allows us to follow phases refracted from 440 km and reflected from 335 and 440 km. On each section we have records documenting intersection of the two branches: $P_{335}P$ and $P_{440}P$ at 2250 km, after which we observed $P_{440}P$ and then $P_{335}P$ as secondary arrivals.

In the case of the WMSR group of events, there is almost the whole range of the useful epicentral distances. The shallow focal depth range, 0–20 km, allows us to put all events

together in one seismic section (Fig. 6). Here we can observe P_{220} as first arrivals at distances from 1700 to 2000 km. At these distances we observe high-amplitude $P_{335}P$ and $P_{440}P$ as secondary arrivals. There is a lack of seismograms in the area where traveltime branches of P_{220} and P_{440} intersect, but after 2200 km we observe P_{440} as first arrivals. Next we interpreted $P_{440}P$ phase followed by the $P_{335}P$ phase.

Figure 7a presents the seismic section for the JMR region. All those earthquakes are located along the rift zone close to the Jan Mayen Island in the northern Atlantic. Their focal depths are from 0 to 10 km. The recorded epicentral distances ranging from 2260 to 2720 km allow us to model 335 and 440 km discontinuities. Taking into account differences in the oceanic and continental lithosphere, we build a 2-D velocity model. We observe P_{440} as first arrivals and $P_{440}P$ as secondary arrivals. The $P_{335}P$ can be also distinguished in this seismic section. Figure 7b illustrates ray-paths for reflection at 335 km depth and refraction and reflection at 440 km discontinuity.

Figure 8 presents all 249 earthquakes with focal-depth traveltime corrections applied in one seismic section. We added values consequent from different focal depths for each group respectively. These values were calculated during the

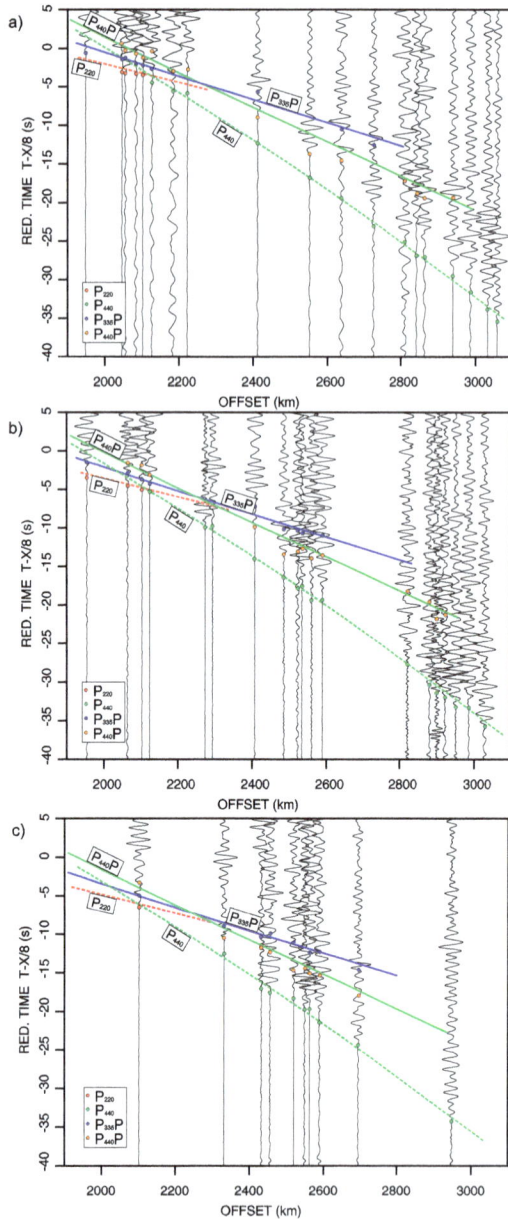

Figure 5. Seismic sections based on the data from the CR group with traveltimes calculated for the MP1-SUW model for three focal depth ranges (**a**) 0–20 km depth, (**b**) 20–50 km depth, and (**c**) > 50 km. Red dots represent P_{200}, green – P_{440}, blue – $P_{335}P$, and orange – $P_{440}P$ phases.

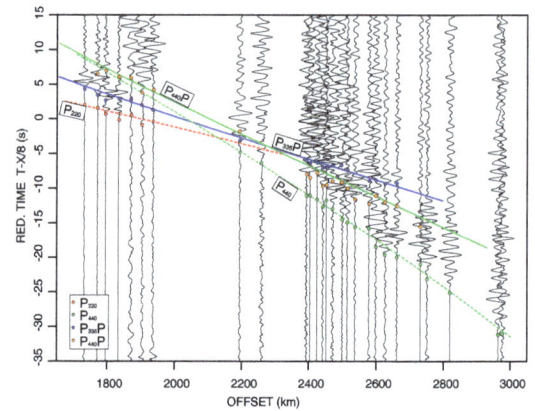

Figure 6. Seismic section based on data from the WMSR region with traveltimes calculated for the MP1-SUW model. Red dots represent P_{200}, green – P_{440}, blue – $P_{335}P$, and orange – $P_{440}P$ phases.

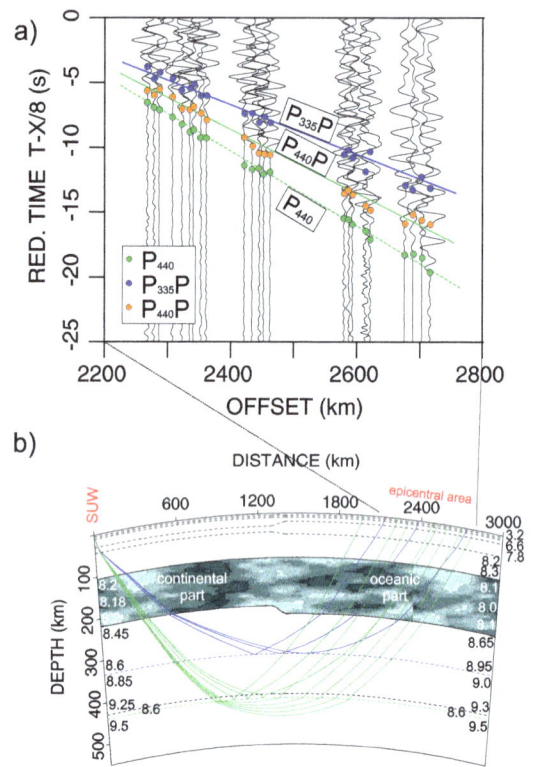

Figure 7. (**a**) Seismic section for the data from the JMR region with traveltimes calculated in the 2-D model shown in (**b**). (**b**) 2-D ray-tracing model and the ray paths refracted and reflected at 335 and 440 km discontinuities. Green dots represent P_{440}, orange – $P_{440}P$ phases, and blue – $P_{335}P$. Green pattern in the LVZ is used to represent heterogeneous nature of this layer modelled as thin low–high-velocity layers.

separation of earthquakes into smaller groups. In order to verify the credibility of the derived models, we calculated synthetic seismograms using the reflectivity method (Fuchs and Muller, 1971; Muller, 1985). There is a good amplitude match between the recorded data (Fig. 8) and the synthetic section (Fig. 9).

There is an interesting feature in the MP1-SUW model presented here (Fig. 4a). Atop the 440 km discontinuity there is a 10 km thick zone with reduced velocities. The insertion

of the LVZ is necessary to explain the separation between P_{440} and $P_{440}P$ branches as shown in Fig. 10. Figure 10a contains part of a section from Fig. 8 with the traveltimes

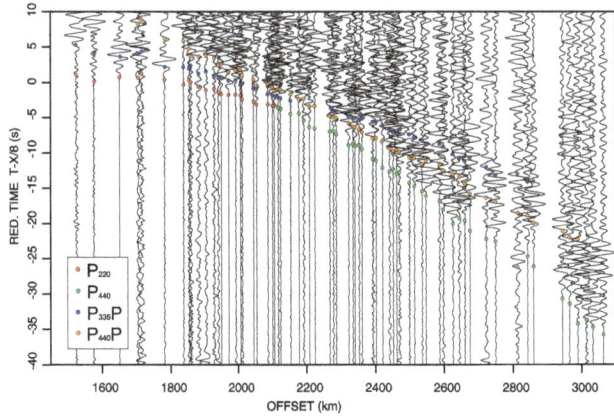

Figure 8. All analysed seismograms recorded at the SUW station with the focal depth corrections applied. Red dots represent P_{200}, green – P_{440}, blue – $P_{335}P$, and orange – $P_{440}P$ phases.

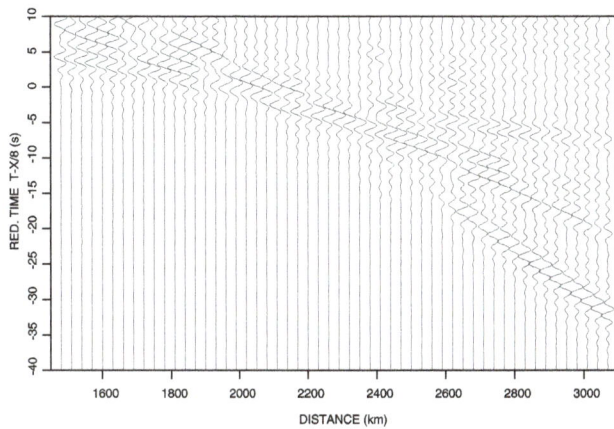

Figure 9. Synthetic seismograms section calculated for the MP1-SUW model using the reflectivity method (0.1–2.0 Hz). Compare with data in Fig. 8.

Table 3. The MP1-SUW P wave velocity model.

Depth (km)	P wave	Depth (km) velocity (km s^{-1})	P wave velocity (km s^{-1})
0	5.60	160	8.20
5	6.00	170	8.20
10	6.05	170	8.18
15	6.10	180	8.18
30	6.20	180	8.20
30	6.60	190	8.18
44	7.20	190	8.20
44	8.15	200	8.20
105	8.35	200	8.18
105	8.20	210	8.18
110	8.20	210	8.20
110	8.18	220	8.20
120	8.18	220	8.45
120	8.20	335	8.60
130	8.20	335	8.85
130	8.18	430	9.05
140	8.18	430	8.60
140	8.20	440	8.60
150	8.20	440	9.60
150	8.18	620	10.10
160	8.18		

model (Fig. 11a–e) and for the reference AK135 model (Kennett et al., 1995) (Fig. 11a'–e').

Figure 11a and 11a' show histograms for all earthquakes used. The adjustment for our model is better than for the AK135. There are smaller errors for the −0.5–0.5 s range. The best fit was obtained for the Jan Mayen group of events (Fig. 11d). The traveltime residual range for that region is ±1 s. We attribute such a good fit to small focal depth uncertainties of the earthquakes from that region. Events with small focal differences are more unique. It could also be a result of the two-dimensional modelling.

The MP1-SUW model predicts two new branches of traveltimes for reflected waves, the $P_{335}P$ and $P_{440}P$. Figure 12 presents how they fit in the records collected at the SUW station. In the case of other models, such as PREM (Preliminary Reference Earth Model; Dziewoński and Anderson, 1981), IASPEI91 (International Association of Seismology and Physics of the Earth's Interior; Kennett and Engdahl, 1991), AK135 (Kennett et al., 1995) or iPREF (Cammarano and Romanowicz, 2008) there is no possibility to make such analysis because they do not reflect regional-scale inhomogeneities.

In Table 4 we summarized the three parameters of the fitting quality: the standard deviation, variance, kurtosis and root mean square for all groups of data. Kurtosis is a measure of the deviation from the normal distribution (the kurtosis of the normal distribution is 3 for the applied formula). Distributions that have higher variation than the normal

calculated for the MP1-SUW model and Fig. 10b shows the same section with the traveltimes calculated for the model without the LVZ but with the same average velocity preserved.

5   Error analysis

We picked all the phases with the accuracy of ±0.1 s. The most significant uncertainties are attributed to the limited accuracy of the event location. The focal depth is the most uncertain hypocentre coordinate in the ISC bulletins, which were used in this analysis.

In order to demonstrate how well the predicted traveltimes fit to the observed data, we present histograms of traveltime residuals ($t_{obs} − t_{calc}$) in Fig. 11. We compare the traveltime residuals of the first arrivals calculated for the MP1-SUW

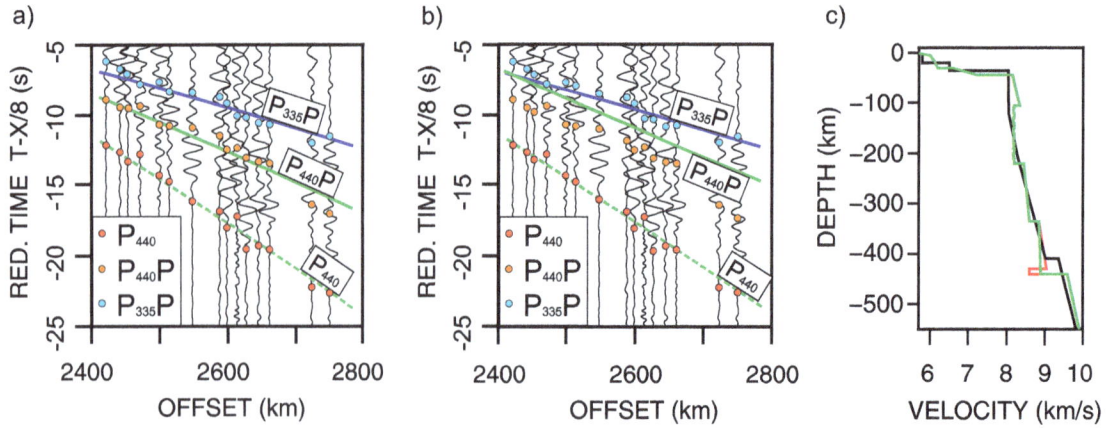

Figure 10. (**a**) Part of the seismic section with the MP1-SUW model traveltimes overlaid. Red dots represent P_{440}, orange – $P_{440}P$, and blue – $P_{335}P$. (**b**) Part of the seismic section with the traveltimes calculated for the model without the LVZ atop the 440 km discontinuity. (**c**) Comparison of 1-D seismic velocity models. Red line corresponds to the MP1-SUW model and green line to the same model without LVZ atop 440 km discontinuity. Black line corresponds to the AK135 model.

Table 4. Statistics of the traveltime residuals for the MP1-SUW (in bold) and the AK135 models (rms – root-mean square).

Region	Number of events	Standard deviation	Variation	Kurtosis	RMS
Greece and Turkey region	103	1.01	1.01	4.46	1.00
		1.99	3.96	3.62	2.25
Greece and Turkey 0–20	53	**0.85**	**0.73**	**3.32**	**0.88**
		1.80	3.25	3.74	3.03
Greece and Turkey 20–50	36	**1.10**	**1.22**	**6.18**	**1.10**
		1.40	1.96	3.76	1.43
Greece and Turkey 50+	14	**1.18**	**1.40**	**2.69**	**1.20**
		1.29	1.65	2.76	1.44
Caucasus region	69	1.50	2.26	3.45	1.46
		1.61	2.59	2.79	1.65
Caucasus 0–20	34	**1.67**	**2.78**	**2.33**	**1.64**
		1.86	3.44	2.21	1.87
Caucasus 20–50	23	**1.58**	**2.51**	**4.38**	**1.55**
		1.55	2.40	3.52	1.59
Caucasus 50+	12	**0.77**	**0.60**	**3.08**	**0.74**
		0.86	0.73	3.65	1.13
Western Mediterranean Sea region	43	1.20	1.44	3.57	1.19
		1.57	2.46	2.15	1.63
Jan Mayen region	34	0.32	0.10	2.22	0.36
		0.27	0.07	2.63	0.29
All events	249	1.08	1.31	3.72	1.07
		1.58	2.79	3.00	1.71
$P_{335}P$	83	**0.75**	**0.56**	**3.91**	**0.75**
$P_{440}P$	76	**0.75**	**0.57**	**4.00**	**0.80**

distribution have a kurtosis greater than 3, distributions with lower variation have a kurtosis less than 3.

The presented summary of the traveltime residuals' statistics shows the fitting parameters with respect to the MP1-SUW model and the AK135 respectively. We subdivided GTR and CR analysis into smaller groups according to their focal depths. The calculated residuals are also smaller for the JMR. The comparison of the data in Table 4 shows that the experimental data is described better by our model than the global reference AK135 model. Although the AK135 model describes first arrivals well, it does not take into account reflected waves and regional discontinuities such as the 300 km discontinuity.

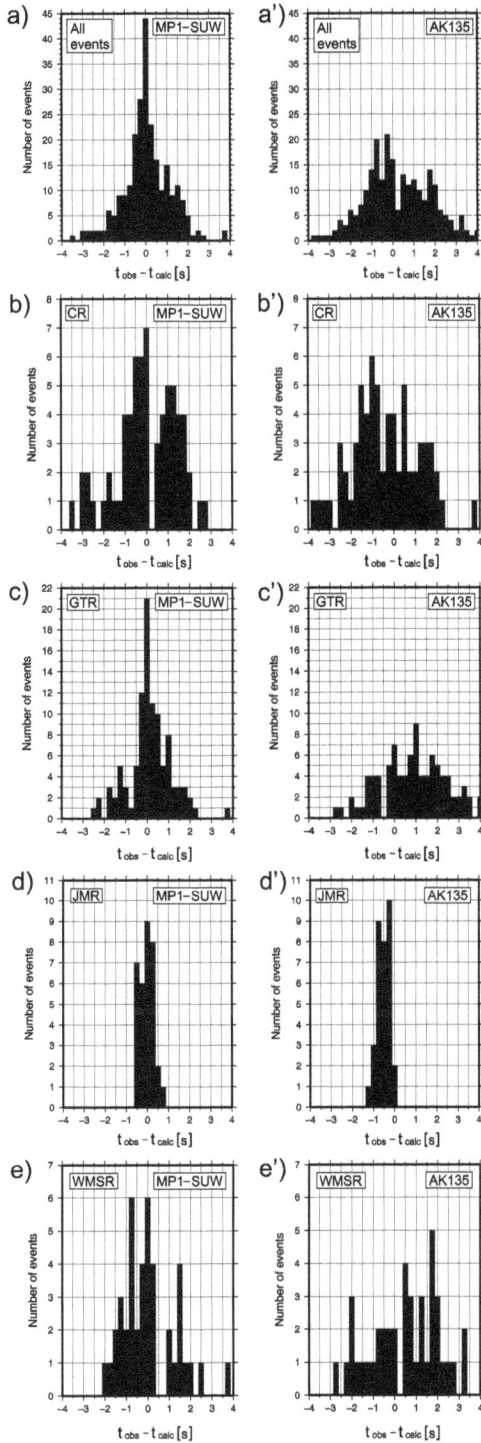

Figure 11. Histograms of the residuals between the observed traveltimes and the traveltimes calculated for the MP1-SUW model (left column) and between the observed traveltimes and the traveltimes calculated for the AK135 model (right column) for the following data: **(a)** **(a')** all events, **(b)** **(b')** events from the CR, **(c)** **(c')** events from the GTR , **(d)** **(d')** events from the JMR, **(e)** **(e')** events from the WMSR.

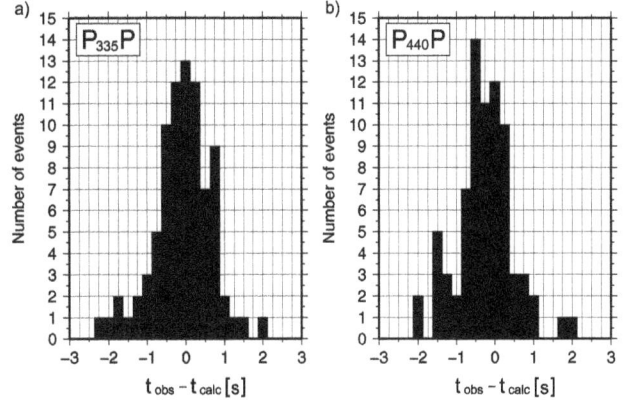

Figure 12. Histograms of the traveltime residuals between the reflected phases $P_{335}P$ (left column), the $P_{440}P$ ones (right column) and the calculated traveltimes for the MP1-SUW model.

Based on the analysis of both the refracted and reflected arrivals, we estimated uncertainty of the velocities in the MP1-SUW model for each layer as $\pm 0.025\,\mathrm{km\,s^{-1}}$. The uncertainty for the depth of each discontinuity is $\pm 10\,\mathrm{km}$.

6 Discussion

The MP1-SUW model is intended to characterize upper mantle structure below the western rim of the EEC. Both 1-D and 2-D models provide a consistent image. The Lehmann 220 km phase, defining the bottom of the low-velocity zone (Lehmann, 1961), is seen as first arrivals for the distance range of 1500–2150 km. A very prominent feature of the MP1-SUW model is the existence of the 300 km discontinuity at the depth of 335 km, which is observed at the 1500–2800 km offset range. The origin of the 300 km discontinuity is generally related to the existence of a subduction zone (Revenaugh and Jordan, 1991), where increased amount of free silica is delivered to the upper mantle by subducted oceanic and continental material. Both coesite–stishovite phase transformation and the thermal anomaly caused by a subducting slab can explain the velocity jump at this discontinuity (Liu et al., 1999). With relation to the mechanical mixture model of Xu et al. (2008), the jump in P wave velocity at approximately 335 km depth would require enrichment in the basaltic component. In this region, the nature of this boundary can be tentatively explained as the traces of the ancient subduction regime related to the closure of the Iapetus and Rheic oceans and the Tornquist Sea (e.g. Torsvik et al., 1996, Torsvik and Rehnström, 2003). This discontinuity, due to its regional scale, is recorded better for the JMR and CR groups of events, although we observe it also for the WMSR and GTR groups. An independent support for the existence of such a discontinuity was recently provided by Knapmeyer-Endrun et al. (2013), who studied the P wave receiver function based on the data from the PASSEQ experiment. There is a clear peak observed in the receiver function section clustered around TTZ at ca. 30 s relative time (see Fig. 7 in Knapmeyer-Endrun et al., 2013), which is roughly equivalent to ca. 300 km depth. However, authors do not comment whether this signal is related to primary or multiple energy.

We modelled 410 km discontinuity at the depth of 440 km. It is observed as first arrivals in the 2150–3100 km offset range. The deeper location of this discontinuity compared to the reference global model (e.g. AK135) relates to the observation of the LVZ atop it. We find reports of such a zone in regional-scale investigations (Thybo et al., 1997). It was also postulated as the global phenomenon (Bercovici and Karato, 2003). In this case it can also be explained as an ancient oceanic structure subducted during the closure of the Tornquist Sea (Torsvik and Rehnström, 2003). We interpret the low-velocity zone above 410 km discontinuity as a result of the dehydration of the subducted plate that brought some residual water in the transition zone. This feature was found and documented also in the western part of the United States (Song et al., 2004), where authors interpreted it as a compositional anomaly. It could be caused by a dense partially melted layer linked to prior subduction of the Farallon Plate and back-arc extension.

Comparing MP1-SUW model with regional tomographic models based on *P* wave traveltimes like the one of Koulakov et al. (2009), we observe similar pattern of the velocity anomalies in the regions analysed in this paper (Fig. 1b), both in horizontal and vertical slices. However the comparison is not straightforward as the tomographic models are inherently smooth and continuous whereas our model contain discontinuities modelled using reflected waves.

7 Conclusions

We derived a new *P* wave velocity model of the upper mantle structure characterizing the western rim of the East European Craton (model MP1-SUW). Analysing seismic record sections from events recorded at the SUW station, we calculated a one-dimensional *P* wave velocity model for the azimuthally differentiated regions: the Western Mediterranean Sea region, the Greece and Turkey region and the Caucasus region. The two-dimensional model for the JMR region was justified by the fact that waves propagating from that source region travel through the oceanic and continental structures.

The MP1-SUW model documents the bottom of the asthenospheric low-velocity zone at the depth of 220 km, 335 km discontinuity and the zone with the reduction of *P* wave velocity atop 410 km discontinuity which is depressed to 440 km depth. The nature of both the 335 and 440 km discontinuities are explained by tracing the ancient subduction regime related to the closure of the Iapetus and Rheic oceans and the Tornquist Sea (Torsvik and Rehnström, 2003). The 335 km discontinuity is a robust feature of the MP1-SUW

model, however we are aware that this feature is not ubiquitous, but linked to the marginal zone of the EEC.

The work presented here shows that even a single station can be a rich source of information when a careful phase identification and modelling are implemented. We hope that the seismological community can benefit from the use of the MP1-SUW model in other regional studies, for example, in receiver function calculation or traveltime tomography.

Acknowledgements. This study has been funded by the National Science Centre. Grant no. DEC-2011/01/N/ST10/07275. We are very grateful to Piotr Środa for fruitful discussions. Some of the figures were prepared using Generic Mapping Tool software (Wessel and Smith, 1995). We thank two anonymous reviewers and the editor for comments that helped us improve our paper.

Edited by: J. Plomerova

References

Abramovitz, T., Thybo, H., and MONA LISA Working Group: Seismic structure across the Caledonian deformation front along MONA LISA profile 1 in the southeastern North Sea, Tectonophys, 288, 153–176, 1998.

Abramovitz, T., Landes, M., Thybo, H., Jacob, A. W. B., and Prodehl, C.: Crustal velocity structure across the Tornquist and Iapetus Suture Zones – a comparison based on MONA LISA and VARNET data, Tectonophys, 314, 69–82, 1999.

Abramovitz, T., Thybo, H., and Perchuc, E.: Tomographic inversion of seismic *P* and *S* wave velocities from the Baltic Shield based on FENNOLORA data, Tectonophys, 358, 151–174, 2002.

BABEL Working Group: Evidence for early Proterozoic plate tectonics from seismic reflection profiles in the Baltic Shield, Nature, 348, 34–38, 1991.

BABEL Working Group: Deep seismic reflection/refraction interpretation of crustal structure along BABEL profiles A and B in the southern Baltic Sea, Geophys. J. Int., 112, 325–343, 1993.

Bastow, I. D.: Relative arrival-time upper-mantle tomography and the elusive background mean, Geophys. J. Int., 190, 1271–1278, 2012.

Bercovici, D. and Karato, S.: Whole-mantle convection and the transition-zone water filter, Nature, 425, 39–44, 2003.

Bock, G., Perchuc, E., Hanka, W., Wiejacz, P., Kind, R., Suchcicki, J., and Wylengalla, K.: Seismic anisotropy beneath the Suwalki station: one year of the activity of the station, Acta Geophys. Pol., 45, 1–12, 1997.

Bogdanova, S. V., Gorbatschev, R., and Stephenson, R. A.: EUROBRIDGE: paleoproterozoic accretion of Fennoscandia and Sarmatia, Tectonophys, 339, vii-x, 2001.

Bogdanova, S. V., Gorbatschev, R., and Garetsky, R. G.: The East European craton, in: Encyclopedia of Geology, edited by: Selley, R. C., Cocks, L. R., and Plimer, I. R., Elsevier, 2, 34–49, 2005.

Cammarano, F. and Romanowicz, B.: Radial profiles of seismic attenuation in the upper mantle based on physical models, Geophys. J. Int., 175, 116–134, 2008.

Chu, R., Schmandt, B. and Helmberger D. V.: Upper mantle P velocity structure beneath the Midwestern United States derived from triplicated waveforms, Geochem. Geophys. Geosyst., 13, 1–21, 2012.

Cocks, L. R. M. and Torsvik, T. H.: European geography in a global context from the Vendian to the end of the Palaeozoic, in: European Lithosphere Dynamics, edited by: Gee, D. G. and Stephenson, R. A., Geological Society, London, Memoirs, 32, 83–95, 2006.

Dziewoński, A. and Anderson, D.: Preliminary reference Earth model, Phys. Earth Planet. In., 25, 297–356, 1981.

EUGENO-S Working Group: Crustal structure and tectonic evolution of the transition between the Baltic Shield and the North German Caledonides (the EUGENO-S Project), Tectonophys, 150, 253–348, 1988.

EUROBRIDGE Seismic Working Group: Seismic velocity structure across the Fennoscandia-Sarmatia suture of the East European craton beneath the EUROBRIDGE profile through Lithuania and Belarus, Tectonophys, 314, 193–217, 1999.

Fuchs, K. and Müller, G.: Computation of synthetic seismograms with the reflectivity method and comparison with observations, Geophys. J. Roy. Astr. S., 23, 417–433, 1971.

Gee, D. G.: Tectonic model for central part of scandinavian caledonides, Am. J. of Sci., A275, 468–515, 1975.

Gorman, A. R.: Ray-theoretical seismic traveltime inversion – modifications for a two-dimensional radially parametrized Earth, Geophys. J. Int., 151, 511–516, 2002.

Grad, M., Keller, G. R., Thybo, H., Guterch, A., and POLONAISE Working Group: Lower lithospheric structure beneath the Trans-European Suture Zone from POLONAISE'97 seismic profiles, Tectonophys, 360, 153–168, 2002.

Grad, M., Tiira, T., and ESC Working Group: The Moho depth map of the European Plate, Geophys. J. Int., 176, 279–292, 2009.

Gregersen, S., Voss, P., Nielsen, L. V., Achauer, U., Busche, H., Rabbel, W., and Shomali, Z. H.: Uniqueness of modeling results from teleseismic P wave tomography in Project Tor, Tectonophys, 481, 99–107, 2010.

Guggisberg, B. and Berthelsen, A.: A two-dimensional velocity model for the lithosphere beneath the Baltic Scield and its possible tectonic significance, Terra Cognita, 7, 631–638, 1987.

Guterch, A., Grad, M., Materzok, R., Perchuc, E., and Toporkiewicz, S.: Results of seismic crustal studies in Poland, Publ. Inst. Geophys. Pol. Acad. Sci., 17, 84–89, 1986.

Guterch, A., Grad, M., Materzok, R., and Perchuc, E.: Deep structure of the Earth's crust in the contact zone of the Paleozoic and Precambrian Platforms in Poland (Tornquist-Teisseyre zone), Tectonophys, 128, 251–279, 1996.

Guterch, A., Grad, M., Thybo, H., Keller, G. R., and POLONAISE Working Group: POLONAISE'97 – international seismic experiment between Precambrian and Variscan Europe in Poland, Tectonophys, 314, 101–121, 1999.

Guterch, A., Grad, M., and Keller, G. R.: Seismologists celebrate the new Millennium with an experiment in Central Europe, EOS Trans. AGU, 82, 529, 534–535, 2001.

Guterch, A., Wybraniec, S., Grad, M., Chadwick, R. A., Krawczyk, C. M., Ziegler, P. A., Thybo, H., and Vos, W. D.: Crustal structure and structural framework, in: Petroleum Geological Atlas of the Southern Permian Basin Area, edited by: Doornenbal, J. C. and Stevenson, A. G., EAGE Publications B. V., 11–23, 2010.

Hoffman, P. F.: United Plates of America, the birth of a craton: early proterozoic assembly and growth of Laurentia, Annu. Rev. Earth Pl. Sc., 16, 543–603, 1988.

ISC Bulletin, available at: http://www.isc.ac.uk/iscbulletin (last access: May 2013), 2013.

Kamiński, W. and Müller, G.: Program LAUFZEIT, University of Karlsruhe, 1979.

Kennett, B. L. N. and Engdahl, E. R.: Travel times for global earthquake location and phase identification, Geophys. J. Int., 105, 429–465, 1991.

Kennett, B. L. N., Engdahl, E. R., and Buland, R.: Constraints on seismic velocities in the earth from travel times, Geophys. J. Int, 122, 108–124, 1995.

Knapmeyer-Endrun, B., Krüger, F., Legendre, C., Geissler, W., and PASSEQ Working Group: Tracing the influence of the Trans-European Suture Zone into the mantle transition zone, Earth Planet. Sc. Lett., Elsevier, 363, 73–87, 2013.

Koulakov, I., Kaban, M. K., Tesauro M., and Cloetingh S.: P- and S-velocity anomalies in the upper mantle beneath Europe from tomographic inversion of ISC data, Geophys. J. Int., 179, 345–366, 2009.

Liu, J., Zhang, J., Flesch, L., Li, B., Weidner, D., and Liebermann, R.: Thermal equation of state of stishovite, Phys. Earth Planet. In., 112, 257–266, 1999.

Lehmann, I.: S and the Structure of the Upper Mantle, Geophys. J. Roy. Astr. S., 4, 124–138, 1961.

Levander, A.: USArray design implications for wavefield imaging in the lithosphere and upper mantle, The Leading Edge, 22, 250–255, 2003.

Müller, G.: The reflectivity method: a tutorial, J. Geophys., 58, 153–174, 1985.

Nita, B., Dobrzhinetskaya, L., Maguire, P., and Perchuc, E.: Age-differentiated subduction regime: An explanation of regional scale upper mantle differences beneath the Alps and the Variscides of Central Europe, Phys. Earth Planet. In., 206–207, 1–1, 2012.

Perchuc, E. and Thybo, H.: A new model of upper mantle P waves velocity below the Baltic Shield, Indication of partial melt in the 95 to 160 km depth range, Tectonophys, 253, 227–245, 1996.

Perchuc, E., Malinowski, M., and Nita, B.: Seismic and petrological properties of the upper mantle between 300 and 400 km depth, AGU Fall Meeting, S14C-01, 2008.

Plomerová, J., Babuška, V., Vecsey, L., and Kouba, D.: Seismic anisotropy of the lithosphere around the Trans-European Suture Zone (TESZ) based on teleseismic body-wave data of the TOR experiment, Tectonophys, 360, 89–114, 2002.

Revenaugh, J. and Jordan, T.: Mantle layering from ScS reverberations: the upper mantle, J. Geophys. Res., 96, 19781–19810, 1991.

Shapiro, N. M. and Ritzwoller, M. H.: Monte-Carlo inversion for a global shear velocity model of the crust and upper mantle, Geophys. J. Int., 151, 1–18, 2002.

Song, T. A., Helmberger, D. V., and Grand, S. P.: Low-velocity zone atop the 410 km seismic discontinuity, Nature, 427, 530–533, 2004.

Somali, Z. H., Roberts, R. G., Pedersen, L. B., and the TOR Working Group: Lithospheric structure of the Tornquist Zone resolved by nonlinear P and S teleseismic tomography along the TOR array, Tectonophys, 416, 133–149, 2006.

Świeczak, M., Grad, M., and TOR and SVEKALAPKO Working Groups: upper mantle seismic discontinuities: topography variations beneath Eastern Europe, Acta Geophys. Pol., 52, 251–270, 2004.

Thybo, H. and Perchuc, E.: The seismic 8° discontinuity and partial melting in continental mantle, Science, 275, 1626–1629, 1997.

Thybo, H., Perchuc, E., and Pavlenkova, N.: Two Reflectors in the 400 km Depth Range Revealed from Peaceful Nuclear Explosion Seismic Sections, Upper Mantle Heterogeneities from Active and Passive Seismology, edited by: Fuchs, K., NATO ASI Series, 97–104, 1997.

Torsvik, T. H., Smethurst, M. A., Meert, J. G., Van der Voo, R., McKerrow, W. S., Brasier, M. D., Sturt, B. A., and Walderhaug. H. J.: Continental break-up and collision in the Neoproterozoic and Palaeozoic: a tale of Baltica and Laurentia, Earth-Sci. Rev., 40, 229–258, 1996.

Torsvik, T. H. and Rehnström, E. F.: The Tornquist Sea and Baltica-Avalonia docking, Tectonophys, 362, 67–82, 2003.

Wessel, P. and Smith, W. H. F.: New version of Generic Mapping Tools released, EOS, 76, 453, 1995.

Wilde-Piórko, M.: Crustal structure from seismic receiver function, Przegląd Geofizyczny, 1–2, 31–45, 2005.

Wilde-Piórko, M., Geissler, W., Plomerová, J., Grad, M., Babuška, V., Brückl, E., Cyziene, J., Czuba, W., England, R., Gaczyński, E., Gazdova, R., Gregersen, S., Guterch, A., Hanka, W., Hegedűs, E., Heuer, B., Jedlička, P., Lazauskiene, J., Keller, G., Kind, R., Klinge, K., Kolinsky, P., Komminaho, K., Kozlovskaya, E., Krüger, F., Larsen, T., Majdański, M., Malek, J., Motuza, G., Novotný, O., Pietrasiak, R., Plenefisch, T., Růžek, B., Sliaupa, S., Środa, P., Świeczak, M., Tiira, T., Voss, P., and Wiejacz, P.: PASSEQ 2006–2008: Passive seismic experiment in Trans-European Suture Zone, Studia Geophysica et Geodetica, 52, 439–448, 2008.

Xu, W., Lithgow-Bertelloni, C., Stixrude, L., and Ritsema, J.: The effect of bulk composition and temperature on mantle seismic structure, Earth Planet. Sci. Lett., 275, 70–79, 2008.

Zelt, C.: ZPLOT – an iteractive plotting and picking program for seismic data, Bullard Lab. Univ. of Cambridge, Cambridge UK, 1994.

Ziegler, P. A. and Dèzes, P.: Crustal configuration of Western and Central Europe, in: European Lithosphere Dynamics, edited by: Gee, D. G. and Stephenson, R. A., Geol. Soc., London, Memoirs 32, 43–56, 2006.

Znosko, J.: Polish Caledonides and their relation to other European Caledonides, Ann. Soc. Geol. Pol., 56, 33–52, 1986.

11

Testing the effects of basic numerical implementations of water migration on models of subduction dynamics

M. E. T. Quinquis[1] **and S. J. H. Buiter**[1,2]

[1]Geodynamics Team, Geological Survey of Norway (NGU), Trondheim, Norway
[2]Centre for Earth Evolution and Dynamics, University of Oslo, Oslo, Norway

Correspondence to: M. E. T. Quinquis (matthieu@quinquis.net)

Abstract. Subduction of oceanic lithosphere brings water into the Earth's upper mantle. Previous numerical studies have shown how slab dehydration and mantle hydration can impact the dynamics of a subduction system by allowing a more vigorous mantle flow and promoting localisation of deformation in the lithosphere and mantle. The depths at which dehydration reactions occur in the hydrated portions of the slab are well constrained in these models by thermodynamic calculations. However, computational models use different numerical schemes to simulate the migration of free water. We aim to show the influence of the numerical scheme of free water migration on the dynamics of the upper mantle and more specifically the mantle wedge. We investigate the following three simple migration schemes with a finite-element model: (1) element-wise vertical migration of free water, occurring independent of the flow of the solid phase; (2) an imposed vertical free water velocity; and (3) a Darcy velocity, where the free water velocity is a function of the pressure gradient caused by the difference in density between water and the surrounding rocks. In addition, the flow of the solid material field also moves the free water in the imposed vertical velocity and Darcy schemes. We first test the influence of the water migration scheme using a simple model that simulates the sinking of a cold, hydrated cylinder into a dry, warm mantle. We find that the free water migration scheme has only a limited impact on the water distribution after 1 Myr in these models. We next investigate slab dehydration and mantle hydration with a thermomechanical subduction model that includes brittle behaviour and viscous water-dependent creep flow laws. Our models demonstrate that the bound water distribution is not greatly influenced by the water migration scheme whereas the free water distribution is.

We find that a bound water-dependent creep flow law results in a broader area of hydration in the mantle wedge, which feeds back to the dynamics of the system by the associated weakening. This finding underlines the importance of using dynamic time evolution models to investigate the effects of (de)hydration. We also show that hydrated material can be transported down to the base of the upper mantle at 670 km. Although (de)hydration processes influence subduction dynamics, we find that the exact numerical implementation of free water migration is not important in the basic schemes we investigated. A simple implementation of water migration could be sufficient for a first-order impression of the effects of water for studies that focus on large-scale features of subduction dynamics.

1 Introduction

Dehydration of subducting lithosphere and the related hydration of the mantle wedge can influence the dynamics of subduction, as water has a weakening effect on viscous and brittle rheologies (e.g. Sibson et al., 1975; Peacock, 1987; Hirschmann, 2006; Connolly, 2005; Gerya et al., 2008). The amount of fluids carried by subducting oceanic lithosphere is debated, but it is thought that water content can reach up to 3 wt% at the surface, decreasing with depth (Rüpke et al., 2004). Water is acquired through near-surface hydration, which is aided by flexure-related extensional fractures, and by hydrothermal activity with circulation of hot water and vapour in the upper section of the oceanic crust (Staudigel, 2003; Rüpke et al., 2004; Faccenda et al., 2008).

The subducting crust carries water in two phases: (1) free fluids that are contained in the porosity of the rock and can percolate along grain boundaries (Stern, 2002; Bercovici and Karato, 2003; Wark et al., 2003; Rüpke et al., 2004; Cheadle et al., 2004) and (2) mineralogically bound fluids in the form of hydroxyl complexes (OH) (Schmidt and Poli, 1998; Hirschmann, 2006). Once a slab starts to subduct, it undergoes dehydration processes due to the increase in pressure and temperature. Most of the water contained in the porosity of the sediments is expelled near the trench through compaction and is not transported into the mantle. This is known as fore arc volatile discharge (Stern, 2002; Rüpke et al., 2004). Hydrated minerals include, among others, amphiboles, chlorite and serpentine. It has been well documented that mineralogically bound water is released when hydrated minerals undergo certain phase transitions (Schmidt and Poli, 1998; Iwamori, 1998; Kerrick and Connolly, 2001; Hacker et al., 2003; Ohtani et al., 2004; Rüpke et al., 2004; Syracus and Abers, 2006; Hirschmann, 2006). At the same time, experimentally determined phase diagrams suggest that mineralogically bound water can be transported to the base of the upper mantle, or perhaps even greater depths (Iwamori, 1998; Schmidt and Poli, 1998; Stern, 2002; Ohtani et al., 2004).

Dehydration processes can influence subduction in multiple ways. For example, the depth at which major dehydration occurs determines the location of volcanic arcs, which are located ca. 110–120 km above the surface of subducted slabs (England et al., 2004; Syracus and Abers, 2006). It is the melting of mantle wedge materials that is thought to lead to arc volcanism. But water released from the subducting slab decreases the pressure and temperature at which melting occurs, thus enhancing mantle wedge melting and causing volcanism. Mantle material that is hydrated by water released from the slab may also form weak, positively buoyant "wet plumes" that rise upwards and efficiently hydrate the mantle wedge (Billen and Gurnis, 2001; Billen, 2008; Gorczyk et al., 2007; Richard and Iwamori, 2010). These fluids can then cause a more vigorous flow in the mantle wedge. Arcay et al. (2005) showed that mantle wedge hydration can result in thermal erosion and softening of the overriding lithosphere.

Once subduction has started, (de)hydration processes may further influence the evolution of subduction by enforcing an asymmetrical geometry of subduction zones, causing subduction to be one-sided (Gerya et al., 2008). Dehydration processes can in addition aid the exhumation of high- and ultrahigh-pressure metamorphic rocks by creating a wide and weak subduction channel through which rocks are exhumed (Gerya et al., 2002). Dehydration of the subducting slab may in turn increase slab strength, but this effect may be overwhelmed by the strong impact of water on the mantle wedge.

The models mentioned above use similar methods to determine the conditions of pressure and temperature at which dehydration processes occur. These are usually based on thermodynamic calculations (de Capitani and Brown, 1987; Holland and Powell, 1998; Powell et al., 1998; Connolly, 2005) or high-pressure experiments (Schmidt and Poli, 1998; Ohtani et al., 2004; Komabayashi et al., 2005; Iwamori, 2007), and the locations of the dehydration fronts during subduction do not greatly vary between models.

Water migration can be described by two-phase flow conservation equations (Spiegelman , 1993a, b; Bercovici et al., 2001). However, these are not routinely used in numerical simulations of subduction zone dynamics at the scale of the upper mantle as it adds a fairly complex set of equations to an already highly non-linear model. Usually simplifications are therefore made to migrate water in large-scale subduction models.

Bound water is advected along the solid flow field. Bound water can in addition be affected by diffusion mechanisms which allow water to migrate through the solid-phase field as a function of the chemical and temperature gradients (Richard et al., 2006). Because we focus on first-order behaviour of free water migration on subduction dynamics, we keep our models relatively simple and neglect the effect of bound water diffusion. Free water can migrate in the interconnected pore space of the solid phase (Stern, 2002; Wark et al., 2003; Cheadle et al., 2004; Rüpke et al., 2004), create its own hydrated channels (Katz et al., 2006), or be absorbed by non-saturated rocks of the mantle wedge (Iwamori, 1998) to be potentially transported with the mantle flow into the lower mantle (Bercovici and Karato, 2003; Iwamori, 2007; Richard and Bercovici, 2009; Fujita and Ogawa, 2013). Free water is also advected by the solid flow field through which it migrates, and this can result in cases where part of the free water migrates up through the mantle wedge while the rest is carried with the solid flow and subducted into the mantle (Cagnioncle et al., 2007).

Numerical studies of hydration of the mantle by slab dehydration have used different simplified numerical approximations for the migration of free water in the mantle:

I Free water migrates vertically in the upper mantle and is not coupled to the solid-phase flow in the mantle wedge (Arcay et al., 2005).

II The migration of free water is vertical but coupled to the mantle flow. The effective migration path of water is therefore no longer purely vertical, but can include a horizontal component. This method has been implemented as an imposed vertical velocity added to the velocity of the solid-phase flow (Gorczyk et al., 2007) or as a dehydration front with an imposed horizontal and vertical velocity (Gerya et al., 2002).

III Free water migrates as a Darcy flow, following the density gradient between the solid phase and the fluid phase in the mantle wedge (Cagnioncle et al., 2007). Darcy flow changes the migration paths of fluids which are now no longer necessarily vertical (Cagnioncle et al.,

2007). Also here the solid flow phase may advect free water in addition to the Darcy mechanism.

Studies that use the above water migration schemes show differences in the spatial distribution of hydrated material in the mantle wedge as subduction evolves (Arcay et al., 2005; Gerya et al., 2002; Gorczyk et al., 2007; Cagnioncle et al., 2007). However, as the numerical setup of the subduction models also differs between these studies, it is difficult to evaluate the possible effects of the numerical implementation of water migration. So far, the influence of the basic numerical implementation of water migration on the dynamics of a subduction model has not been investigated.

We aim to investigate the effects of the three numerical water migration schemes described above (schemes I, II and III) on the dynamics of a subducting slab and its overlying mantle wedge. These models are kept simple, allowing us to focus on the first-order effects of dehydration and water migration. Therefore our models do not include melting, shear heating or adiabatic heating. We first illustrate the effects of dehydration and water migration for a simple model of a cold and hydrated cylinder sinking in a warm mantle. Our second series of models examines the effects of (de)hydration and water migration on a thermo-mechanical subduction model at the scale of the upper mantle.

2 Modelling approach

2.1 Thermo-mechanical equations

We solve the equations for conservation of mass (assuming incompressibility) (Eq. 1), momentum (Eq. 2) and energy (Eq. 3):

$$\nabla \cdot \boldsymbol{v} = 0, \tag{1}$$

$$-\nabla P + \nabla \cdot \bar{\sigma}' + \rho \boldsymbol{g} = 0, \tag{2}$$

$$\rho C_p \frac{\partial T}{\partial t} = k \nabla^2 T - \rho C_p \boldsymbol{v} \cdot \nabla T + H; \tag{3}$$

$\boldsymbol{v}$ is the velocity vector, ρ density, t time, P pressure (mean stress), $\bar{\sigma}'$ the deviatoric stress tensor, $\boldsymbol{g}$ gravitational acceleration ($g_x = 0$ and $g_y = -9.81\,\mathrm{m\,s^{-2}}$), C_p specific heat, T temperature, k thermal conductivity, and H radioactive heat production per unit volume. In the subduction models, the Boussinesq approximation is assumed, i.e. $\frac{\partial \rho}{\partial t} = 0$ but $\rho = \rho_0(1 - \alpha(T - T_0))$, where ρ_0 is the reference density at $T = T_0$ and α is the volumetric thermal expansion coefficient.

Materials are either viscous or brittle. Our viscous rheologies are linear or pressure- and temperature-dependent:

$$\eta_{\mathrm{df,ds}} = \frac{1}{2}\left(\frac{d^p}{A\,C_{\mathrm{OH}}^r}\right)^{\frac{1}{n}} \dot{\varepsilon}_e'^{\frac{1-n}{n}} e^{\left(\frac{Q+PV}{nRT}\right)}; \tag{4}$$

$\dot{\varepsilon}_e'$ is the effective deviatoric strain rate ($\dot{\varepsilon}_e' = (\frac{1}{2}\dot{\varepsilon}_{ij}'\dot{\varepsilon}_{ij}')^{\frac{1}{2}}$), A a material constant, n the power law stress exponent, d grain

size, p grain size exponent, C_{OH} water content in ppm, r the water content exponent, Q activation energy, V activation volume and R the molar gas constant. df and ds refer to deformation by diffusion creep ($p > 0$ and $n = 1$) and dislocation creep ($p = 0$ and $n > 1$) respectively. Diffusion and dislocation creep are assumed to act in parallel in all materials, resulting in a composite viscosity (η_{comp}) (Karato and Li, 1992; van den Berg et al., 1993):

$$\eta_{\mathrm{comp}} = \left(\frac{1}{\eta_{\mathrm{ds}}} + \frac{1}{\eta_{\mathrm{df}}}\right)^{-1}. \tag{5}$$

In our models, only bound water influences the viscosity and we only consider the impact of water on sub-crustal materials. A water content of $0.4\,\mathrm{wt\,\%}$ results in a viscosity decrease by ca. 2 orders of magnitude when using the dislocation or diffusion creep flow law for wet olivine from Hirth and Kohlstedt (2003). This can result in viscosities that are lower than the minimum viscosity of $10^{18}\,\mathrm{Pa\,s}$ which is imposed in our models. We therefore assume that viscosity no longer decreases further once water content exceeds $0.4\,\mathrm{wt\,\%}$.

Brittle behaviour in the subduction models follows a Drüker–Prager criterion (Handin, 1969; Jaeger and Cook, 1976; Twiss and Moores, 1992):

$$\sigma_e' = P \sin\phi + C \cos\phi; \tag{6}$$

σ_e' is the effective deviatoric shear stress ($\sigma_e' = (\frac{1}{2}\sigma_{ij}'\sigma_{ij}')^{\frac{1}{2}}$), ϕ the angle of internal friction, and C the cohesion. ϕ undergoes a linear decrease with total effective plastic strain (measured as the square root of the second invariant of the strain tensor) to simulate strain weakening. Such strain weakening is thought to result from a reduction in fault rock grain size (Handy et al., 2007), mineral transformations (White and Knipe, 1978; Tingle et al., 1993) or the development of foliation or high fluid pressures (Hubbert and Rubey, 1959; Sibson, 1977). The effective viscosity for plastic flow is (Lemiale et al., 2008)

$$\eta_p = \frac{P \sin\phi + C \cos\phi}{2\dot{\varepsilon}_e}. \tag{7}$$

In our thermo-mechanical subduction models, we use a minimum viscosity value of $10^{18}\,\mathrm{Pa\,s}$ and a maximum cut-off of $10^{24}\,\mathrm{Pa\,s}$. These values ensure efficient convergence of our mechanical solution, while allowing for viscosity contrasts of 6 orders of magnitude. The minimum viscosity of $10^{18}\,\mathrm{Pa\,s}$ is low enough to capture almost all viscosity variations in our model. The cut-off value is only reached very locally in the mantle wedge. We solve the thermal and mechanical equations with a 2-D version of SULEC, which is an arbitrary Lagrangian–Eulerian (ALE) finite-element code (Buiter and Ellis, 2012). The mesh consists of quadrilateral elements which have linear continuous velocity and constant discontinuous pressure fields. Particles are used to track the

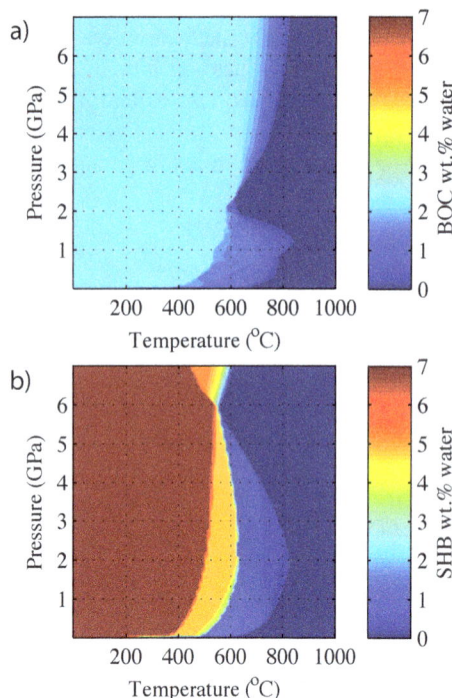

Figure 1. Water content as a function of pressure and temperature calculated using Perple_X for (**A**) bulk oceanic crust, and (**B**) serpentinised harzburgite lithologies (Chemia et al., 2010). Bulk compositions are given in Table 1.

Table 1. Bulk compositions in percent for the lithological oceanic lithosphere model (Chemia et al., 2010).

Oxides	BOC	SHB
SiO_2	47.32	41.023
TiO_2	0.63	0.075
Al_2O_3	16.11	1.114
FeO	7.21	7.66
MgO	9.27	42.298
CaO	12.17	1.029
Na_2O	1.65	–
H_2O	2.68	6.8
CO_2	2.95	–
Total	99.99	99.999

material field (through a material identifier) and properties such as particle strain, stress and water content. The particles are advected with the solid flow field using a second-order Runge–Kutta scheme. We use harmonic viscosity averaging and arithmetic density averaging schemes from particles to elements (Schmeling et al., 2008). The subduction models have a free surface and we use the surface stabilisation algorithm of Kaus et al. (2010) and Quinquis et al. (2011).

2.2 Calculation of water content

Our subduction model includes three lithologies: a bulk oceanic crust (BOC), serpentinised harzburgite (SHB) for the lithospheric mantle (Chemia et al., 2010), and pyrolite for the sub-SHB mantle (Schmidt and Poli, 1998). Water content is determined in wt% as a function of pressure, temperature and bulk composition (i.e. the average chemical composition of each lithology, Table 1). The water contents of BOC and SHB are calculated using Perple_X (Connolly, 2005) by Chemia et al. (2010) (Fig. 1). Perple_X is a thermodynamic code that minimises the Gibbs free energy of a chemical system to determine the stability fields of the phases which compose the mineral assemblages. Once these stability fields are calculated, it is possible to determine the maximum allowed water content of each phase as a function of pressure and temperature. The thermodynamic calculations

of Perple_X are valid up to pressures of 7 GPa and temperatures of 1300 °C and therefore do not cover upper-mantle conditions. We base our water contents in the upper mantle on Schmidt and Poli (1998), who experimentally determined water storage capacity for pyrolite up to 8 GPa and 1100 °C. We extrapolate these data to 11 GPa and 1400 °C by linearly continuing the Clapeyron slopes of the stability fields, similar to Arcay et al. (2005) (Fig. 2). Serpentinisation of the mantle wedge can locally increase the water storage capacity up to 7 wt% H_2O (Iwamori, 1998; Rüpke et al., 2004; Connolly, 2005; Férot and Bolfan-Casanova, 2012). We furthermore assume that the water storage capacity in the sublithospheric mantle (i.e. the pyrolytic material) does not converge to zero as shown by the phase diagrams determined by Schmidt and Poli (1998) because this would not allow the transport of bound water into the upper mantle. We impose the minimum storage capacity of the mantle to be 0.2 wt%, following Bercovici and Karato (2003) (Fig. 2).

2.3 Water migration schemes

Mineralogically bound water is advected along the solid-phase flow field. In our models, each particle therefore carries not only a material identifier (which determines its material properties) but also the amount of bound and free water it contains. Free water migrates following one of these imposed migration schemes: (I) one element vertically up per time step (also referred to as "elemental"), (II) imposed vertical velocity, or (III) Darcy flow velocity. All water migration schemes follow three steps: (1) determine the maximum water storage capacity of each particle and the amounts of free and bound water; (2) determine the migration path for the free water, if present and; (3) distribute the free water along the migration path. For every particle we calculate the maximum bound water that the particle can contain using a standard bilinear interpolation in pressure and temperature of wt% H_2O on gridded versions of Figs. 1 and 2. If the mineralogically bound water is less than the water storage

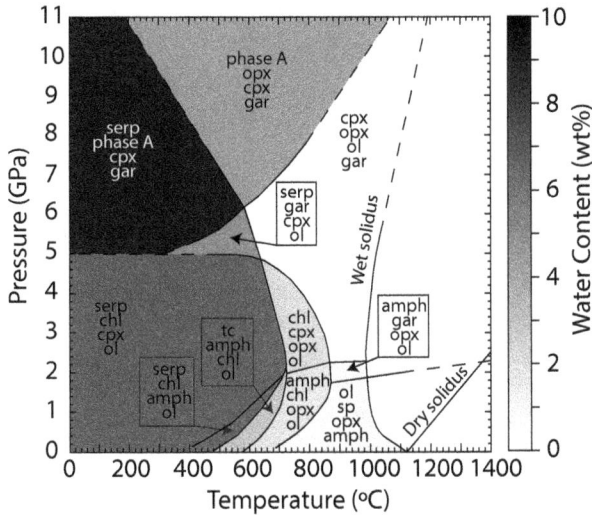

Figure 2. Water content as a function of pressure and temperature for mantle material of pyrolytic composition, modified from Schmidt and Poli (1998). The solid lines are experimentally determined Clapeyron slopes (Schmidt and Poli, 1998), while the dashed lines have been extrapolated in a similar manner to Arcay et al. (2005). amph: amphibole; chl: chlorite; cpx: clinopyroxene; gar: garnet; opx: ortopyroxene; ol: olivine; serp: serpentine; sp: spinel; and tc: talc.

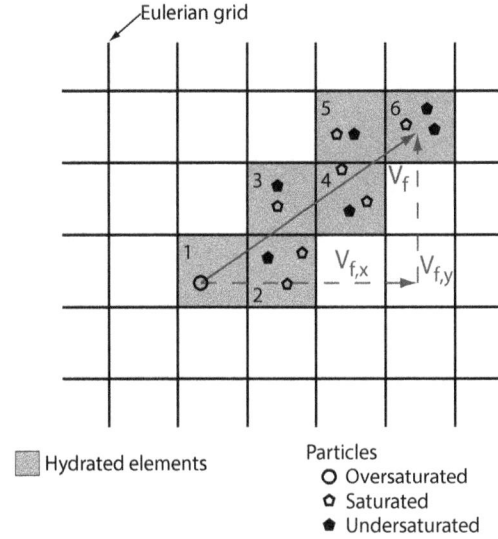

Figure 3. Schematic migration of free water and its distribution along a prescribed path for our velocity controlled water migration schemes II and III (see text for further explanation). V_f is water velocity.

capacity, the particle is undersaturated and no free water is produced. If the mineralogically bound water of the particle exceeds the water storage capacity, it is oversaturated in water and dehydration occurs. The amount of free water is the difference between the mineralogically bound water of the particle and the maximum allowed water. The free water migrates through the model following one of the schemes we are investigating.

The first migration scheme (scheme I) assumes that free water moves vertically upwards owing to its negative buoyancy, with one element per time step (Δt), and is not affected by the solid-phase flow (Arcay et al., 2005). This implies that the water migration velocity is purely vertical and is imposed as the local vertical grid size divided by the time step. A model using a variable grid size would not have a constant free water velocity, and therefore this should be avoided. If free water is present, undersaturated particles in the current element are hydrated first. If free water remains after this first step, it migrates to the element above; there it hydrates the undersaturated particles of that element from the bottom up. If all particles are saturated and free water is still present, it is evenly distributed over all particles of the element, waiting for the next time step for further upward migration.

The second migration scheme (scheme II) imposes the velocity for free water ($v_{f,x}$ and $v_{f,y}$) (Gorczyk et al., 2007). This method reduces the grid dependence of the migration scheme, though does not eliminate it completely as the migration path itself is grid-dependent. When a particle is over-

saturated, it releases water distributed along the path defined by $v_f \times \Delta t$ (Fig. 3). The horizontal and vertical components of the solid flow phase are added to the respective components of the free water velocity. Migration of free water is therefore no longer necessarily vertical. As in the case of the migration scheme I, the first step is to hydrate undersaturated particles in the current element. If free water is still present after this step, the remaining free water migrates to the next element, saturating the undersaturated particles from the bottom up, and so on. If all particles along the migration path of this time step are saturated (i.e. all particles in elements 1 to 6 of Fig. 3), the remaining water is distributed evenly over all particles of the last element. The motivation for element-wise distribution is that water migration paths are likely irregular and a linear path for free water would be unlikely (Rüpke et al., 2004; Katz et al., 2006). We only show examples with an imposed vertical velocity and $v_{f,x} = 0$.

The third migration scheme (scheme III) follows a simplification of Darcy flow where the fluid follows the pressure gradient caused by the difference in density between the fluid and the solid it is percolating through (Turcotte and Schubert, 2002):

$$q = \frac{(\rho_s - \rho_f)g\kappa}{\eta_f}, \tag{8}$$

where q is the Darcy velocity, ρ_s and ρ_f the density of the solid and fluid respectively, g the gravitational acceleration, η_f the viscosity of the fluid, and k the permeability. The permeability follows the empirical definition of Wark et al.

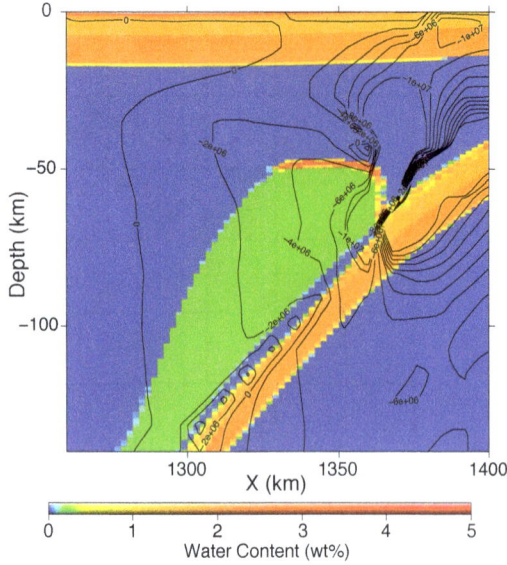

Figure 4. A close-up of the mantle wedge region at 5 Myr in a subduction model that uses free water migration scheme I in which viscosity decreases with bound water content. Shown is the elemental bound water content with contours at every 2 MPa of the horizontal pressure difference, $\frac{dP}{dx}$. Horizontal pressure variations are low in the region with highest bound water content.

(2003):

$$\kappa = \frac{d^2 \Phi^3}{270}, \tag{9}$$

where Φ is the volume fraction of fluid and d the grain size (same as in Eq. 4). The fluid velocity is the sum of the Darcy velocity relative to the volume fraction of fluid and the solid velocity:

$$v_f = \frac{q}{\Phi} + v_s, \tag{10}$$

where v_f is the fluid velocity and v_s the solid-phase velocity.

The water migration and distribution then follow scheme II. We assume, however, that water migration in scheme III is vertical. This assumption (also previously made by (Cagnioncle et al., 2007)) is reasonable, as the horizontal pressure gradient in our models in the mantle wedge is much smaller than the vertical pressure gradient (Fig. 4). Φ, the volume fraction of fluid, is determined from the initial water content and the grids for pressure, temperature and wt% H_2O. However, this assumes that all the free water that is present in interconnected channels is used to calculate the fluid velocity (Eqs. 8–10). This would result in unnaturally high fluid velocities. We therefore introduce an efficiency factor, ω, that corresponds to the percentage of interconnected channels of the network through which water can migrate. The effective permeability in Eq. 8 then becomes $\kappa_e = \omega \kappa$. This reduces the effective fluid velocity as water can only migrate through the interconnected network.

The Darcy water velocity is calculated for every particle of the model. The water velocity is not constant throughout the model, and areas with higher water content have higher water migration velocities.

We use the following output values to quantify the influence of the water migration schemes on the water distribution and the dynamic evolution of the models: (1) the water distribution is described by tracking the depths of the top-most, lower-most, and left- and rightmost particles of hydrated material. The top hydrated particle gives insight into the rate of advancement of the hydration front, which is not necessarily the same as the imposed water migration velocity. For example, assuming a high water migration velocity and a low amount of free water, free water will first saturate undersaturated materials, resulting in an effectively lower velocity of the hydration front. (2) The root-mean-square water contents (OH_{rms}) for slab or cylinder and the mantle provide information on the rate of dehydration of the slab or cylinder and the rate of hydration of the mantle. This gives the average water content over various domains in the model (e.g. in the mantle or in the slab)

$$OH_{rms} = \frac{1}{A} \sqrt{\int\limits_y \int\limits_x (OH_f^2 + OH_b^2) \, dx \, dy}, \tag{11}$$

where OH_b is the bound water content, OH_f is the free water content, and A is the area of computation. (3) For sinking cylinder models in which viscosity depends on water content, the bottom-most particle of the subducting cylinder is also tracked. This shows the influence of water content on the dynamic evolution of the model.

3 The effects of (de)hydration on a simple model of a sinking wet cylinder

3.1 Simple model of a sinking cylinder

We first investigate the effects of (de)hydration and water migration with a simple model which simulates the subduction of a detached piece of lithosphere by the sinking of a cold, hydrated cylinder into a warm, dry mantle. These experiments are based on simple Stokes flow and use linearly viscous rheologies. We solve for the advection and conduction of temperature in addition to the mechanical flow, but the temperature does not play a role in the mechanical flow as viscosities are linear viscous.

The model domain is $300 \, \text{km} \times 300 \, \text{km}$ and has a uniform Eulerian resolution of $1 \, \text{km} \times 1 \, \text{km}$ elements. The initial particle density is 25 particles per element. In this series of experiments no particle injection or deletion scheme is used. The cold cylinder has a radius of 20 km and is centred on the coordinates $x = 150 \, \text{km}$, $y = -130 \, \text{km}$ (Fig. 5). A strong viscosity contrast between the cylinder and the mantle avoids deformation of the cylinder (Table 2). The mechanical boundary

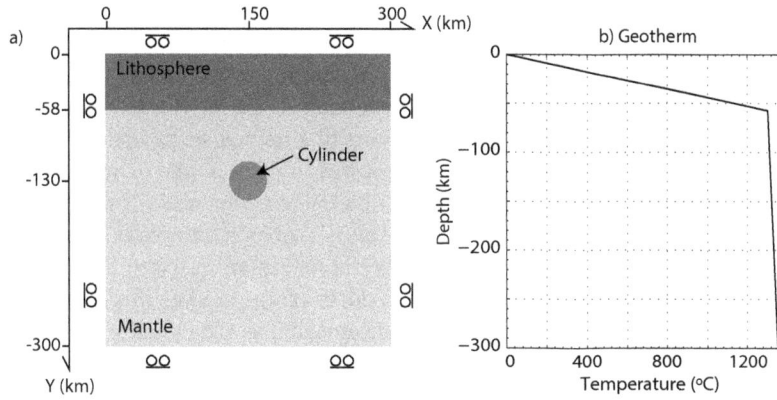

Figure 5. (**A**) Model setup for a cold, hydrated cylinder sinking in a warm, dry mantle. All materials are linear viscous, and their rheological parameters are given in Table 2. (**B**) Initial geotherm for model in (**A**).

Table 2. Input parameters for the linear viscous sinking cylinder model. For all materials thermal expansivity $\alpha = 0$ and heat production $H = 0$.

Lithology	Parameter	Symbol	Unit	Value
Dry lithosphere	Reference density	ρ_0	$kg\,m^{-3}$	3200
	Viscosity	η	Pa s	10^{23}
	Thermal conductivity	k	$W\,m^{-1}\,K^{-1}$	4.5
	Specific heat	C_p	$J\,kg^{-1}\,K^{-1}$	750
Dry mantle	Reference density	ρ_0	$kg\,m^{-3}$	3200
	Viscosity	η	Pa s	10^{20}
	Thermal conductivity	k	$W\,m^{-1}\,K^{-1}$	105
	Specific heat	C_p	$J\,kg^{-1}\,K^{-1}$	1250
Hydrated cylinder	Reference density	ρ_0	$kg\,m^{-3}$	3250
	Viscosity	η	Pa s	10^{23}
	Thermal conductivity	k	$W\,m^{-1}\,K^{-1}$	4.5
	Specific heat	C_p	$J\,kg^{-1}\,K^{-1}$	1250

conditions are free-slip on all sides (i.e. the velocity component parallel to the boundary is free, whereas the velocity component perpendicular to the boundary is zero). A 20 Myr old lithosphere, 58 km thick, overlies the mantle. The initial thermal condition of this oceanic lithosphere is determined from the plate-cooling model (e.g. Turcotte and Schubert, 2002) for a surface temperature of $0\,°C$, a temperature of $1300\,°C$ at 58 km, and a thermal diffusivity of $10^{-6}\,m^2\,s^{-1}$. The surface temperature is held at $0\,°C$ and the bottom temperature (at $y = -300\,km$) at $1360.5\,°C$ throughout model evolution, while the lateral sides are insulated (zero heat flux). A high conductivity ($k = 105\,W\,m^{-1}\,K^{-1}$) is used in the mantle to enforce a mantle adiabat of $0.25\,°C\,km^{-1}$ (Pysklywec and Beaumont, 2004). The initial temperature of the cylinder is $400\,°C$, which gradually increases as the cylinder warms up. The mantle is of pyrolytic composition, while the lithosphere and the cylinder are composed of SHB. The material and thermal properties are in Table 2. The initial water content of the hydrated cylinder is imposed at 0.2 wt %.

This model setup is run using the three different migration schemes (schemes I, II and III; see Sect. 2.3). First, we use an elemental vertical migration of free water. The water migration velocity is the vertical element size divided by the time step: $20\,cm\,yr^{-1}$. Second, we use migration scheme II, imposing vertical free water velocities of 10, 20 and $60\,cm\,yr^{-1}$. These velocities are in line with free water velocities reached in previous models (Gorczyk et al., 2007; Gerya et al., 2002). As we use a uniform grid resolution, the $20\,cm\,yr^{-1}$ model should be similar to the scheme I. However, the schemes are not identical because the vertically imposed velocity scheme assumes that free water is also displaced by the solid flow phase field. Finally, we use migration scheme III, where the vertical water velocity is calculated from the Darcy equation (Eqs. 8–10).

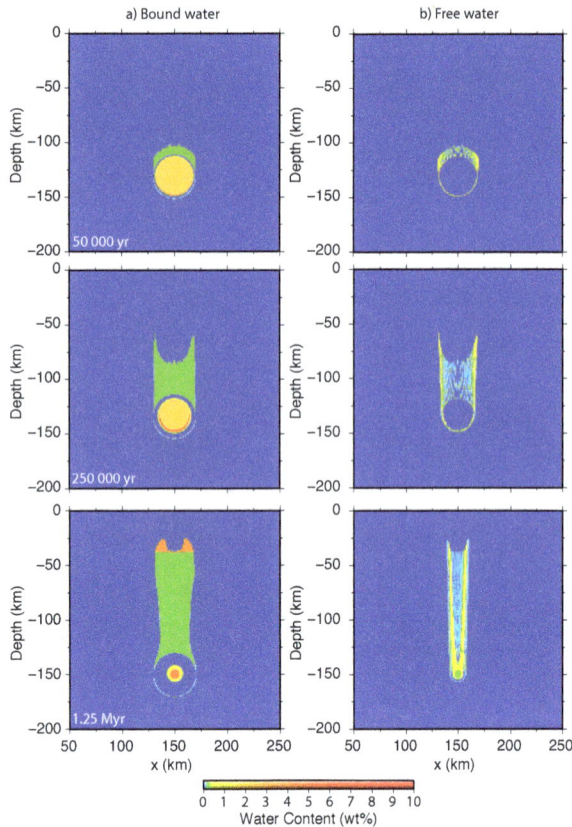

Figure 6. Elemental water content for the Stokes flow models using scheme I ($20 \, \mathrm{cm \, yr^{-1}}$ for a grid resolution of $1 \, \mathrm{km} \times 1 \, \mathrm{km}$) showing (**A**) bound water and (**B**) free water.

3.2   Sinking cylinder results

We first examine models in which water content does not influence the linear viscosity. The thermal and mechanical evolution of these models is therefore identical. These models are used to test cylinder dehydration and mantle hydration for the three free water migration schemes.

As only the water migration schemes are changed, the evolution of the cylinders dehydration is identical in all cases. Dehydration of the cylinder occurs due to the increase in temperature of the cylinder from the outer rim inwards (Fig. 6a). Dehydration processes are assumed to be nearly instantaneous, whereas water migration velocities are more than an order of magnitude larger than the flow velocities of the mantle. Therefore, the influence of pressure on dehydration is negligible as the cylinder does not sink as quickly as it dehydrates. The hydrated cylinder is initially undersaturated as the initial water content is $0.2 \, \mathrm{wt\%}$ (but it could contain up to $6.8 \, \mathrm{wt\%}$ at its initial pressure and temperature). The interior of the cylinder is therefore hydrated by water from the lower rim, which undergoes dehydration. This explains the differences observed in cylinder $\mathrm{OH_{rms}}$ during the first 1 Myr of model evolution (Fig. 7). Once the centre of the subducting

cylinder becomes saturated, dehydration in all models converges (Fig. 7c, f, i).

Hydration of the mantle initiates at the lateral sides of the cylinder, as the interior is being hydrated (Fig. 6b). This results in a horned-shape area of hydrated mantle, which progresses inwards as the cylinder dehydrates. The spatial distribution of bound water is affected by the flow of the mantle. This is visible in the area of hydrated mantle above the cylinder: a minimum in width of the hydrated domain occurs at a depth of ca. 125 km (Fig. 6a, last stage). This thinning is accentuated as the cylinder sinks towards the bottom of the model domain. The spatial distribution of free water is not affected by the mantle flow as water migration velocities are much higher than the mantle flow velocities.

The final bound water distribution in the model is independent of the water migration scheme, due to the limited amount of initial water, but water migration schemes do influence the evolution of the water distribution in the mantle (Fig. 7d). Due to the limited amount of water that mantle material can absorb, the mantle is rapidly saturated by water released from the dehydrating cylinder. Therefore the faster the vertical migration of water, the faster the mantle hydrates (Fig. 7d). Due to the much higher water saturation values in the lithosphere, the rate of advancement of the hydration front in the overlying lithosphere is greatly reduced and differences in migration schemes are negligible there. Increasing the efficiency factor in the Darcy flow models (ω, in Sect. 2.3) by an order of magnitude can locally increase the water velocity by an order of magnitude. However, this has a limited influence on the distribution of water as the average water velocity in the Darcy model stays close to $10 \, \mathrm{cm \, yr^{-1}}$ (Fig. 7g).

To simulate the effect of water on viscosity in these linear viscous models, a linear decrease in the lithospheric and mantle viscosity of 2 orders of magnitude is used over a water content of 0 to $0.4 \, \mathrm{wt}, \%$. This is of the same order of magnitude as obtained by including $0.4 \, \mathrm{wt\%}$ in the Hirth and Kohlstedt (2003) dislocation or diffusion creep flow law for wet olivine. Including the influence of water on mantle viscosity does not change the overall water distribution of the models (Fig. 7). It does, however, have a small effect on the mechanical evolution of the model (Fig. 8). The lower viscosity values of the hydrated mantle above the cylinder increase the mantle flow velocity in this region, which results in an increase in the sinking velocity of the cylinder. However, this effect is limited as it corresponds to a difference in velocity of $0.1 \, \mathrm{cm \, yr^{-1}}$. This is because hydration only occurs directly above the sinking cylinder. The sinking velocity of the cylinder is less sensitive to lowering the viscosity above the cylinder than it would be to changing the mantle viscosity below the cylinder.

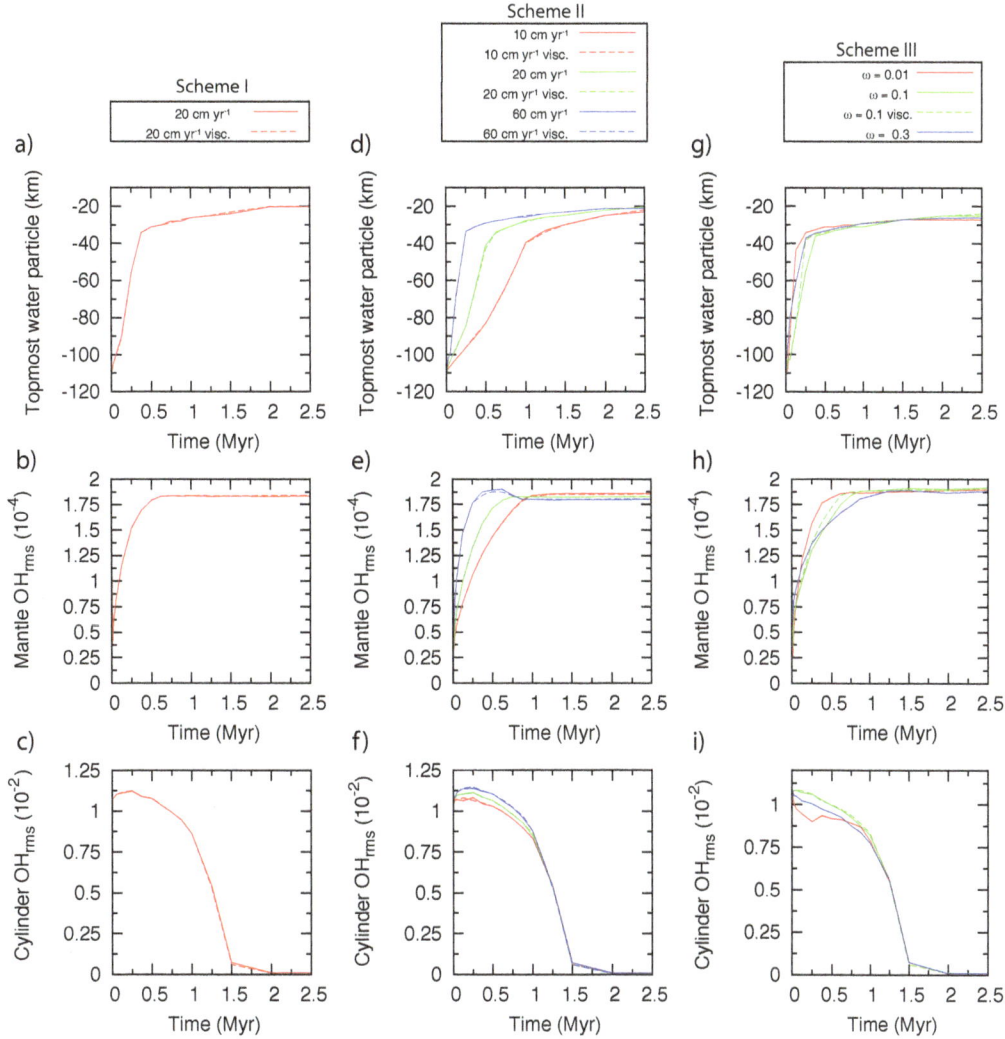

Figure 7. Evolution with time of the depth of the top-most particle of hydrated material (top row), mantle OH_{rms} (middle row) and cylinder OH_{rms} (bottom row) for our models of a wet sinking cylinder. The first column show results for scheme I (water migration velocity of $20\,cm\,yr^{-1}$), the second for scheme II with different imposed velocities ($v_{f,y} = 10, 20$ and $60\,cm\,yr^{-1}$) and the third for scheme III, in which the water migration velocity is calculated using the Darcy law with efficiency factor $\omega = 0.01, 0.1$ or 0.3 (Sect. 2.3). Models labelled "visc" have a decrease in viscosity with increasing water content.

3.3 Sinking cylinder discussion

Our models of cylinder dehydration and mantle hydration show that different numerical water migration schemes do not result in large differences in the distribution of bound water in the mantle. The sinking cylinder induces a vertical solid flow field in the area of mantle hydration. The free water migration velocities are also vertical. The combined fluid migration velocity is therefore vertical and the difference between the schemes lies mainly in the rate of free water migration. The water migration velocities for our three migration schemes are within the same order of magnitude, resulting in a similar bound and free water distribution. However, changing the magnitude of the water migration veloc-

ity does initially increase the hydration rate of the mantle and the hydration front rises at different velocities (Fig. 7d). This is a transient phenomenon as the models converge after ca. 2.4 Myr. The low water absorption capability of the mantle material results in rapid effective migration velocities of bound water in the mantle, confirming the results of Arcay et al. (2005). Models that simulate a decrease in mantle viscosity with water content show an increase in mantle velocities and a slightly faster sinking of the cylinder. Our simple models of a sinking wet cylinder in a dry mantle indicate that the three schemes we investigated for numerical implementation of water migration lead to very similar results. Within these schemes, the exact implementation might be secondary to the first-order effect that water could have

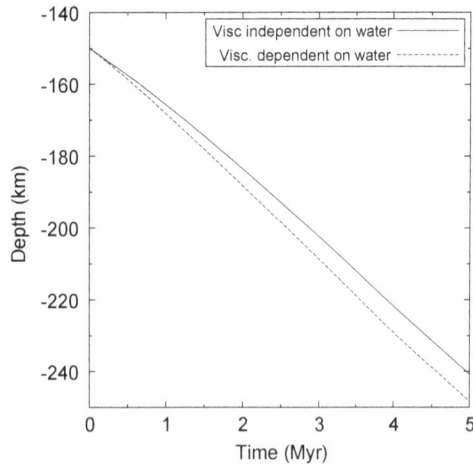

Figure 8. Depth of the bottom of the subducting cylinder versus time for linear viscous models with or without including the effects of water on viscosity. This figure is for models that follow free water migration scheme I (Fig. 7a–c).

on the system. We will test this in the next section with thermo-mechanical models that also include horizontal mantle (wedge) flow components.

Figure 6 shows a one-element-thick ring of hydrated mantle around the subducting cylinder, on the lateral and bottom sides. This is an artefact of the free water migration scheme. Once a particle dehydrates, undersaturated particles of the element are hydrated first. When the contact between cylinder and mantle lies within that element, mantle particles can absorb the released water, resulting in the hydrated ring around the cylinder.

4   The effects of (de)hydration on subduction

4.1   Subduction model setup

We investigate the effects of slab dehydration, water migration, and mantle wedge hydration using a model of a 70 Myr old oceanic lithosphere subducting under a 40 Myr oceanic lithosphere. The model without water follows Quinquis et al. (2013) and has been tested with a number of different numerical codes.

The oceanic plates are composed of two layers: (1) a 7 and 8 km crustal layer for the overriding and subducting lithospheres, respectively, composed of bulk oceanic crust (BOC) (Chemia et al., 2010), and (2) a 32 km thick serpentinised harzburgite (SHB) layer (Chemia et al., 2010). The rheological and thermal parameters are given in Tables 3 and 4. We assume that the upper 16 km of the oceanic plates is hydrated to a certain degree through fractures in the oceanic crust. The upper kilometre of BOC is fully hydrated, resulting in an initial water content of 2.68 wt% H_2O. Few faults exceed 1 km

depth, resulting in an undersaturation of the remaining BOC. The initial water content of the BOC below 1 km depth is set at 1.5 wt% H_2O. Finally, between 7 (or 8) and 15 km (or 16 km) depth, the overriding and subducting SHB is also undersaturated at an initial water content of 2 wt% H_2O. These values follow Rüpke et al. (2004) and Faccenda et al. (2012).

The model domain is 3000 km wide and 670 km deep (Fig. 9) and has the highest horizontal and vertical Eulerian resolution at the trench (1 km per element). The total number of elements is 473×269 (horizontal $\times$ vertical), and 16 particles per element are used initially. Due to the variable grid resolution, this series of models requires injection and deletion of particles to maintain elemental particle density between 12 and 36. The particle deletion scheme helps keep the code memory requirements reasonable. To maintain the overall water content of the model, particles are injected dry, whereas the bound or free water content of a deleted particle is distributed evenly over all other particles of the element. Subduction is initialised by a "weak seed" located at the interplate boundary. The weak seed simulates a pre-existing shear zone in the oceanic lithosphere separating the overriding and subducting plates. It is 14 km thick (in the direction perpendicular to the dip angle), extends to a depth of 82 km and has a 35° dip angle. The top mechanical boundary is a true free surface (both v_x and v_y are free), whereas the bottom and left boundaries of the model are free-slip. Balanced material in- and outflow is defined on the right boundary of the model domain (the boundary parallel component is again free). To avoid strong shearing at the transition between in- and outflow, a linear velocity gradient from in- to outflow is defined over a 20 km depth interval. The inflow velocity is 5 cm yr^{-1} over the thermal thickness of the 70 Myr old lithosphere. The outflow velocity is imposed from a depth of -130 km to the bottom of the model domain at -670 km.

The initial thermal conditions of the 40 Myr (from $x = 0$ to $x = -1500$ km) and 70 Myr (from $x = 1500$ to $x = 3000$ km) old oceanic lithospheres are determined from the plate-cooling model (Turcotte and Schubert, 2002) for a surface temperature of 0 °C, a mantle temperature of 1300 °C at 82 and 110 km depth respectively, and a thermal diffusivity of 10^{-6} m^2 s^{-1}. The initial step in temperatures at $x = 1500$ km is rapidly diffused. During model evolution, the surface temperature is held at 0 °C and the bottom temperature (at $y = -670$ km) at 1440 °C, while the lateral sides are insulated (zero heat flux). A high conductivity ($k = 183.33$ W m^{-1} K^{-1}) is defined for the mantle to enforce the mantle adiabat of 0.25 °C km^{-1} (Pysklywec and Beaumont, 2004).

The subduction models are run using the three water migration schemes. Because vertical grid size varies (coarsening downwards from 1 to 7 km per element), water migration velocities are no longer constant in the elemental scheme and vary between 10 and 70 cm yr^{-1}. However, the bulk of the dehydration processes occurs in the mantle wedge, and there our model has a constant mesh resolution and therefore a

Table 3. Subduction model parameters. The viscosity ranges from a minimum of 10^{18} to a maximum of 10^{24} Pa s. Heat production $H = 0$.

Parameter	Symbol	Unit	Mantle	Sublithospheric mantle	Overriding plate BOC	Overriding plate SHB	Subducting plate BOC	Subducting plate SHB	Weak seed
Olivine rheology			Dry	Dry	Wet	Wet	Wet	Wet	Wet
Angle of friction[a]	ϕ		20°/10°	20°/10°	20°/10°	20°/10°	5°/2°	10°/5°	2°
Cohesion	C	MPa	20	20	15	15	5	15	5
Reference density	ρ_0	kg m^{-3}	3200	3200	3000	3250	3000	3250	3200
Reference temperature	T_0	°C	1300	1300	200	200	200	200	200
Conductivity	k	W m^{-1} K^{-1}	183.33	2.5	2.5	2.5	2.5	2.5	2.5
Heat capacity	C_p	J kg^{-1} K^{-1}	750	750	750	750	750	750	750
Thermal expansivity	α	10^{-5} K^{-1}	2.5	2.5	2.5	2.5	2.5	2.5	2.5
Initial water content	–	wt% H$_2$O	0	0	1.5–2.68	2	1.5–2.68	2	2.68

[a] Angle of internal friction (ϕ) softens from first to second value over an effective strain interval of 0.5 to 1.5.

Table 4. Flow law parameters from Hirth and Kohlstedt (2003).

Parameter	Unit	Dry olivine Diffusion	Dry olivine Dislocation	Wet olivine Diffusion	Wet olivine Dislocation
A[a]	Pa^{-n} m^p s^{-1} H(10^6Si)$^{-1}$	2.25×10^{-15}	6.514×10^{-16}	1.5×10^{-18}	5.3301×10^{-19}
n	–	1	3.5	1	3.5
Q	kJ mol^{-1}	375	530	335	480
V	10^{-6} m^3 mol^{-1}	4	14	4	11
d	mm	5	–	5	–
p	–	3	–	3	–
C_{OH}	H(10^6Si)$^{-1}$	–	–	1000	1000
r	–	–	–	1	1.2

[a] A is given for a general state of stress, and was converted from a uni-axial stress (Ranalli, 1995).

constant water migration velocity of 10 cm yr^{-1}. All schemes are used to investigate the evolution of models with or without the effect of bound water content on viscosity (Eq. 4). The sinking cylinder models showed that an efficiency factor ω of 0.1 in scheme III (Sect. 2.3) provides realistic water migration velocities. We therefore also use an efficiency factor ω of 0.1 in scheme III in the subduction models. For scheme II, two water velocities are investigated: $v_{f,y} = 5$ cm yr^{-1} and 10 cm yr^{-1}.

4.2 Subduction model results

Subduction is initiated at the weak seed by pushing the 70 Myr oceanic plate inwards, resulting in ca. 15 km of advance of the interplate contact until brittle failure helps localise deformation at the trench. The slab subducts at a fairly steep angle and with a more-or-less constant sinking velocity. The hydrated portion of the slab is limited to the top 16 km, and most of the slab is dry and therefore stiff. As we discuss below, the evolution of this stiff subducting slab is affected by (de)hydration processes, but not significantly.

Main dehydration occurs at two locations in the slab: at ca. 150 km depth, where dehydration occurs at the phase tran-

sition of blueschists to eclogite, and 210 km depth, where dehydration of chlorite occurs (Figs. 10 and 11). The mantle wedge is hydrated by these dehydration reactions. Hydrated mantle wedge material is entrained by the downwards flow above the subducting slab. This can bring bound water down to the transition zone. Figure 10 shows that the horizontal width of hydrated mantle above the slab decreases with depth. In the wedge, i.e, above 200 km depth, the mantle can be hydrated up to a distance of 100 km away from the surface of the slab. As the slab deepens, the distance of hydrated mantle from the slab surface decreases, from 50 km at ca. 200 km depth to less than 10 km at the slab tip (Fig. 10). Including the effects of bound water on viscosity increases the width of the hydrated mantle, BOC and SHB regions (Fig. 12e, g, h). The difference is, however, limited as the width of the hydrated regions increases by ca. 40 km towards the overriding plate.

Due to inflow of hydrated oceanic lithosphere during model evolution, the OH$_{rms}$ of the subducting slab increases as the model evolves (Fig. 12j). After 3 Myr, the OH$_{rms}$ of the slab decreases slightly. This represents the onset of dehydration and is synchronous to the increase in OH$_{rms}$ of the mantle (Fig. 12i). The numerical water migration schemes cause

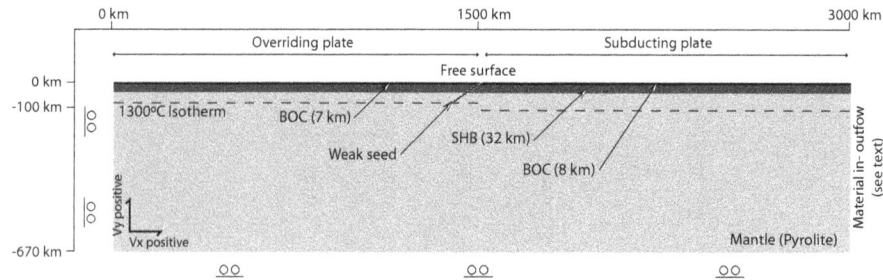

Figure 9. Model setup for subduction of a 70 Myr old oceanic plate under a 40 Myr old oceanic plate. The top boundary is free. The bottom and left boundary are free-slip, while the right boundary condition includes material in- and outflow. SHB: serpentinised harzburgite; BOC: bulk oceanic crust.

small differences in the distribution of bound water in the mantle. The differences in mantle OH_{rms} between the three water migration schemes is small initially, but it increases to ca. 1×10^{-5} after 6 Myr and then remains constant (Fig. 12i). The difference corresponds to variations in the lateral distribution of hydrated mantle material (Figs. 10 and 12). Including the effects of water on the viscosity does not change the OH_{rms} of the slab (Fig. 12j), but it does increase the OH_{rms} of hydrated mantle (Fig. 12i). The effects on the distribution of free water are more substantial (Fig. 11). The distribution of free water for the elemental and Darcy schemes is similar, but the free water domain is somewhat broader, at ca. 40 km (Fig. 11) in the Darcy models. This is caused by the horizontal component of the mantle flow that is added to the free water velocity. Larger variations occur in the free water distribution for scheme II and can result in locally large quantities of free water of up to $4 \, \mathrm{wt\%} \, H_2O$ (Fig. 11). However, in our models, free water has no effect on rheology and therefore no influence on the dynamics of the system.

Introducing the effect of water content on viscosity does not have a strong impact on the large-scale mechanical evolution of the model. This is because the evolution of the subducting slab is controlled by its stiffness. Due to the relatively little amount of water present in the slab (which is initially hydrated up to 16 km depth), dehydration processes will not greatly affect evolution of the already stiff slab. We do find that water weakening of viscosity increases flow in the mantle wedge and slightly reduces the curvature of the slab (Figs. 14 and 15). Corner flow is more pronounced in the models that have water-dependent viscosity, as a large part of the mantle above the slab is hydrated and thus weakened, promoting stronger mantle flow.

4.3 Subduction model discussion

Our models show a similar slab dehydration evolution to Rüpke et al. (2004), Arcay et al. (2005) and Cagnioncle et al. (2007). This is because a similar method for determining the locations of dehydration reactions are used in these experiments. The depths at which dehydration reactions occur

are slightly different between these studies because of differences in the thermal structure of the slab in the models.

We find that the overall dynamics of our subduction model are not strongly influenced by the viscosity decrease of mantle materials due to increase in water content. We suggest that this may be caused by the subducting slab being fairly stiff. Our slab is largely dry and the flow law of Hirth and Kohlstedt (2003) results in an average viscosity of ca. $5 \times 10^{23} \, \mathrm{Pa \, s}$, which is up to 5 orders of magnitude above the viscosity of the mantle. The evolution of this strong slab is not greatly influenced by a further viscosity increase caused by dehydration processes. Our models do show an increase in the corner flow (Fig. 14) but not on the same scale as that observed in the models of Arcay et al. (2005). The slab in the models of Arcay et al. (2005) dips at a shallower angle, which could focus corner flow and cause a larger effect of water on the flow field.

The numerical implementation of water migration has a significant effect on the distribution of free water in the mantle wedge (Fig. 11). In our models, this effect is not visible in the overall evolution of the model because free water does not affect the rheology of the mantle materials. However, the distribution of free water in the mantle wedge could influence the dynamics of subduction by changing the pore pressure, thereby changing the stress, and the viscosity. Similarly, the pore pressure effect of free water could reduce the plastic yield stress, thereby reducing effective viscosities in the slab and brittle parts of the mantle wedge. During fluid flow, compaction and dilation of the solid matrix may occur, related to pore pressure effects. This could locally change the pressure field and thus in return effect water migration paths, changing the hydration patterns in the solid matrix.

To obtain a first-order assessment of the potential impact of free water on the rheology of our models, we calculated a scheme II water migration model with $v_{f,y} = 5 \, \mathrm{cm \, yr^{-1}}$ in which the free water content was added to the bound water content in the calculation of the creep viscosity (Eq. 4). As bound water in the model can locally already reach values of 2–3 wt % (Fig. 11), we increased the maximum value that is allowed for C_{OH} in the viscosity calculation from 0.4 to

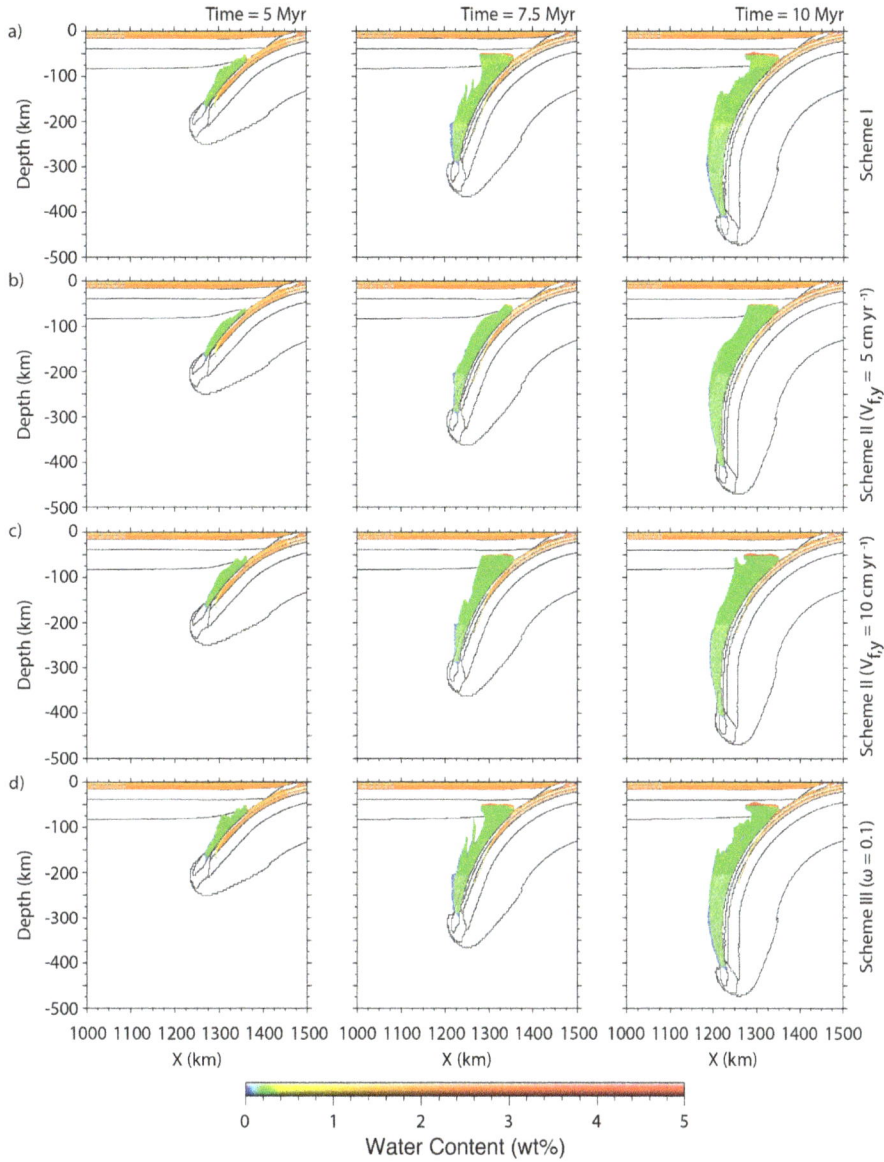

Figure 10. Bound water distribution for subduction models after 5, 7.5 and 10 Myr for the following free water migration schemes: (**A**) elemental water migration scheme I (velocity $v_{f,y}$ increases from $10\,\text{cm}\,\text{yr}^{-1}$ just below the lithosphere to $70\,\text{cm}\,\text{yr}^{-1}$ at the base of the model where the grid is coarsest), (**B**) imposed vertical velocity $v_{f,y} = 5\,\text{cm}\,\text{yr}^{-1}$ (scheme II), (**C**) imposed vertical velocity $v_{f,y} = 10\,\text{cm}\,\text{yr}^{-1}$ and (**D**) Darcy velocity with efficiency factor $\omega = 0.1$ (scheme III). Bound water moves with the solid-phase flow. Viscosity of all materials changes with water content following the flow law of Eq. 4.

$1\,\text{wt}\,\%$. In the initial stages of subduction, where only small amounts of free water are present, the difference in viscosity between the bound water viscosity model and the bound and free water viscosity model is small. However, as the model evolves and the amount of free water increases, the viscosity field changes (Fig. 16). Low viscosity values are localised in the mantle wedge. The amount of hydrated mantle material that reaches the bottom of the model domain is greatly reduced. This is because mantle flow is localised higher up in the mantle wedge. We emphasise that this model is a preliminary result, as the effect of free water (as opposed to bound water) on creep flow laws is not established.

In our simplified models of water migration we did not include melts. Melts are sinks that are thought to absorb most of the excess free water. This would then decrease the potential effects of free water on pore pressures. However, the resulting melt would build up pore pressure instead of a pore pressure increase that would have been caused by the now dissolved free water. Therefore, a potential weakening effect of free water on viscosity could remain. Water also decreases the temperature at which melting can occur,

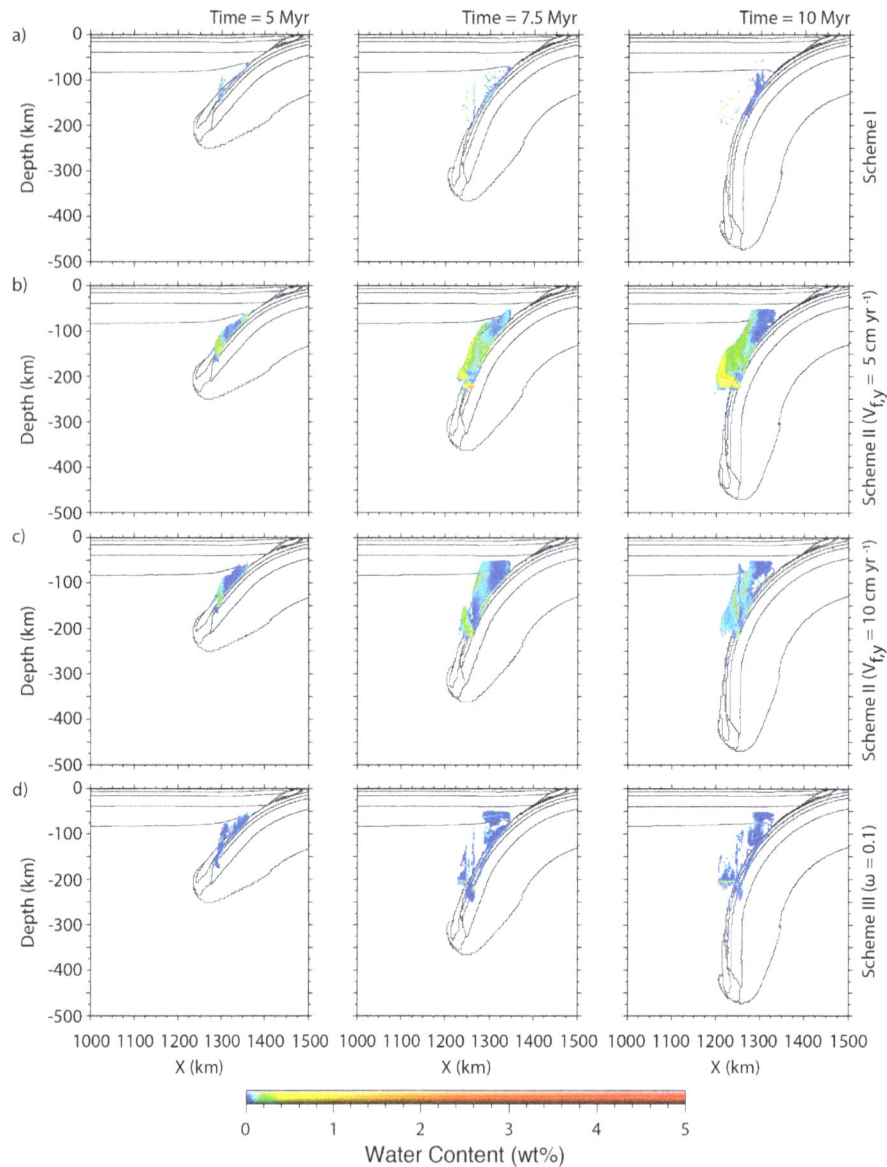

Figure 11. Free water distribution for subduction models after 5, 7.5 and 10 Myr for the following water migration schemes: (**A**) elemental water migration scheme I (velocity $v_{f,y}$ varies between 10 and 70 cm yr^{-1}), (**B**) imposed vertical water migration scheme II with $v_{f,y} = 5$ cm yr^{-1}, (**C**) Imposed vertical water migration scheme II with $v_{f,y} = 10$ cm yr^{-1}, and (**D**) Darcy water migration scheme III with efficiency factor $\omega = 0.1$. Viscosity of all materials changes with water content following flow law of Eq. 4.

encouraging melting, and because melts have a low viscosity, this could impact subduction dynamics. A logical next step would therefore be to include the effects of melts in our models of subduction with (de)hydration processes.

Including a decrease of viscosity with water content in the models influences the bound water distribution in the mantle close to the surface of the subducting slab (Fig. 14). Due to the limited water absorption capabilities of the mantle and the weak mantle wedge, the flow in the mantle wedge is increased, causing a further increase in the area over which the mantle is hydrated and weak. This suggests that subduction and mantle wedge studies that investigate (de)hydration pro-

cesses should preferably include the dynamic effects of water on viscosity and thus mantle flow during model evolution.

We assumed that the mantle material in our models can contain up to 0.2 wt % water at upper-mantle pressures and temperatures (Bercovici and Karato, 2003). We find that hydrated mantle material up to 0.2 wt % is entrained by the flow caused by the subducting slab down to the bottom of the model domain (Figs. 10 and 13). This therefore supports the initial assumptions used in the studies of slab dehydration at the transition zone (i.e. between 670 and 410 km) of Richard et al. (2006, 2007). It also agrees with experimentally determined phase diagrams that suggest water could be

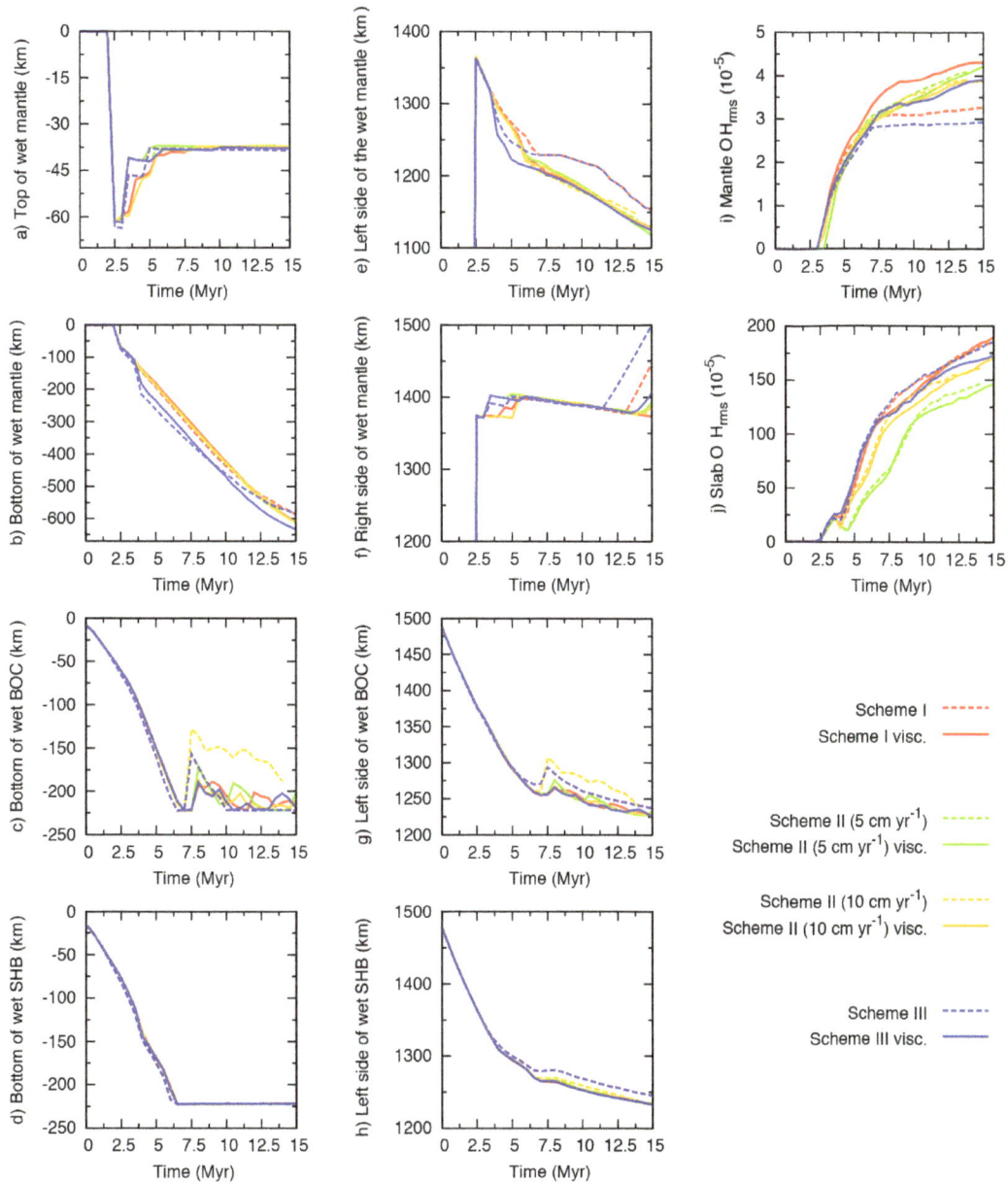

Figure 12. Evolution of: (**A**) depth of the topmost particle of hydrated mantle, (**B**) depth of the lowermost particle of hydrated mantle, (**C**) depth of the lowermost particle of hydrated BOC, (**D**) depth of the lowermost particle of hydrated SHB for the subduction models, (**E**) horizontal coordinate of the leftmost particle of hydrated mantle, (**F**) horizontal coordinate of the rightmost particle of hydrated mantle, (**G**) horizontal coordinate of the leftmost particle of hydrated BOC, (**H**) horizontal coordinate of the leftmost particle of hydrated SHB, (**I**) mantle OH$_{rms}$, and (**J**) slab OH$_{rms}$. "visc" indicates models that have a decrease in viscosity with increasing water content.

present in the transition zone or deeper (Ohtani et al., 2004; Komabayashi et al., 2005). The amount of water reaching the transition zone could be greatly increased by including other chemical reactions in the mantle wedge, such as serpentinisation (Iwamori, 1998).

We show that including the weakening effect of bound water on viscosity increases the amount of water brought down to the bottom of the model domain (Fig. 14). This is because the viscosity reduction causes a stronger corner flow that en-

trains more hydrated mantle material in the downward flow above the subducting slab.

5   Conclusions

We have used a linear viscous model of a wet cylinder sinking in a dry mantle and a thermo-mechanical subduction model to investigate the effects of the numerical implementation of basic schemes of free water migration. We find that

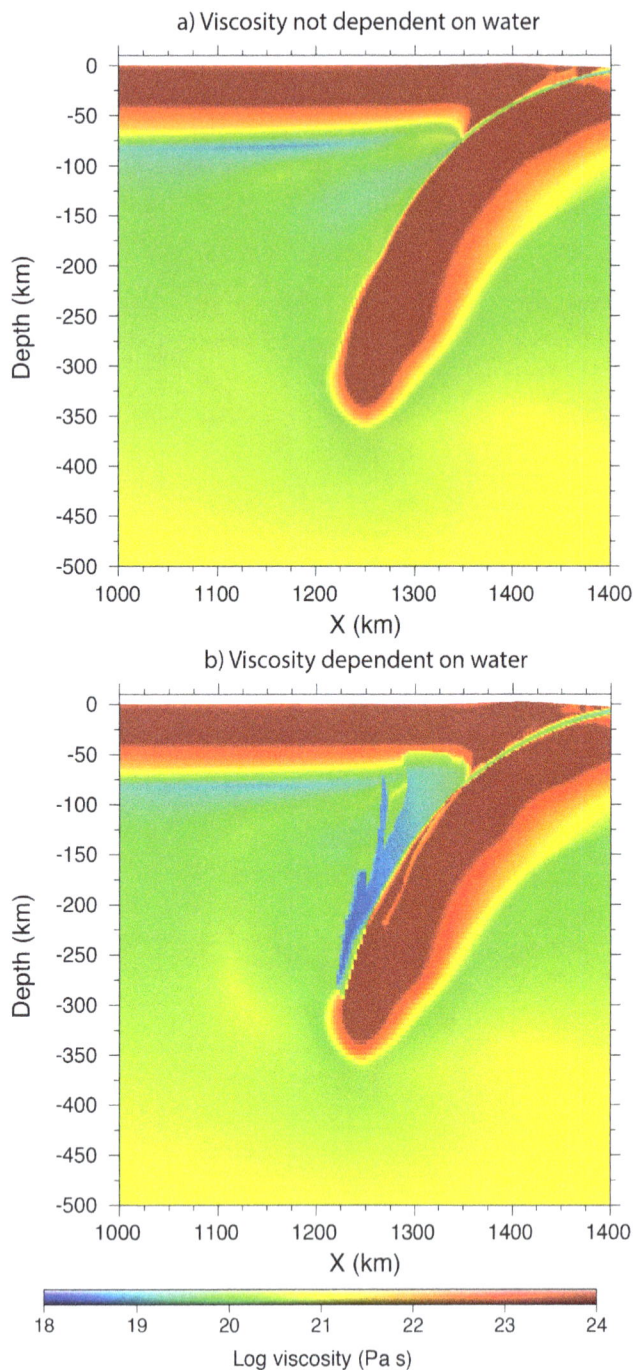

a) Viscosity not dependent on water

b) Viscosity dependent on water

Log viscosity (Pa s)

Figure 13. Viscosity field for models that follow the Darcy free water migration scheme III with efficiency factor $\omega = 0.1$ after 7.5 Myr for (**A**) viscosity not dependent on water content and (**B**) viscosity decrease with water content.

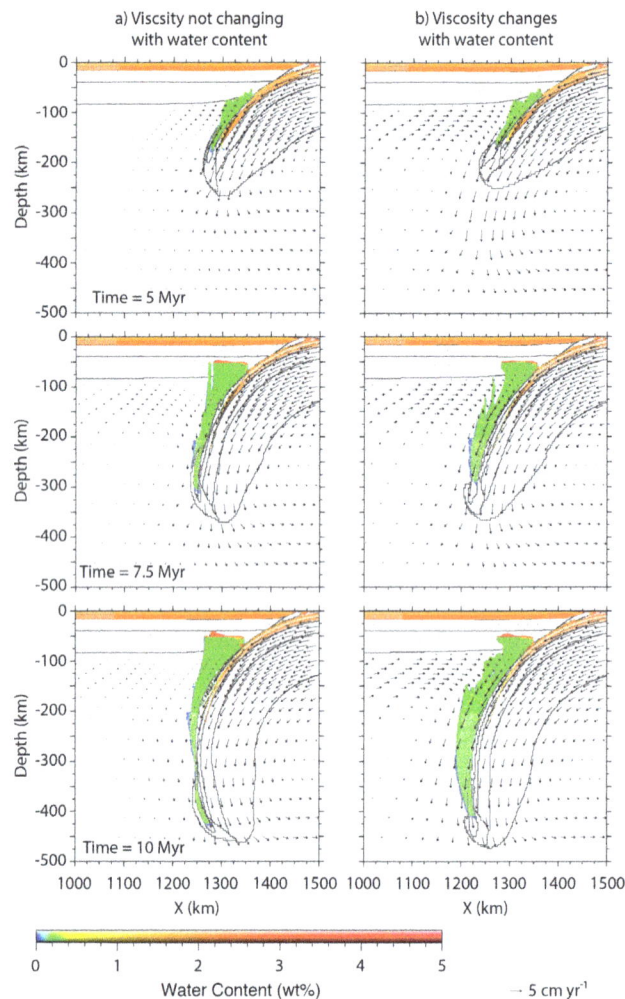

a) Viscsity not changing with water content

b) Viscosity changes with water content

Water Content (wt%)   → 5 cm yr^{-1}

Figure 14. Bound water distribution for subduction models using Darcy free water migration scheme III (with efficiency factor $\omega = 0.1$) at 5, 7.5 and 10 Myr. (**A**) Water content does not change viscosity; (**B**) water content decreases the viscosity (following flow law Eq. 4).

(de)hydration influences our models but that the exact manner of water migration is not that important. We suggest that studies of especially large-scale subduction dynamics may use a simple implementation of free water migration to capture the first-order effects of (de)hydration, as, for example, an imposed velocity or simple Darcy flow water migration scheme. Elemental water migration (simply moving free water up one element per time step) can be used as long as grid resolution is constant and it is realised that grid size determines the water migration velocity.

We find that the different water migration schemes influence the distribution of free water in subduction models. Elemental water migration results in a localised distribution of free water, while an imposed water migration velocity and Darcy flow result in a broader distribution of free water in the mantle wedge. This effect is caused by the free water being moved by the mantle flow in the latter migration schemes. Free water could affect pore pressure and thus material strength, but melting could decrease the amount of free water, and therefore the effect of free water on a subduction system requires future study.

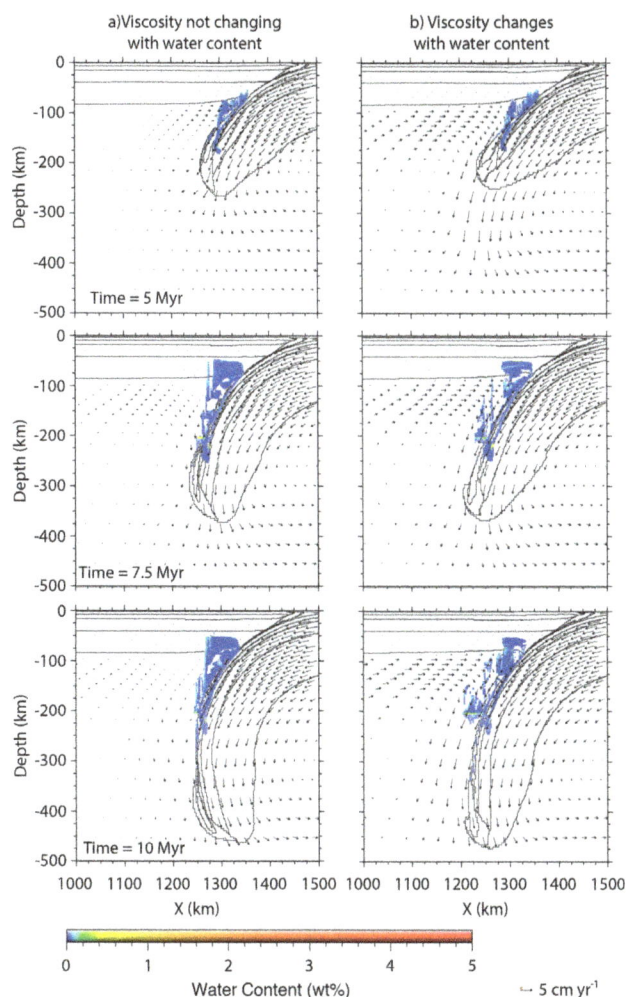

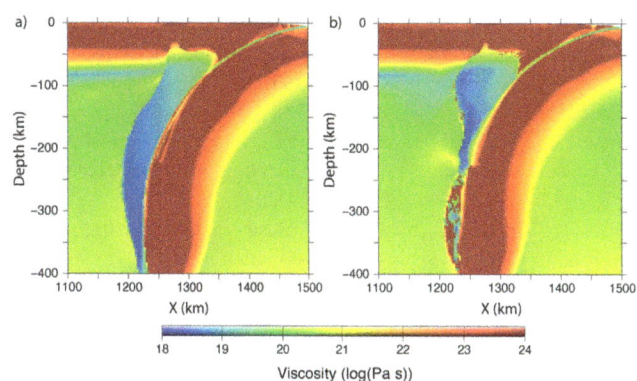

Figure 16. Viscosity fields in the mantle wedge for subduction models calculated using the imposed free water migration scheme II with $v_{f,y} = 5\,\mathrm{cm\,yr}^{-1}$. **(A)** Only the effects of bound water are taken into account in the viscosity calculations (using a maximum cut-off for C_{OH} of 0.4 wt % in Eq. 4). This is the same model as Fig. 10b. **(B)** The effects of both bound and free water are taken into account in the viscosity calculations (using a maximum cut-off for C_{OH} of 1 wt % in Eq. 4).

Figure 15. Free water distribution for subduction models using the Darcy water migration scheme (with $\omega = 0.1$) at 5, 7.5 and 10 Myr. **(A)** Water content does not change viscosity; **(B)** water content decreases the viscosity (following flow law Eq. 4).

Including the effects of bound water content on viscosity does not strongly impact the overall evolution of the subducting slab as it is controlled by the slab stiffness. We do find that the decrease in mantle viscosity with increasing water content as slab dehydration continues causes a more vigorous corner flow. If we assume that the upper mantle contains an average of 0.2 wt % water, this allows saturated hydrated mantle material to be transported down to the base of the upper mantle, supporting previous assumptions of hydrated material residing in the mantle transition zone.

Acknowledgements. We acknowledge financial support from the European Commission through Marie-Curie Research Training Network "c2c" (contract MRTN-CT-2006-035957). SULEC is developed jointly by Susanne Buiter and Susan Ellis. We warmly thank Zurab Chemia, Mark Caddick and James Connolly for help with Perple_X calculations and discussions on water behaviour, and Joya Tetreault for many numerical discussions. This manuscript benefitted from the constructive reviews by Manuele Faccenda, Tobias Keller and Guillaume Richard.

Edited by: T. Gerya

References

Arcay, D., Tric, E., and Doin, M.-P.: Numerical simulations of subduction zones – effect of slab dehydration on the mantle wedge dynamics, Phys. Earth Planet. In., 149, 133–153, 2005.

Bercovici, D. and Karato, S.-I.: Whole mantle convection and the transition-zone water filter, Nature, 425, 39–44, 2003.

Bercovici, D., Ricard, Y, and Schubert, G.: A two-phase model for compaction and damage 1. General theory, J. geophys. Res., 106, 8887–8906. 2001

Billen, M. I.: Modeling the dynamics of subducting slabs, Annu. Rev. Earth Pl. Sc., 36, 325–356, doi:10.1146/annurev.earth.36.031207.124129, 2008.

Billen, M. I. and Gurnis, M.: A low viscosity wedge in subduction zones, Earth Planet. Sc. Lett., 193, 227–236, 2001.

Buiter, S. J. H. and Ellis, S. M.: Benchmarking a new ALE Finite-Element code, Geophys. Res. Abstracts, 14, EGU2012-7528, 2012.

Caddick, M. J. and Thompson, A. B.: Quantifiying the tectonometamorphic evolution of Pelitis rocks from a wide range of tectonic settings: mineral compositions in equilibrium, Contrib. Mineral. Petr., 156, 177–195, doi:10.1007/s00410-008-0280-6, 2008.

Cagnioncle, A.-M., Parmentier, E., and Elkins-Tanton, L.: Effect of solid flow above a subducting slab on water distribution and melting at convergent plate boundaries, J. Geophys. Res., 112, B09402, doi:10.1029/2007JB004934, 2007.

Cheadle, M. J., Elliott, M. T., and McKenzie, D.: Percolation threshold and permeabiliy of crystallizing igneous rocks: the importance of textural equilibrium, Geology, 32, 757–760, doi:10.1130/G20495.1, 2004.

Chemia, Z., Dolejš, D., and Steinle-Neumann, G.: Thermal effects of metamorphic reactions in a three component slab, Geophys. Res. Abstracts 12, EGU2010-5768, 2010.

Connolly, J. A. D.: Computation of phase equilibria by linear programming: a tool for geodynamic modeling and its application to subduction zone decarbonation, Earth Planet. Sc. Lett., 236, 524–541, doi:10.1016/j.epsl.2005.04.033, 2005.

de Capitani, C. and Brown, T. H.: The computation of chemical equilibrium in complex systems containing non-ideal solutions, Geochim. Cosmochim. Ac., 51, 2639–2652, 1987.

England, P., Engdahl, R., and Thatcher, W.: Systematic variation in the depths of slabs beneath arc volcanoes, Geophys. J. Int., 156, 337–408, doi:10.1111/j.1365-246X.2003.02132.x, 2004.

Faccenda, M., Burlini, L., Gerya, T. V., and Mainprice, D.: Fault-induced seismic anisotropy by hydration in subducting oceanic plates, Nature, 455, 1097–1100, doi:10.1038/nature07376, 2008.

Faccenda, M., Gerya, T. V., Mancktelow, N. S., and Moresi, L.: Fluid flow during slab unbending and dehydration: implications for intermediate-depth seismicity, slab weakening and deep water recycling, Geochem. Geophy. Geosy., 13, Q01010, doi:10.1029/2011GC003860, 2012.

Férot, A., and Bolfan-Casanova, N.: Water storage capacity in olivine and pyroxene to 14 GPa: Implications for the water content of the Earth's upper mantle and nature of seismic discontinuities , Earth Planet. Sc. Lett., 349–350, 218–230, doi:10.1016/j.epsl.2012.06.022, 2012.

Fujita, K. and Ogawa, M.: A preliminary numerical study on water-circulation in convecting mantle with magmatism and tectonic plates, Phys. Earth Planet. Int., 216, 1–11, doi:10.1016/j.pepi.2012.12.003, 2013.

Gerya, T. V., Stöckhert, B., and Perchuk, A. L.: Exhumation of high pressure metamorphic rocks in a subduction channel: a numerical simulation, Tectonics, 21, 6-1–6-19, doi:10.1029/2002TC001406, 2002.

Gerya, T. V., Connolly, J. A. D., and Yuen, D. A.: Why is terrestrial subduction one-sided?, Geology, 36, 43–46, doi:10.1130/G24060A.1, 2008.

Gorczyk, W., Gerya, T. V., Connolly, J. A. D., and Yuen, D. A.: Growth and mixing dynamics of the mantle wedge plumes, Geology, 35, 587–590, 2007.

Hacker, B. R., Abers, G. A., and Peacock, S. M.: Subduction factory: 1. Theoretical mineralogy, density, seismic wave speeds, and H_2O content, J. Geophys. Res., 108(B1), 2029, doi:10.1029/2001JB001127,2003.

Handin, J.: On the Coulomb-Mohr failure criterion, J. Geophys. Res., 74, 5343–5348, 1969.

Handy, M., Hirth, G., and Bürgmann, R.: Continental fault structure and rheology from the frictional-to-viscous transition downward, in: Tectonic Faults: Agents of Change on a Dynamic Earth, edited by: Handy, M. R. et al., MIT Press, Cambridge, MA, 139–181, 2007.

Hirschmann, M. M.: Water, melting, and the deep earth H_2O cycle, Annu. Rev. Earth Pl. Sc., 34, 629–653, doi:10.1146/annurev.earth.34.031405.125211, 2006.

Hirth, G. and Kohlstedt, D.: Rheology of the upper mantle and the mantle wedge: a view from the experimentalists, in: Inside the Subduction Factory, edited by: Eiler, J., vol. 138 of Geophys. Monogr. Seer., AGU, Washington, D.C., 83–105, 2003.

Holland, T. J. B. H. and Powell, R.: An internally consistent thermodynamic data set for phases of petrological interest, J. Metamorph. Geol., 16, 309–343, 1998.

Hubbert, M. K. and Rubey, W. W.: Role of fluid pressure in mechanics of overthrust faulting: I. Mechanics of fluid-filled porous solids and its application to overthrust faulting, Geol. Soc. Am. Bull., 70, 115–166, 1959.

Iwamori, H.: Transportation of H_2O and melting in subduction zone, Earth Planet. Sc. Lett., 160, 65–80, 1998.

Iwamori, H.: Seismic evidence for deep-water transportation in the mantle, Chem. Geol., 239, 182–198, 2007.

Jaeger, J. C. and Cook, N. G. W.: Fundamentals of Rock Mechanics, Chapman and Hall, New York, 488 pp., 1976.

Karato, S.-I. and Li, P.: Diffusion creep in perovskite and the linear rheology of the Earth's lower mantle, Science, 255, 1238–1240, 1992.

Katz, R. F., Spiegelman, M., and Holtzman, B.: The dynamics of melt and shear localization in partially molten aggregates, Nature, 442, 676–679, doi:10.1038/nature05039, 2006.

Kaus, B. J. P., Mühlhaus, H.-B., and May, D. A.: A stabilization algorithm for geodynamic numerical simulations with a free surface, Phys. Earth Planet. In., 181, 12–20, doi:10.1016/j.pepi.2010.04.007, 2010.

Kerrick, D. M. and Connolly, J. A. D.: Metamorphic devolatilization of subducted oceanic metabasalts: implications for seismicity, arc magmatism and volatile recycling, Earth Planet. Sc. Lett., 189, 19–29, 2001.

Komabayashi, T., Hirose, K., Funakoshi, K.-I., and Takafuji, N.: Stability of phase A in antigorite (serpentine) composition determined by in situ X-ray pressure observations, Phys. Earth Planet. In., 151, 276–289, 2005.

Lemiale, V., Mühlhaus, H.-B., Moresi, L., and Stafford, J.: Shear banding analysis of plastic models formulated for incompressible viscous flows, Phys. Earth Planet. Int., 171, 177–186, doi:10.1016/j.pepi.2008.07.038, 2008.

Ohtani, E., Litasov, K., Hosoya, T., Kubo, T., and Kondo, T.: Water transport into the deep mantle and formation of a hydrous transition zone, Phys. Earth Planet. In., 143–144, 255–269, doi:10.1016/j.pepi.2003.09.015, 2004.

Peacock, S.: Thermal effects of metamorphic fluids in subduction zones, Geology, 15, 1057–1060, 1987.

Powell, R., Holland, T. J. B. H., and Worley, B.: Calculating phase diagrams involving solid solutions via non-linear equations, with examples using THERMOCALC, J. Metamorph. Geol., 16, 577–588, 1998.

Pysklywec, R. N. and Beaumont, C.: Intraplate tectonics: feedback between radioactive thermal weakening and crustal deformation driven by mantle lithosphere instabilities, Earth Planet. Sc. Lett., 221, 275–292, 2004.

Quinquis, M. E. T., Buiter, S. J. H., and Ellis, S.: The role of boundary conditions in numerical models of subduction zone dynamics, Tectonophysics, 497, 57–70, doi:10.1016/j.tecto.2010.11.001, 2011.

Quinquis, M. E. T., Buiter, S. J. H., Tosi, N., Thieulot, C., Maierová, P., and Quinteros, J.: A numerical model setup for subduction: from linear viscous to thermo-mechanical rheologies, Geophys. Res. Abstracts, 15, EGU2013-7255-1, 2013.

Ranalli, G.: Rheology of the Earth, 2nd edn., Chapman & Hall, London, UK, 392 pp., 1995.

Regenauer-Lieb, K., Yuen, D. A., and Branlund, J. M.: The initiation of subduction: criticality by addition of water?, Science, 294, 578–580, doi:10.1126/science.1063891, 2001.

Richard, G. C. and Bercovici, D.: Water-induced convection in the Earth's mantle transition zone, J. Geophys. Res., 114, B01205, doi:10.1029/2008JB005734, 2009.

Richard, G. C. and Iwamori, H.: Stagnant slab, wet plumes and Cenozoic volcanism in East Asia, Phys. Earth Planet. In., 183, 280–287, doi:10.1016/j.pepi.2010.02.009, 2010.

Richard, G. C., Bercovici, D., and Karato, S.-I.: Slab dehydration in the Earth's mantle transition zone, Earth Planet. Sc. Lett., 251, 156–167, doi:10.1016/j.epsl.2006.09.006, 2006.

Richard, G. C., Monnereau, M., and Rabinowicz, M.: Slab dehydration and fluid migration at the base of the upper mantle: implications for deep earthquake mechanisms, Geophys. J. Int., 168, 1291–1304, doi:10.1111/j.1365-246X.2006.03244.x, 2007.

Rüpke, L. H., Phipps Morgan, J., Hort, M., and Connolly, J. A. D.: Serpentine and the subduction zone water cycle, Earth Planet. Sc. Lett., 223, 17–34, doi:10.1016/j.epsl.2004.04.018, 2004.

Schmeling, H., Babeyko, A. Y., Enns, A., Faccenna, C., Funiciello, F., Gerya, T. V., Golabek, G. J., Grigull, S., Schmalholz, S., and van Hunen, J.: A benchmark comparison of spontaneous subduction models – towards a free surface, Phys. Earth Planet. Int., 171, 198–223, doi:10.1016/j.pepi.2008.06.028, 2008.

Schmidt, M. and Poli, S.: Experimentally based water budgets for dehydrating slabs and consequences for arc magma generation, Earth Planet. Sc. Lett., 163, 361–379, 1998.

Sibson, R. H.: Fault rocks and fault mechanisms, J. Geol. Soc. London, 133, 191–213, 1977.

Sibson, R. H., Moore, J. M., and Rankin, A. H.: Seismic pumping – a hydrothermal fluid transport mechanism, J. Geol. Soc. London, 131, 653–659, 1975.

Spiegelman, M.: Flow in deformable porous media. Part 1. Simple analysis, J. Fluid. Mech., 247, 17–38. 1993a.

Spiegelman, M.: Flow in deformable porous media. Part 2. Numerical analysis - the relationship between shock waves and solitary waves, J. Fluid. Mech., 247, 39–63, 1993b.

Staudigel, H.: Hydrothermal alteration processes in the Oceanic Crust, in: Treatise of Geochemistry, vol. 3, chap. 15, edited by: Holland, H. and Turekian, K., 511–537, 2003.

Stern, R.: Subduction zones, Rev. Geophys., 40, 3-1–3-38, doi:10.1029/2001RG000108, 2002.

Syracus, E. M. and Abers, G. A.: Global compilations of variations in slab depth beneath arc volcanoes and implications, Geochem. Geophy. Geosy., 7, Q05017, doi:10.1029/2005GC001045, 2006.

Tingle, T., Green, H. W., Scholz, C. H., and Koczynski, T. A.: The rheology of faults triggered by the olivine-spinel transformation in Mg_2GeO_4 and its implications for the mechanism of deep-focus earthquakes, J. Struct. Geol., 15, 1249–1256, 1993.

Turcotte, D. L. and Schubert, G.: Geodynamics, 2nd edn., Cambridge Univ. Press, 186 pp., 2002.

Twiss, R. J. and Moores, E. M.: Structural Geology, W. H. Freeman and Company, New York, 532 pp., 1992.

van den Berg, A. P., van Keken, P. E., and Yuen, D. A.: The effects of a composite non-Newtonian and Newtonian rheology on mantle convection, Geophys. J. Int., 115, 62–78, 1993.

Wark, D. A., Williams, C. A., Watson, E. B., and Price, J. D.: Reassessment of pore shapes in microstructurally equilibrated rocks, with implications for permeability of the upper mantle., J. Geophys. Res., 108, B1, doi:10.1029/2001JB001575, 2003.

White, S. H. and Knipe, R. J.: Transformation-and reaction-enhanced ductility in rocks, J. Geol. Soc. London, 135, 513–516, 1978.

12

The sensitivity of GNSS measurements in Fennoscandia to distinct three-dimensional upper-mantle structures

H. Steffen[1] and P. Wu[2]

[1] Lantmäteriet, Lantmäterigatan 2c, 80182 Gävle, Sweden
[2] Department of Earth Sciences, The University of Hong Kong, Pokfulam Road, Hong Kong

Correspondence to: H. Steffen (holger-soren.steffen@lm.se)

Abstract. The sensitivity of global navigation satellite system (GNSS) measurements in Fennoscandia to nearby viscosity variations in the upper mantle is investigated using a 3-D finite element model of glacial isostatic adjustment (GIA). Based on the lateral viscosity structure inferred from seismic tomography and the location of the ice margin at the last glacial maximum (LGM), the GIA earth model is subdivided into four layers, where each of them contains an amalgamation of about 20 blocks of different shapes in the central area. The sensitivity kernels of the three velocity components at 10 selected GNSS stations are then computed for all the blocks.

We find that GNSS stations within the formerly glaciated area are most sensitive to mantle viscosities below and in its near proximity, i.e., within about 250 km in general. However, this can be as large as 1000 km if the stations lie near the center of uplift. The sensitivity of all stations to regions outside the ice margin during the LGM is generally negligible. In addition, it is shown that prominent structures in the second (250–450 km depth) and third layers (450–550 km depth) of the upper mantle may be readily detected by GNSS measurements, while the viscosity in the first mantle layer below the lithosphere (70–250 km depth) along the Norwegian coast, which is related to lateral lithospheric thickness variation there, can also be detected but with limited sensitivity.

For future investigations on the lateral viscosity structure, preference should be on GNSS stations within the LGM ice margin. But these stations can be grouped into clusters to improve the inference of viscosity in a specific area. However, the GNSS measurements used in such inversion should be weighted according to their sensitivity. Such weighting should also be applied when they are used in combination with other GIA data (e.g., relative sea-level and gravity data) for the inference of mantle viscosity.

1 Introduction

It is well known that observations of the glacial isostatic adjustment (GIA) process allow us to determine the earth's viscosity structure, especially that beneath formerly glaciated areas such as Fennoscandia and North America. So far, the most frequently employed GIA data are relative sea-levels, global navigation satellite system (GNSS) measurements and the gravity-rate-of-change data from the Gravity Recovery and Climate Experiment (GRACE) satellite mission (e.g., Peltier, 2004; Wu et al., 2013). These data are commonly used to infer lithospheric thickness and radial variations of mantle viscosity. Owing to recent improvement in modeling techniques and advances in computational power, lateral variations of both lithospheric thickness and mantle viscosity can also be inferred. In view of that, it is important to understand the capability of the many GIA observations for the determination of lateral lithospheric and mantle variations. This study will analyze how sensitive class "A" GNSS stations of the EUREF Permanent Network (EPN, Bruyninx et al., 2013) are to distinct areas in the upper mantle beneath Fennoscandia. (Note that class "A" stations are the best and well-maintained stations of the EPN – they have position accuracy of 1 cm at all epochs of the time span of the used observations (Bruyninx et al., 2013).)

Sensitivity (or Fréchet) kernels of GIA observations at a specific GNSS station show how sensitive they are to viscosity variations in a specific region of the mantle in comparison to another region. The methodology in computing sensitivity kernels for a laterally homogeneous earth can be found in Mitrovica and Peltier (1991) and Peltier (1998), while that for a laterally heterogeneous earth is described in Wu (2006). The fundamentals of Fréchet kernel sensitivity are also developed in seismology (see Dahlen and Tromp, 1998, for a review).

Sensitivity kernels of GIA observations to radial changes in viscosity have been calculated in Mitrovica and Peltier (1991, 1993, 1995), Peltier and Jiang (1996a, b), and Peltier (1998). These studies showed that sensitivity is generally higher in the upper mantle than in the lower mantle. This is especially true for data in Fennoscandia, where the resolving power of GIA observations is too low to provide accurate inference of lower-mantle viscosity (Mitrovica and Peltier, 1993; Steffen and Wu, 2011). However, the data in North America can see deeper because the load is wider and the sensitivity is higher down to the shallow part of the lower mantle (Mitrovica and Peltier, 1995).

The sensitivity kernels for selected stations of the BIFROST (Baseline Inferences for Fennoscandian Rebound Observations, Sea Level, and Tectonics) project to radial viscosity variations have been studied by Milne et al. (2004). Interestingly, they found sufficiently high sensitivities for the lower mantle. This was not supported by Steffen et al. (2006), who showed with a 3-D model, that lateral variations in lower-mantle viscosity do not affect the GNSS velocity field in Fennoscandia. As pointed out in Wu (2006), the sensitivity of the Fennoscandian data to the lower mantle may actually be due to Laurentia.

When lateral viscosity is included in the earth model, the normal mode formulation of Mitrovica and Peltier (1991) and Peltier (1998) no longer applies or becomes impractical to apply due to mode coupling (Wu, 2002). To overcome this, Wu (2006) showed that the sensitivity kernel can equivalently be computed from the difference in response between a model with a small but fixed perturbed viscosity in a single mantle block (or layer) at the location of interest and the response of the reference model without the perturbation. In Wu (2006), an axisymmetric (2-D) earth model and simplistic ice load, with size comparable to the Laurentian ice load, are used to study the sensitivity of global GIA data. Later, Steffen et al. (2007) employed a 3-D earth model with a realistic (4-D) ice model to study the sensitivity of GNSS stations in Fennoscandia. An advantage of the latter study is that all three horizontal components can be investigated. In addition, the spatial resolution is much higher, with element block sizes of 600 km × 600 km or 1000 km × 1000 km. In any case, both studies showed that sensitivity is highest within the formerly glaciated area. The radial or vertical variation of sensitivity kernels for uplift rate peaks around 300–450 km depth but becomes small below the upper mantle. Also, both the load distribution and the deglaciation history strongly affect the magnitude of sensitivity (Steffen et al., 2007). Furthermore, if one is interested in the lateral viscosity outside

the former ice margin, then the horizontal velocities should be analyzed, although that also depends on the size of the perturbed mantle region.

It should be noted that in the models of Wu (2006) and Steffen et al. (2007), the perturbed viscosity has a fixed magnitude and lies in a single rectangular block in an otherwise laterally homogeneous mantle. When lateral heterogeneity in the mantle is taken into account, the shape of the blocks must be modified to reflect the shape of the lateral heterogeneity. Also, the magnitude of the viscosity in the block must reflect the true viscosity value there. This study will include such changes for a 3-D viscosity model and a realistic ice load history will also be used.

The GNSS stations where sensitivity kernels are computed in this study (Fig. 1) belong to the class "A" stations, except for station Vaasa, which is of class "B" (positions with an accuracy of 1 cm at the epoch of minimal variance of the station). The selected stations are also used in BIFROST investigations to GIA (see Lidberg et al., 2010).

The aims of this study are (i) to investigate the sensitivity of velocity fields at selected GNSS stations to certain regions of the mantle with similar viscosity and location relative to the former ice margin; and (ii) outline where (future) GNSS stations in Fennoscandia would be helpful to identify lateral viscosity changes. The next section describes the model in more detail. This is followed by the presentation and discussion of the results. Finally, the conclusion is presented in Sect. 5.

2 Modeling

The finite-element method is used to model the GIA process in Fennoscandia. The earth model used is flat with isotropic, compressible, Maxwell-viscoelastic layers, but the finite-element model allows both vertical and lateral variations to be taken into account. This model is described in Steffen et al. (2006) and is based on the approach of Wu (2004) which has been used successfully in many GIA investigations in North America (e.g., Wu, 2005), Fennoscandia (e.g., Steffen et al., 2006), the Barents Sea (e.g., Kaufmann and Wu., 1998), Antarctica (e.g., Kaufmann et al., 2005) and Iceland (e.g., Schmidt et al., 2012). The model consists of a central area of 3000 km × 3000 km size, where each element has a horizontal length of 100 km. The ice-load history model FBKS8 (Lambeck et al., 1998) is applied to the surface in the central area. Surrounding the central area is a 60 000 km wide peripheral area, that is connected to infinity horizontally with semi-infinite elements.

In this preliminary study, where we are not interested in the sensitivity of small-scale features in the mantle, we continue to use the laterally heterogeneous model U3L1_V1 in Steffen et al. (2006) to define the viscosity and shape of the blocks. This model has a uniform 70 km thick lithosphere on top of a laterally heterogeneous mantle which consists of

EPN/BIFROST stations used

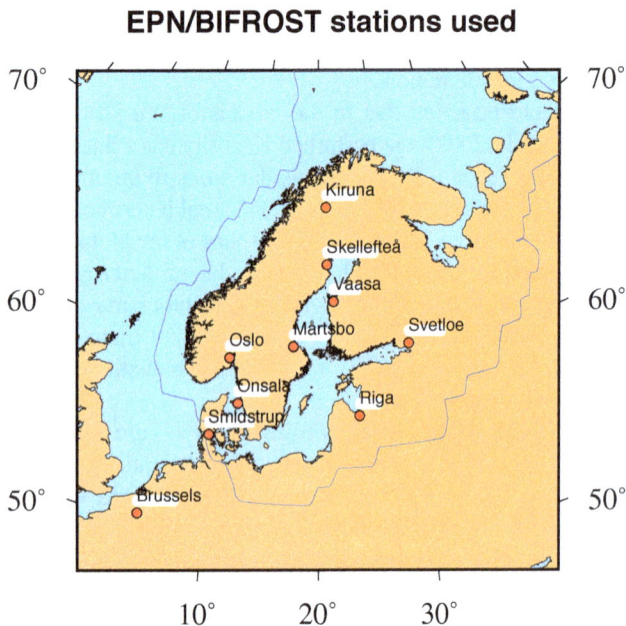

Figure 1. Location of EPN/BIFROST station used in our investigation. The blue line marks the location of the ice margin at the Last Glacial Maximum.

four layers in the upper mantle and another four in the lower mantle. The lateral viscosity variations in each layer of the upper mantle are converted from the SH shear wave tomography model S20A (Ekström and Dziewonski, 1998) using the scaling relationship derived from Ivins and Sammis (1995), but modified by Steffen et al. (2006). The viscosity structure within the four upper-mantle layers is shown in Fig. 2 with solid black contour lines. Model U1L1_V1 of Steffen et al. (2006) is used as the reference model, on which the blocks with lateral viscosity are superposed. It has an upper-mantle viscosity value of 4×10^{20} Pa s, which is a good average value of upper-mantle viscosity beneath Fennoscandia (Steffen and Wu, 2011). Lower-mantle viscosity is set to 2×10^{22} Pa s (Steffen and Kaufmann, 2005). Elastic parameters (density ρ, shear modulus μ and bulk modulus κ) of the model are obtained by volume-averaging the values in the Preliminary Reference Earth Model (PREM; Dziewonski and Anderson, 1981) and are shown in Table 1 of Steffen et al. (2006).

To investigate the sensitivity of lateral heterogeneity in the mantle, finite elements with similar viscosities (i.e., within one order of magnitude in the first mantle layer and within half an order of magnitude in the three other layers) are grouped together to form blocks, so that the blocks reflect the lateral viscosity structure within each layer (see red lines in Fig. 2). These blocks of similar viscosity are further subdivided into groups of blocks that lie inside the former ice margin and those lying outside in order to study the effect of location relative to the former ice margin. In addition, we design three groups of blocks in the center of uplift that have

the same shape in all four layers. This will help us investigate the sensitivity as a function of depth without the shape-effects of the blocks. The number of blocks in the first mantle layer is 22, while that in the second and bottom layers is 19 and the third layer has 18 blocks. That means 78 models are generated – where each model has a different block of lateral heterogeneity included in Model U1L1_V1. One should note that different ice-load models (e.g., ICE-5G; Peltier, 2004) or different seismic tomography models (e.g., Grand et al., 1997) will give different shapes of the subdivisions. But, that should not significantly change the major conclusions of this investigation which is quite general in nature.

3 Results

The normalized sensitivity kernels $K_{lj}(r_i)$ of block j in layer i at location l is computed based on the approach of Wu (2006), which is modified from Peltier (1998):

$$K_{lj}(r_i) = \frac{\delta p_l}{\delta m_j(r_i) \, \Delta V_j(r_i) \{V_{\max}(r_i)\}}, \tag{1}$$

where δp_l is the difference between the prediction $p_l^{3-\mathrm{D}}$ of the perturbed 3-D model and the prediction $p_l^{1-\mathrm{D}}$ of the reference model U1L1_V1 at GNSS location l. (Here, the prediction p_l is one of the three velocity components.) $\delta m_j(r_i)$ is the (dimensionless) viscosity perturbation of block j in layer i (i.e., the difference between the log of the viscosity of the block in model U3L1_V1 and that in model U1L1_V1 which is log of 4×10^{20} Pa s). Also, $\Delta V_j(r_i)$ is the fractional volume of this particular block. The latter is given by

$$\Delta V_j(r_i) = \frac{V_j(r_i)}{V_{\mathrm{model}}}, \tag{2}$$

where $V_j(r_i)$ is the block volume, and V_{model} the volume of the entire central area in the model, which includes the upper and lower mantle. (For example, $V_1(r_2)$ refers to the volume of block 1 in the second layer shown in Fig. 2.) Normalization by this term is useful as we are only interested in the relative amplitude of the sensitivity kernels, i.e., which viscosity block is comparatively more sensitive to the particular measurement at a GNSS station. Unlike the approach of Wu (2006), we introduce the extra dimensionless normalization factor $\{V_{\max}(r_i)\}$, which is the value of the maximum fractional volume of the four layers investigated. This normalization is introduced here mainly for plotting purpose.

The kernels for the three velocity components are calculated for the location of each of the 10 selected EPN stations. Thus, we are able to analyze the relative sensitivity of the station to every block. Figure 3 presents two typical sensitivity kernels for the three velocity components (EW, NS, Z) to all the different viscosity blocks in the model at the two stations Kiruna and Brussels. Kiruna (Fig. 3a) is located above block 1 and also not too far away from the center of rebound.

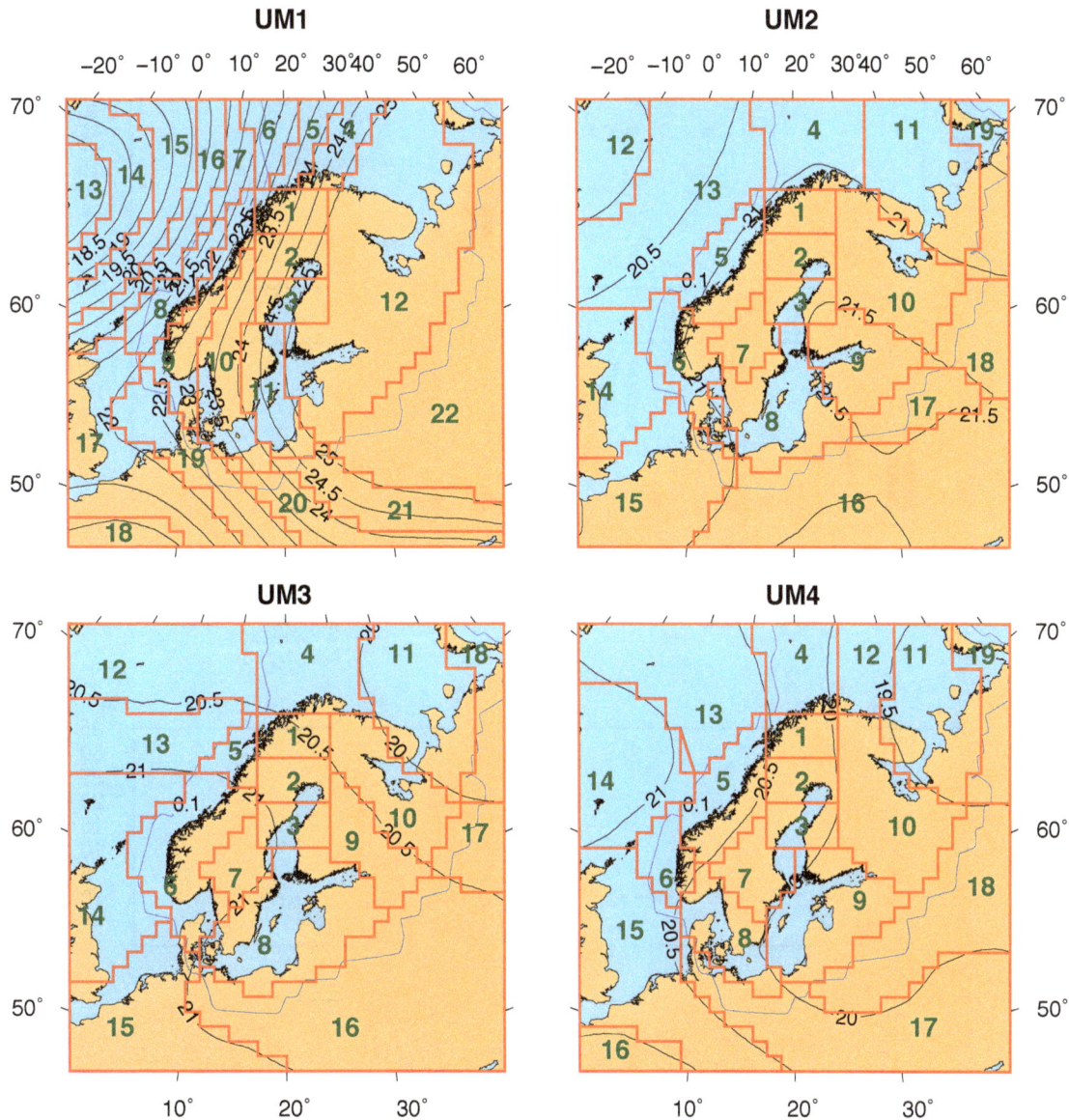

Figure 2. Distribution of viscosity blocks (red lines, green numbers) based on the lateral viscosity structure (black lines) as calculated from seismic tomography model S20A (Ekström and Dziewonski, 1998). Viscosity blocks are additionally subdivided into blocks inside and outside the former glaciated area (blue line, based on ice model FBKS8; Lambeck et al., 1998). Depth ranges: UM1 70–250 km, UM2 250–450 km, UM3 450–550 km, UM4 550–670 km.

Figure 3 clearly shows that sensitivity of any velocity component to a block in the first mantle layer (70–250 km depth) is small. However, sensitivity is highest for the blocks right underneath the station or close to it, provided that the blocks lie within the former ice margin. The largest sensitivity for the vertical velocity is in block 1 of the second mantle layer (250–450 km depth). This block also has the highest sensitivity for any velocity component at any one of the 10 selected stations. Sensitivity is smaller in the two neighboring blocks in the north (block 4) and south (block 2), and become almost negligible for all other blocks in layer 2. In blocks 1, 2 and 5 in the third mantle layer (450–550 km depth) as well

as blocks 1 and 4 in the fourth mantle layer (550–670 km depth), sensitivity of the vertical component is also larger than in other blocks of that layer. Sensitivities for horizontal velocities are generally smaller, but they also peak in the second layer and their amplitudes decrease with depth. The exception is the EW component in block 5of the third layer, which shows the largest sensitivity for a horizontal component of all stations.

For the station in Brussels (Fig. 3b), which is located outside the former ice margin, sensitivities of all velocities to any block in any layer become almost negligible. This is true even for the viscosity blocks directly underneath the station.

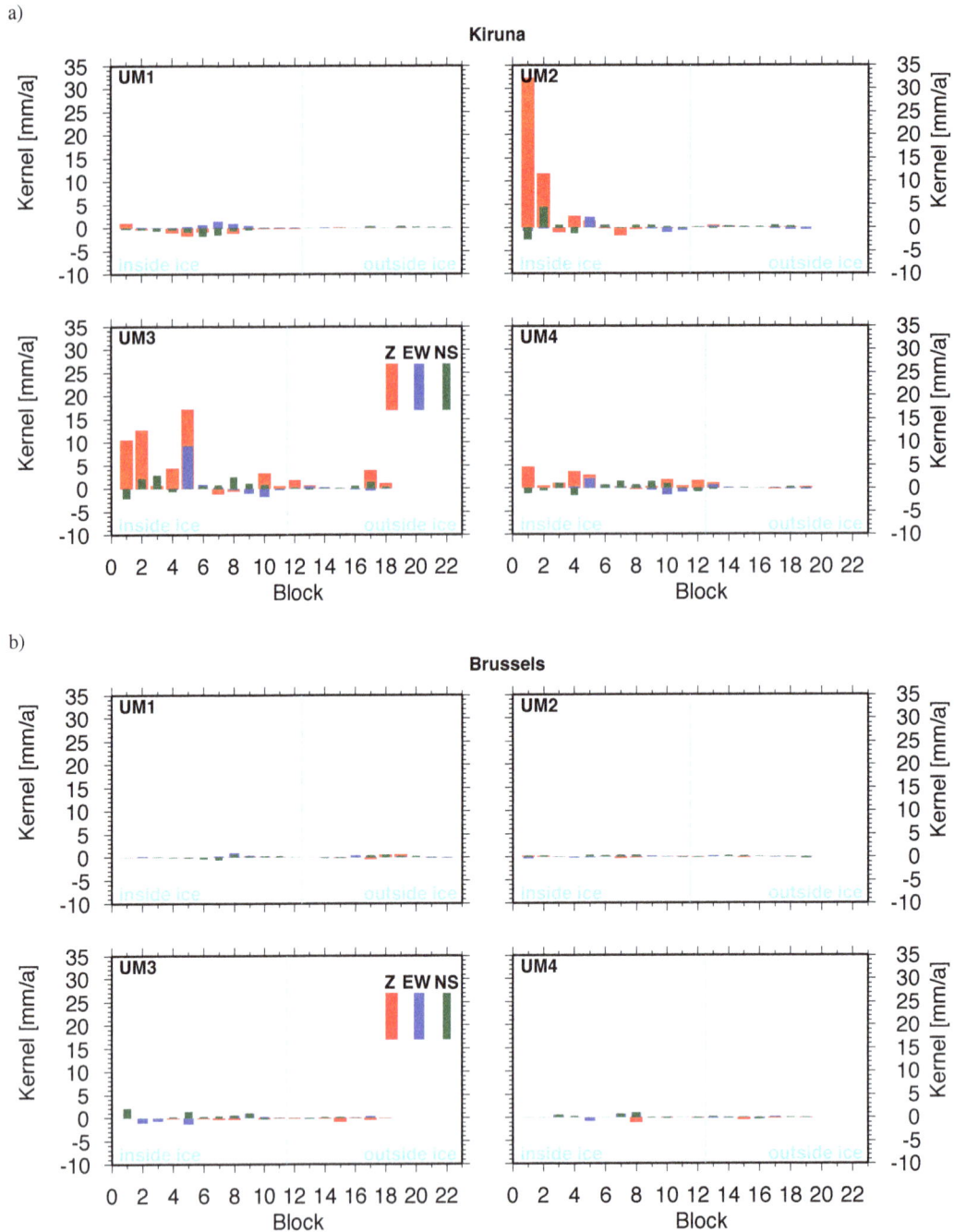

Figure 3. Sensitivity kernels shown as bars for vertical (red) and horizontal velocities (EW: blue, NS: green) at the stations of **(a)** Kiruna, and **(b)** Brussels to viscosity blocks in the four layers of the upper mantle (UM1–UM4). Depth ranges: UM1 70–250 km, UM2 250–450 km, UM3 450–550 km, UM4 550–670 km.

Any sensitivity that shows up marginally is related to horizontal velocities and they are located within the former ice margin.

The presentation of kernels as in Fig. 3 does not allow us to see visually where the blocks with significant sensitivity lie in the map. To overcome this, we set an arbitrary threshold value for the normalized sensitivity and plot only the blocks with sensitivity above the threshold value in Figs. 4–13. (Likewise, in the rest of the paper, when we say the data is sensitive to a block, we mean that the sensitivity of the block is above the threshold.) Among the 10 stations, Kiruna shows the largest kernel values, and Brussels has the lowest. After testing, we find a threshold of $3\,\mathrm{mm\,yr^{-1}}$ for the vertical velocity. A threshold of $1\,\mathrm{mm\,yr^{-1}}$ is found for

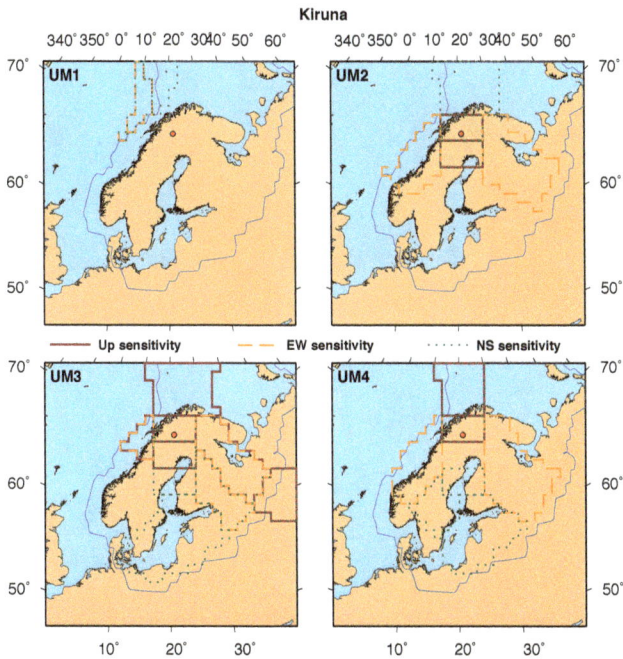

Figure 4. Location of viscosity blocks in each layer where sensitivity kernels of the velocities in Kiruna (red dot) lie above the selected threshold. Red solid line: vertical velocity. Orange dashed line: EW velocity. Green dotted line: NS velocity. Blue line: former glaciated area based on ice model FBKS8. Depth ranges: UM1 70–250 km, UM2 250–450 km, UM3 450–550 km, UM4 550–670 km.

Figure 5. Same as Fig. 4, but for Skellefteå.

the horizontal velocities. (Note that these threshold values as well as the normalized sensitivity kernels should NOT be compared to actually observed velocities of GNSS measurements because they are physically different quantities.) Figures 4–13 show the blocks where sensitivity of the velocities observed at the stations exceed the threshold values.

For example, GNSS velocities measured at Kiruna (Fig. 4) have sensitivity to several blocks in each layer. However, not all components of the velocity field are sensitive to the same viscosity block. In the first mantle layer, vertical velocity is insensitive to any viscosity block but horizontal velocities show sensitivity to blocks west of the station at the edge of the former ice sheet. In the second mantle layer, the vertical and NS components have sensitivity to the block below the station as well as to the block south of it. In addition, the NS component is also sensitive to the block north of it while blocks in the west and east are detectable by the EW component. This shows that horizontal velocities may provide information for adjacent blocks. In the third layer, there are more viscosity blocks with sufficiently large sensitivity than in the second layer. The vertical component is sensitive to the underlying block and blocks surrounding it, the EW component to blocks west and east of the station, and the NS component to the underlying block and all three blocks south of the station within the former ice margin. Both NS and EW component are also sensitive to a block southeast of

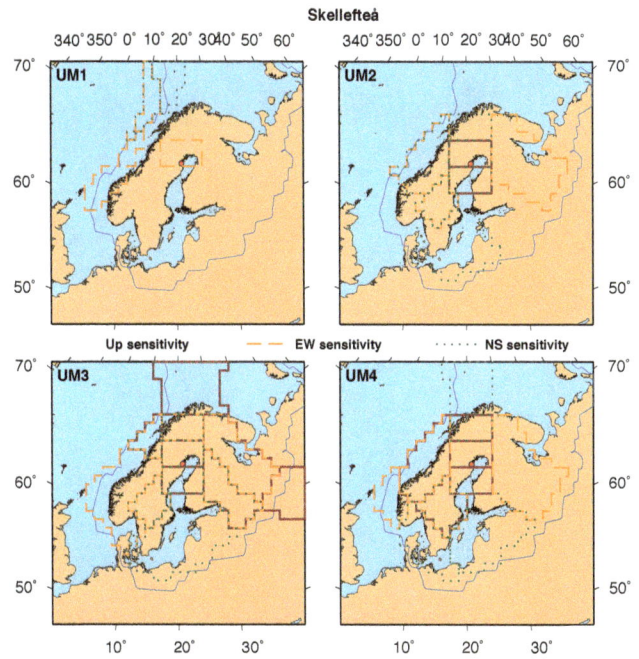

the station. At the bottom of the upper-mantle layer, the underlying block and the one north of it influence the vertical component. Several blocks have sensitivity in the horizontal components, which have similar characteristics as found for the third mantle layer.

The reader can continue to look at how sensitivity varies with depth at each station in Figs. 5–13. However, it is more profitable to compare the block sensitivities for all stations at a certain layer of the upper mantle. This is what we will do next.

4 Discussion

Let us begin with the first mantle layer (70–250 km depth). Horizontal components of all stations except Brussels are sensitive to at least one block of the area along the Norwegian Atlantic coast (blocks 6–8 in UM1 in Fig. 2), where there is a strong gradient in the viscosity of the first mantle layer. Since lateral thickening of the lithosphere from west to east under Fennoscandia (Steffen and Wu, 2011) can appear as a strong viscosity gradient in the first mantle layer, sensitivity to viscosity blocks in the first mantle layer along the Norwegian Atlantic coast also implies sensitivity to lithospheric thickness variations there. The station of Vaasa is the only one where the vertical component is sensitive to a block (the underlying one). Therefore, the vertical component does most likely not contain sufficient information of the first mantle layer.

For the second mantle layer (250–450 km depth), the GNSS stations in the center of rebound or in the southern

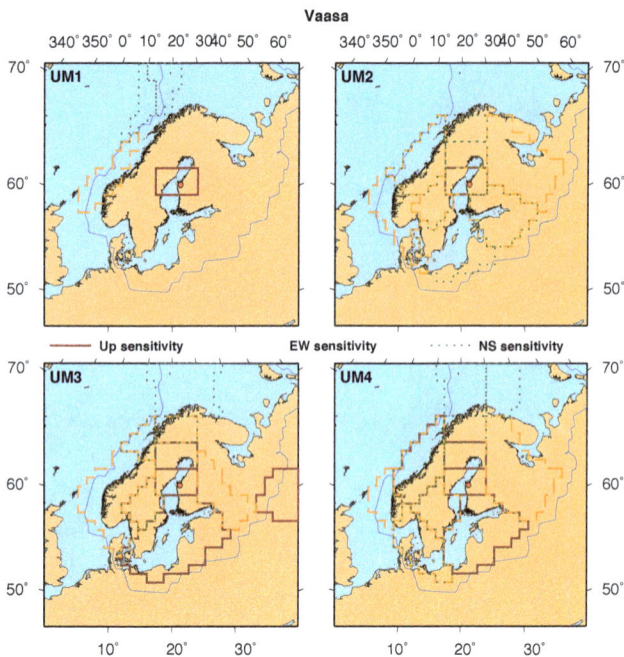

Figure 6. Same as Fig. 4, but for Vaasa.

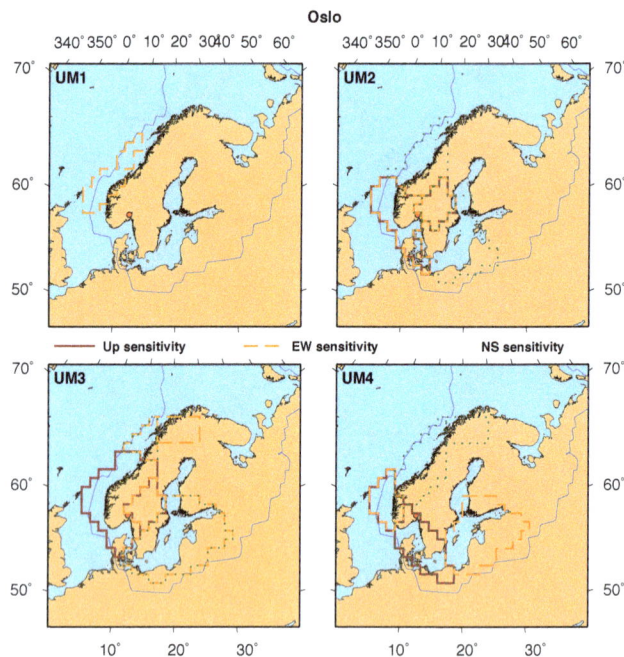

Figure 8. Same as Fig. 4, but for Oslo.

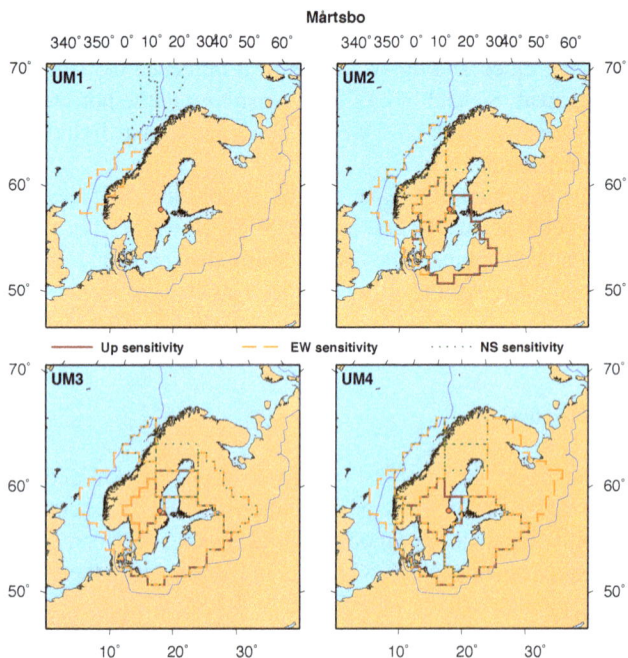

Figure 7. Same as Fig. 4, but for Mårtsbo.

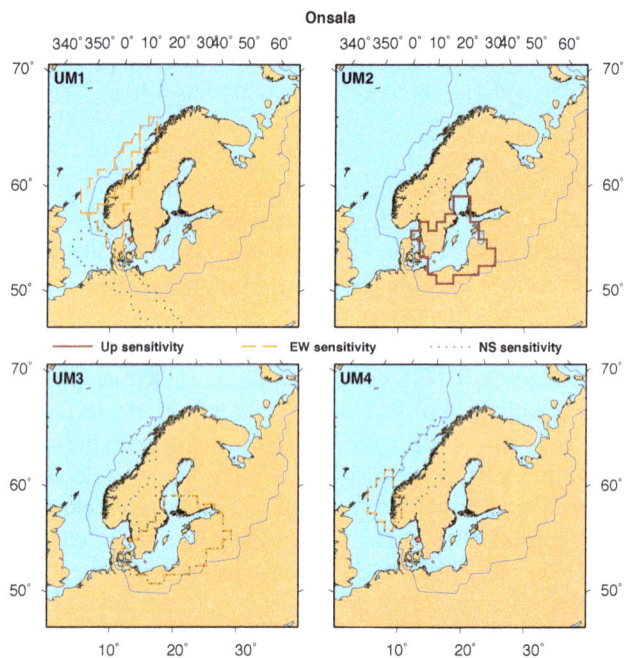

Figure 9. Same as Fig. 4, but for Onsala.

part of the Scandinavian Peninsula (Figs. 4–10) are sensitive to the underlying block in this layer where maximum sensitivity for a particular station is usually found. Significant sensitivities for the vertical component are also found for the surrounding blocks. For the horizontal components, blocks located in the north or south of the station generally show sensitivity in the NS component, while blocks in the east or west of the station reflect sensitivity in the EW component. The two stations near the center of rebound, Skellefteå and Vaasa, are sensitive to the largest number of blocks in the second layer. The number of blocks in the second layer is reduced as the station moves away from the center of rebound. Also, these blocks are generally located in the vicinity of the GNSS station, i.e., within a radius of about 500 km. Svetloe

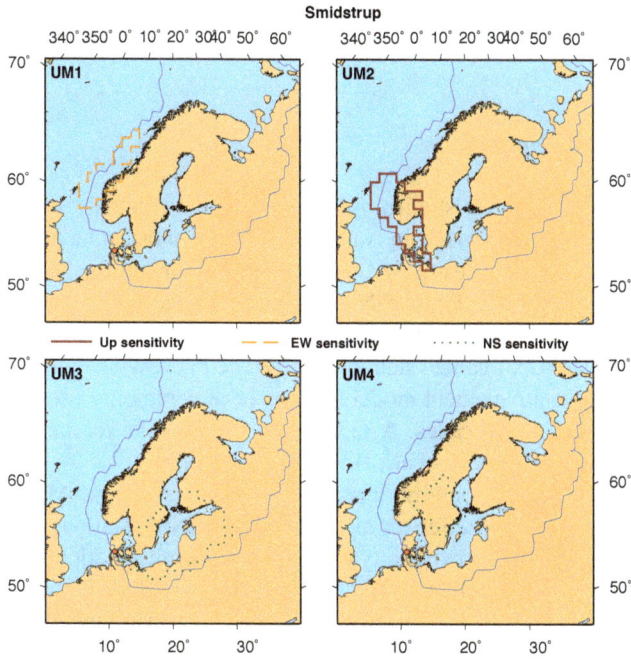

Figure 10. Same as Fig. 4, but for Smidstrup.

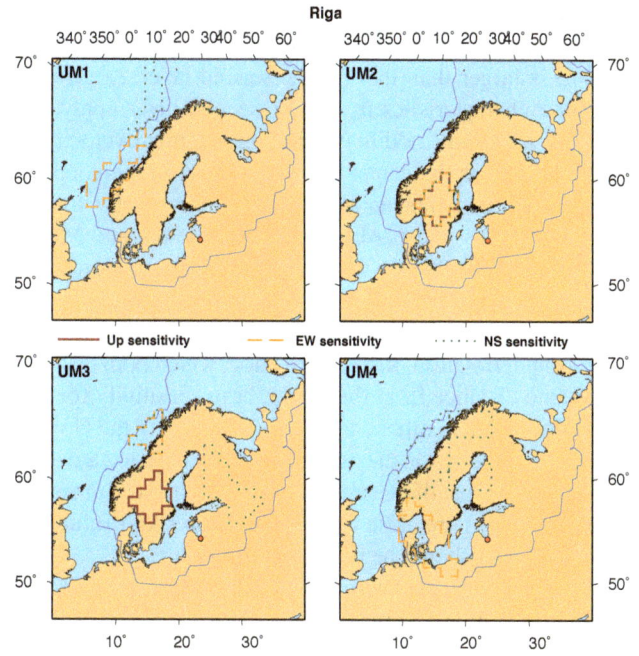

Figure 12. Same as Fig. 4, but for Riga.

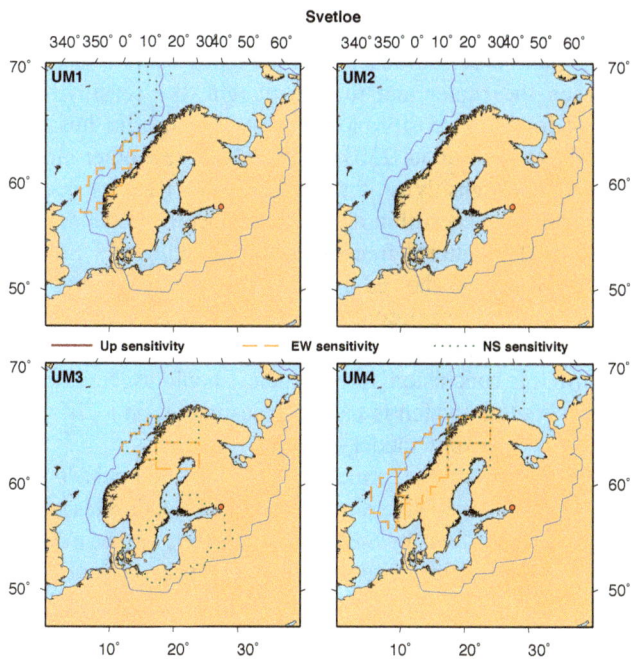

Figure 11. Same as Fig. 4, but for Svetloe.

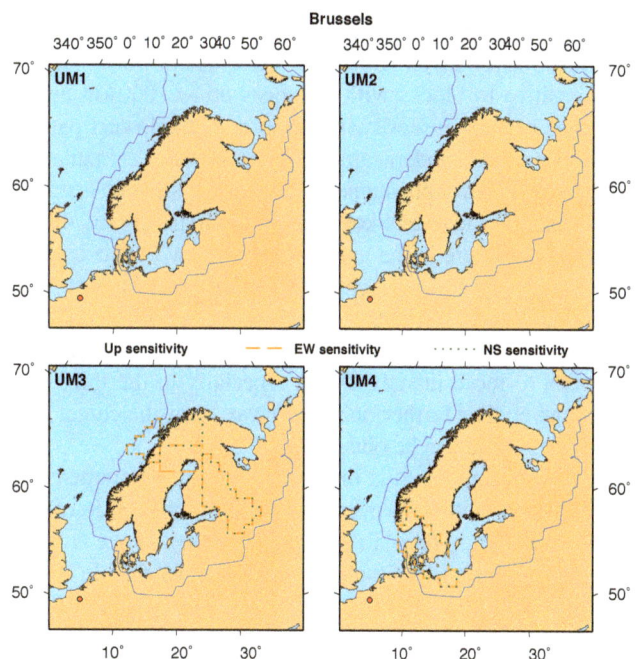

Figure 13. Same as Fig. 4, but for Brussels.

and Brussels show no sensitivity to any block in this layer, while Riga has sensitivity (for all components) to block number 7 in the second layer (Fig. 2).

Block number 7, which lies below the southern part of the Scandinavian Peninsula, is of special interest. Many stations show sensitivity of at least one velocity component to this particular block. In other words, the viscosity of this block affects the measurements of these stations within the former ice margin, and strongly affects the value of viscosity inverted from such GNSS measurements. Also, by comparing the viscosity inverted from the vertical component of GNSS stations within the former ice margin and the viscosity inverted without the stations in the center, one should get a feeling of the accuracy of the viscosity inverted for this area.

For the third mantle layer (450–550 km depth), the number of blocks with enough sensitivity (for any velocity component) is larger than that for the second layer. In general, the same characteristics as for the second layer apply, but there are also some additional findings. Vertical velocity of the stations near the center of rebound and in the southern part of the Scandinavian Peninsula has sensitivity to surrounding blocks and partly (Kiruna, Skellefteå and Vaasa) to blocks about 1000 km away from the station. Horizontal components at the stations Svetloe and Brussels are now sensitive to certain blocks which are located more than 800 km away from the station. Riga and Svetloe, which both lie at intermediate distance from the center, have sensitivity for one of the horizontal components to one block along the Norwegian coast, which is on the opposite side of the glaciated area. However, note that the sensitivity values for most of these additional findings are close to the chosen threshold value and so these results should be interpreted with care.

The sensitivity of the fourth mantle layer (550–670 km depth) has similar characteristics as that of the third mantle layer, but the vertical component with enough sensitivity is only found in the underlying block and/or blocks within 400 km distance from the station. The stations in Riga and Svetloe still have sensitivity to the Norwegian coast area. Brussels has sensitivity to an area closer to the station than in the third mantle layer. Horizontal velocities of most stations are sensitive to blocks with distances up to 1000 km away, and thus provide viscosity information of the lowest part of the upper mantle underneath the Scandinavian Peninsula.

Our results strongly support the usage of stations near the center of rebound (e.g., Skellefteå or Vaasa) to investigate the viscosity structure in the upper mantle below Fennoscandia. The vertical component gives information of the viscosity structure in an area of 500 to 1000 km around the station from about 250 to 670 km depth. Horizontal velocities may enlarge this area to more than 1000 km, especially in the third and fourth layer. The farther one goes away from the center, the less information can be obtained.

An interesting result is that the horizontal components at many GNSS stations – even those on the other side of the former ice margin (e.g., in Riga and Svetloe) – have enough sensitivity in almost all the layers in the upper mantle below the Norwegian Atlantic coast. Thus, thorough analysis of the horizontal velocities in Fennoscandia can probably result in better estimation of viscosity or lithospheric thickness variations there.

Stations outside the former glaciated area are of limited value, e.g., sensitivities found for Brussels are very close to the threshold value and thus are quite small compared to values found for stations in the center of rebound.

Oceanic areas far off the coast, i.e., the ones that were never affected by ice load on top, do not show any significant sensitivity at any GNSS station. Similarly, blocks in the southwest do not have enough sensitivity. However, if ice load on the British Isles is investigated, then the situation

may be different. Future investigations with British GNSS stations should analyze their potential sensitivity for the area.

5 Conclusions

Unlike previous studies, this paper includes realistic structures of lateral viscosity variation under Fennoscandia to investigate the sensitivity of GNSS measurements in 10 selected stations. These GNSS stations are backbones of the EPN and the BIFROST project and thus represent excellent and well-maintained stations of high accuracy. We employed a 3-D finite element model that has been commonly used in the last two decades. A realistic ice load of the ice model FBKS8 (Lambeck et al., 1998) was also applied.

Our results confirm previous findings (see e.g., Milne et al., 2004; Steffen et al., 2007) that GNSS stations are most sensitive to viscosity changes underneath a station, but mainly at a depth between 250 and 550 km. Both horizontal and vertical velocities show significant sensitivities provided that the GNSS station and the block are located within the former ice margin. The depth of sensitivity depends on the ice thickness – thinner ice gives less information on the fourth layer, which confirms the resolving power of GIA data in general.

For stations closer to the center of rebound or mid-distance between the center and the ice margin, the sensitivity is largest for the viscosity blocks right underneath (thus to a lateral extent of about 250 km), but also a few other blocks nearby have sensitivity if these blocks or parts of such a block are located within a lateral distance of about 500 to 1000 km. The latter is in contrast to the findings by Steffen et al. (2007), who showed that the sensitivity of neighboring blocks is mainly negligible. This difference is related to the regular block structure used in Steffen et al. (2007). To test or confirm this conclusion, future studies should use a different block structure, which is based, for example, on a different seismic tomography model.

Stations outside the former glaciated area do not have sufficient sensitivity to viscosity directly underneath. This is different to the findings by Steffen et al. (2007), who found that horizontal velocities of such stations might be helpful. This is probably due to the approach of averaging the kernels of a block they used. This may have increased the kernel value for blocks that covered glaciated and non-glaciated areas. It should be noted that Steffen et al. (2007) already suggested using a more realistic viscosity block structure for a sophisticated analysis to find out if their result is correct or not. In turn, it is indicated that horizontal velocities at stations outside the former glaciated area have sensitivity to certain regions within the former glaciated area at 450–670 km depth, which should be further investigated in the future as well.

Regarding the planning of future GNSS stations for GIA research, Wu et al. (2010) investigated optimal locations in Fennoscandia and suggested more stations in the Baltic

States and NW Russia. Our results clearly support this argument as both regions are located within the former glaciated area. Furthermore, the dense network installed in the countries of Norway, Sweden, Finland and Denmark will be further densified in the next few years. In Sweden, for example, the network will consist of 400 stations by 2020 (Lantmäteriet, 2011). Recently, 20 stations of the existing network have been proposed as new EPN stations (Engberg et al., 2013), which means that the quality of the observed data will increase. Together with the existing network of GNSS stations, they should allow a thorough investigation of lateral viscosity structure under Fennoscandia.

The results from this study are helpful in future investigations on lateral variations of mantle viscosity and lithospheric thickness. We recommend a careful grouping of GNSS velocity measurements from selected areas, e.g., from the uplift center or the Baltic States, to investigate the vertical viscosity profile underneath the center of rebound or the viscosity structure of the Norwegian Atlantic coast, respectively. In this or in combined analyses with other GIA observations, such as relative sea-level data or gravity on ground and in space (Steffen et al., 2012, 2014), the observations should be properly weighted according to their sensitivity to a specific region.

Acknowledgements. We thank Kurt Lambeck for providing the FBKS8 ice model and two anonymous reviewers for their useful comments and suggestions. Figures are prepared using GMT software (Wessel and Smith, 1998).

Special Issue: "Lithosphere-cryosphere interactions"
Edited by: M. Poutanen, B. Vermeersen, V. Klemann, and C. Pascal

References

Bruyninx, C., Altamimi, Z., Caporali, A., Kenyeres, A., Lidberg, M., Stangl, G., and Torres, J. A.: Guidelines for EUREF densifications. IAG sub-commission for the European Reference Frame – EUREF, available at: ftp://epncb.oma.be/pub/general/Guidelines_for_EUREF_Densifications.pdf, 2013.

Dahlen, F. A. and Tromp J.: Theoretical Global Seismology. Princeton University Press, Princeton, NJ, 1998.

Dziewonski, A. M. and Anderson D. L.: Preliminary reference Earth model, Phys. Earth Planet. Inter., 25, 297–356, doi:10.1016/0031-9201(81)90046-7, 1981.

Ekström, G. and Dziewonski A. M.: The unique anisotropy of the Pacific upper mantle, Nature, 394, 168–172, doi:10.1038/28148, 1998.

Engberg, L. E., Engfeldt, A., Jivall, L., Kempe, C., Lidberg, M., Lilje, C., Lilje, M., Norin, D., Steffen, H., Wiklund, P., and Ågren, J.: National Report of Sweden to the EUREF 2013 Symposium – activities at Lantmäteriet, Report, 8 pp., available at: http://euref2013.fomi.hu/Download/Session_6/Sweden_NationalReport2013_paper.pdf, 2013.

Grand, S. P., Van Der Hilst, R. D., and Widiyantoro, S.: Global seismic tomography: A snapshot of convection in the earth, GSA Today, 7, 1–7, 1997.

Ivins, E. R. and Sammis, C. G.: On lateral viscosity contrast in the mantle and the rheology of low frequency geodynamics, Geophys. J. Int., 123, 305–322, doi:10.1111/j.1365-246X.1995.tb06856.x, 1995.

Kaufmann, G. and Wu, P.: Lateral asthenospheric viscosity variations and postglacial rebound: A case study for the Barents Sea, Geophys. Res. Lett., 25, 1963–1966, doi:10.1029/98GL51505, 1998.

Kaufmann, G., Wu, P., and Ivins, E. R.: Lateral viscosity variations beneath Antarctica and their implications on regional rebound motions and seismotectonics, J. Geodyn., 39, 165–181, doi:10.1016/j.jog.2004.08.009, 2005.

Lambeck K., Smither, C., and Johnston, P.: Sea-level change, glacial rebound and mantle viscosity for northern Europe, Geophys. J. Int., 134, 102–144, doi:10.1046/j.1365-246x.1998.00541.x, 1998.

Lantmäteriet: Geodesy 2010 – A strategic plan for Lantmäteriet's geodetic activities 2011–2020, Lantmäteriet, Gävle, Sweden, available at: http://www.lantmateriet.se/Global/Kartor%20och%20geografisk%20information/GPS%20och%20m%c3%a4tning/Geodesi/Rapporter_publikationer/Publikationer/Geodesy_2010.pdf, 2011.

Lidberg, M., Johansson, J. M., Scherneck, H.-G., and Milne, G. A.: Recent results based on continuous GPS observations of the GIA process in Fennoscandia from BIFROST, J. Geodyn., 50, 8–18, doi:10.1016/j.jog.2009.11.010, 2010.

Milne, G. A., Mitrovica, J. X., Scherneck, H.-G., Davis, J. L., Johansson, J. M., Koivula, H., and Vermeer, M.: Continuous GPS measurements of postglacial adjustment in Fennoscandia: 2. Modeling results, J. Geophys. Res., 109, doi:10.1029/2003JB002619, 2004.

Mitrovica, J. X. and Peltier, W. R.: A complete formalism for the inversion of post glacial rebound data: Resolving power analysis, Geophys. J. Int., 104, 267–288, doi:10.1111/j.1365-246X.1991.tb02511.x, 1991.

Mitrovica, J. X. and Peltier, W. R.: The inference of mantle viscosity from an inversion of the Fennoscandian relaxation spectrum, Geophys. J. Int., 114, 45–62, doi:10.1111/j.1365-246X.1993.tb01465.x, 1993.

Mitrovica, J. X. and Peltier, W. R.: Constraints on mantle viscosity based upon the inversion of post-glacial uplift data from the Hudson Bay region, Geophys. J. Int., 122, 353–376, doi:10.1111/j.1365-246X.1995.tb07002.x, 1995.

Peltier, W. R.: The inverse problem for mantle viscosity, Inverse Problems, 14, 441–478, doi:10.1088/0266-5611/14/3/006, 1998.

Peltier, W. R.: Global glacial isostasy and the surface of the Ice-Age Earth: the ICE-5G (VM2) model and GRACE, Annu. Rev. Earth Pl. Sc., 32, 111–149, doi:10.1146/annurev.earth.32.082503.144359, 2004.

Peltier, W. R. and Jiang, X.: Glacial isostatic adjustment and earth rotation: refined constraints on the viscosity of the deepest mantle, J. Geophys. Res., 101, 3269–3290, doi:10.1029/95JB01963, 1996a.

Peltier, W. R. and Jiang, X.: Mantle viscosity from the simultaneous inversion of multiple data sets pertaining to postglacial rebound,

Geophys. Res. Lett., 23, 503–506, doi:10.1029/96GL00512, 1996b.

Schmidt, P., Lund, B., Árnadóttir, T., and Schmeling, H.: Glacial isostatic adjustment constrains dehydration stiffening beneath Iceland, Earth Planet. Sci. Lett., 359–360, 152–161, doi:10.1016/j.epsl.2012.10.015, 2012.

Steffen, H. and Kaufmann, G.: Glacial isostatic adjustment of Scandinavia and northwestern Europe and the radial viscosity structure of the Earth's mantle, Geophys. J. Int., 163, 801–812, doi:10.1111/j.1365-246X.2005.02740.x, 2005.

Steffen, H. and Wu, P.: Glacial isostatic adjustment in Fennoscandia – A review of data and modeling, J. Geodyn., 52, 169–204, doi:10.1016/j.jog.2011.03.002, 2011.

Steffen, H., Kaufmann, G., and Wu, P.: Three-dimensional finite-element modelling of the glacial isostatic adjustment in Fennoscandia, Earth Planet. Sci. Lett., 250, 358–375, doi:10.1016/j.epsl.2006.08.003, 2006.

Steffen, H., Wu, P., and Kaufmann, G.: Sensitivity of crustal velocities in Fennoscandia to radial and lateral viscosity variations in the mantle, Earth Planet. Sci. Lett., 257, 474–485, doi:10.1016/j.epsl.2007.03.002, 2007.

Steffen, H., Wu, P., and Wang, H. S: Optimal locations for absolute gravity measurements and sensitivity of GRACE observations for constraining glacial isostatic adjustment on the northern hemisphere, Geophys. J. Int., 190, 1483–1494, doi:10.1111/j.1365-246X.2012.05563.x, 2012.

Steffen, H., Wu, P., and Wang, H. S: Optimal locations of sea-level indicators in glacial isostatic adjustment investigations, Solid Earth, 5, 511–521, doi:10.5194/se-5-511-2014, 2014.

Wessel, P. and Smith, W. H. F.: New, improved version of generic mapping tools released, EOS, 79, 579, doi:10.1029/98EO00426, 1998.

Wu, P.: Mode coupling in a viscoelastic self-gravitating spherical earth induced by axisymmetric loads and lateral viscosity variations, Earth Planet. Sci. Lett., 202, 49–60, doi:10.1016/S0012-821X(02)00750-1, 2002.

Wu, P.: Using commercial Finite element packages for the study of earth deformations, sea levels and the state of stress, Geophys. J. Int., 158, 401–408, doi:10.1111/j.1365-246X.2004.02338.x, 2004.

Wu, P.: Effects of lateral variations in lithospheric thickness and mantle viscosity on glacially induced surface motion in Laurentia, Earth Planet. Sci. Lett., 235, 549–563, doi:10.1016/j.epsl.2005.04.038, 2005.

Wu, P.: Sensitivity of relative sea levels and crustal velocities in Laurentide to radial and lateral viscosity variations in the mantle, Geophys. J. Int., 165, 401–413, doi:10.1111/j.1365-246X.2006.02960.x, 2006.

Wu, P., Steffen, H., and Wang, H. S.: Optimal locations for GPS measurements in North America and northern Europe for constraining Glacial Isostatic Adjustment, Geophys. J. Int., 181, 653–664, doi:10.1111/j.1365-246X.2010.04545.x, 2010.

Wu, P., Wang, H. S., and Steffen, H.: The role of thermal effect on mantle seismic anomalies under Laurentia and Fennoscandia from observations of Glacial Isostatic Adjustment, Geophys. J. Int., 192, 7–17, doi:10.1093/gji/ggs009, 2013.

13

Future Antarctic bed topography and its implications for ice sheet dynamics

S. Adhikari[1,2], **E. R. Ivins**[1], **E. Larour**[1], **H. Seroussi**[1], **M. Morlighem**[3], **and S. Nowicki**[4]

[1] Jet Propulsion Laboratory, California Institute of Technology, 4800 Oak Grove Dr., Pasadena, CA 91109, USA
[2] Division of Geological and Planetary Sciences, California Institute of Technology, 1200 E. California Blvd., Pasadena, CA 91125, USA
[3] Department of Earth System Science, University of California – Irvine, 3200 Croul Hall, Irvine, CA 92697, USA
[4] Code 615, NASA Goddard Space Flight Center, Greenbelt, Maryland, USA

Correspondence to: S. Adhikari (surendra.adhikari@jpl.nasa.gov)

Abstract. The Antarctic bedrock is evolving as the solid Earth responds to the past and ongoing evolution of the ice sheet. A recently improved ice loading history suggests that the Antarctic Ice Sheet (AIS) has generally been losing its mass since the Last Glacial Maximum. In a sustained warming climate, the AIS is predicted to retreat at a greater pace, primarily via melting beneath the ice shelves. We employ the glacial isostatic adjustment (GIA) capability of the Ice Sheet System Model (ISSM) to combine these past and future ice loadings and provide the new solid Earth computations for the AIS. We find that past loading is relatively less important than future loading for the evolution of the future bed topography. Our computations predict that the West Antarctic Ice Sheet (WAIS) may uplift by a few meters and a few tens of meters at years AD 2100 and 2500, respectively, and that the East Antarctic Ice Sheet is likely to remain unchanged or subside minimally except around the Amery Ice Shelf. The Amundsen Sea Sector in particular is predicted to rise at the greatest rate; one hundred years of ice evolution in this region, for example, predicts that the coastline of Pine Island Bay will approach roughly 45 mm yr^{-1} in viscoelastic vertical motion. Of particular importance, we systematically demonstrate that the effect of a pervasive and large GIA uplift in the WAIS is generally associated with the flattening of reverse bed slope, reduction of local sea depth, and thus the extension of grounding line (GL) towards the continental shelf. Using the 3-D higher-order ice flow capability of ISSM, such a migration of GL is shown to inhibit the ice flow. This negative feedback between the ice sheet and the solid Earth may promote stability in marine portions of the ice sheet in the future.

1 Introduction

Projecting the evolution of the Antarctic Ice Sheet (AIS) into the next few centuries relies on simulating a complex and non-linear coupled Earth system. A recent survey of experts by Bamber and Aspinall (2013) reveals that projections for AIS contribution to the rate of sea level rise at the year AD 2100 are generally rather moderate (~ 1.7 mm yr^{-1}) and that the upper end of the spectrum of projections would be about 7 times this value, mainly owing to intensification of dynamics of the West Antarctic Ice Sheet (WAIS). However, projections beyond AD 2100 are much more uncertain (Bindschadler et al., 2013) and are mainly limited by the poor knowledge of the physics involved in the grounding line (GL) migration and ice shelf melting and calving (e.g., Vaughan and Arthern, 2007; Walker et al., 2013). Furthermore, there is strong evidence that over the past four million years, during times of increased global atmospheric temperatures by 2–3 °C, the marine WAIS collapsed, and possibly some portions of the larger East Antarctic Ice Sheet (EAIS) as well (Naish et al., 2009; Raymo et al., 2011; Cook et al., 2013).

If we are to improve our abilities to assess the risk of the catastrophic consequences of partial collapse of AIS marine-based ice currently locked to the Antarctic continent, steps must be taken to assess fully the role of solid Earth deformation over tens to thousands of years, during which time gravitational viscoelastic flow of the underlying mantle may act to change the stability of the AIS. Past assessments have been that isostatic uplift following the ice sheet retreat promoted stability of the WAIS near the end of the last deglaciation during the mid-Holocene (Thomas, 1976). More recently, the promotion of stability by glacial isostatic adjustment (GIA) has been shown with increasingly sophisticated computations (Gomez et al., 2010, 2013). It is also now recognized that past AIS recession and ice flow direction are plausibly explained by strong interaction of ice loading with the solid Earth (Siegert et al., 2013). It is this ice-sheet/solid-Earth (IS/SE) interaction that we now explore in this paper for the AIS as a whole, using the vertical surface motions of bedrock GIA (Ivins and James, 1999) and the 3-D thermomechanical ice flow (Pattyn, 2003) capability of the Ice Sheet System Model (ISSM) (Larour et al., 2012b).

1.1   IS/SE feedback: GL migration

Perhaps the most important IS/SE feedback is associated with the GL migration (e.g., Lingle and Clark, 1985). For equilibrium sea level, any change in vertical bedrock elevation also changes the local sea depth. This perturbs buoyant forces in the regional ocean water and may promote the migration of the GL. Uplift of the seabed occurs, for example, in response to thinning of the inland ice sheet, and it causes local sea depth to decrease. If the GL is in initial equilibrium but sea depth decreases due to bed uplift, the GL tends to advance towards the continental shelf (e.g., Weertman, 1974; Schoof, 2007). The conceptual illustration of this negative feedback is shown in Fig. 1a. Lingle and Clark (1985) explored the effect of GIA-related seabed uplift on GL migration for Ice Stream E, now known as the MacAyeal Ice Stream, in the WAIS. The modeled GIA uplift, caused by thinning of the ice stream catchment area, delays the onset of GL retreat, thus reducing the rate of retreat during Holocene sea level rise. Notably, it was argued that the regional advance of Ross Sea GL may have occurred over the past three millennia. The gravitational attraction effects on local sea level developed later by Gomez et al. (2010) act to amplify this negative feedback during ice sheet retreat, since the diminished mass behind the GL has less mutual gravitational attraction with adjacent sea water, thus causing local sea level to drop.

The pace and magnitude of GL migration are also dictated by the bedrock slope. If sea depth decreases (due to bed uplift and sea level drop associated with the thinning of inland ice) at the initially stable GL on a reverse bed slope, for example, the GL tends to advance further on a relatively flat bed. Therefore the GLs associated with the Ross and Ronne ice shelves in the WAIS having relatively flat reverse bed slopes are sensitive to this effect (Conway et al., 1999). Capturing the dynamics of such GL sensitivity demands proper understanding of interactions between the ice, the ocean and

the solid Earth, and is indeed the key to successful modeling of the WAIS (e.g., Vaughan and Arthern, 2007; Katz and Worster, 2010).

1.2   Additional IS/SE feedbacks

There are other important feedback mechanisms that the solid Earth deformation provides to ice sheets. The GIA uplift can be important in providing basal resistance to ice flow and buttressing the ice sheet by raising bedrock pinning points (e.g., Favier et al., 2012; Siegert et al., 2013). This concept is captured in Fig. 1b. The GIA-induced changes in surface elevation and regional slope of the ice sheet may affect the gravitational driving stress, as well as some processes at the ice–atmosphere interface (e.g., surface mass balance). These perturb the momentum balance and affect the englacial velocity field (e.g., Cuffey and Paterson, 2010; Winkelmann et al., 2012), which in turn may impact the ice thermodynamics via changes in strain heating. Spatially varying bed uplift also affects the hydraulic potential field and hence the subglacial hydrology and the sliding rate of the ice sheet. There is additionally the complication of bulge migration, a broad-scale phenomenon involving bending of the elastic lithosphere and mantle lateral flow. Due to the lateral motion of this topographic bulge, local crustal motions (and slopes) may change sign during GIA (e.g., Fjeldskaar, 1994).

These mechanisms are extremely difficult to isolate and quantify, and it is therefore not obvious whether (and under what circumstances) each of these acts to accelerate or inhibit the ice flow. As long as the thermomechanical ice sheet model and other companion models (e.g., surface mass balance model and the hydrological model) are dynamically coupled to a comprehensive solid Earth model, however, most of these feedbacks are intrinsically taken into account.

1.3   New solid Earth computations

For large timescale (millennia) simulations, most of the ice sheet models (e.g., Pollard and DeConto, 2009; Hughes et al., 2011) capture the solid Earth physics of varying degrees of complexity (cf. Le Meur and Huybrechts, 1996). With the exception of the recent work of Gomez et al. (2013), none of these studies provides the explicit assessment of effects of GIA uplift on several aspects of ice sheet dynamics (e.g., GL migration and gravitational driving stress). Le Meur and Huybrechts (2001) explicitly pointed out the need for the more complete coupling that could be found in the wavelength-dependent relaxation spectra of viscoelastic solid Earth models. In this paper, we quantify two distinct IS/SE feedback mechanisms applied to the AIS on centennial timescales using multiple wavelength-dependent decay spectra. Assuming the equilibrium sea level, we first evaluate how the future bed uplift (governed by the past and future evolution of AIS; cf. Sect. 2) controls the GL migration. Second, we assess the role of bed uplift in the gravitational driving

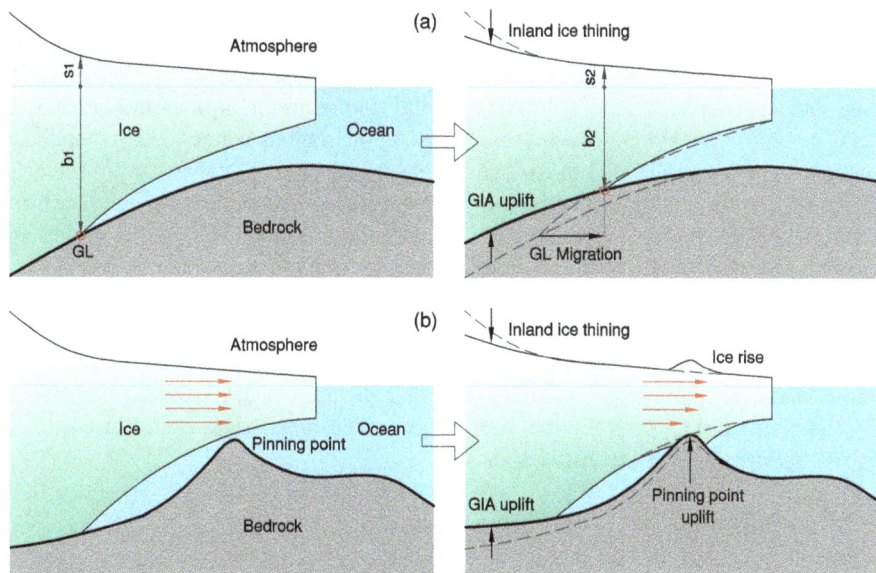

Figure 1. Schematic of IS/SE feedbacks. **(a)** GL migration controlled by the local sea depth. For equilibrium sea level, the GIA uplift due to thinning of the inland ice promotes the GL migration towards the continental shelf. The hydrostatic equilibrium requires that $s_1/b_1 = s_2/b_2 = -(\rho_w - \rho)/\rho$, where s_i and b_i are the surface and bedrock elevations (subscript $i = 1, 2$ represents the initial and final configurations, respectively), and ρ_w and ρ are the water and ice densities. **(b)** Pinning point, raised by GIA uplift due to thinning of the inland ice, provides basal resistance and buttressing to the ice sheet. Red arrows depict the velocity profiles. Dashed lines in the final configurations (second column) represent the initial geometry. Note that the destabilizing effects of GIA during the periods of inland ice thickening that causes the bedrock to subside can be conceptualized by reverting the direction of mid-arrows.

stress. The overall influence of solid Earth deformation (via changes in GL and driving stress) on the ice surface velocities is also quantified. These assessments, based on reliable models and data, help us to understand whether the GIA effects are significant in controlling the future evolution of AIS on centennial timescales.

The paper layout is as follows. In Sect. 2, we introduce the solid Earth model employed in this research, discuss how the change in ice thickness is translated into the ice load height, provide a detailed account of model tuning procedure, and describe all of the required model input data. In Sect. 3, we present modeling results of the future bedrock topography, and assess the relative importance of the past and future ice load changes. In Sect. 4, we quantify the influence of predicted change in Antarctic bed topography on several aspects of ice sheet dynamics. Finally, in Sect. 5, we summarize key conclusions of broad interest.

2   Model and data

Ivins et al. (2013) presented a much improved ice loading history for the AIS since the Last Glacial Maximum (LGM). As for the future, the recently concluded SeaRISE (Sea-level Response to Ice Sheet Evolution) project (Bindschadler et al., 2013; Nowicki et al., 2013) provided quantitative projections of the evolution of AIS under ongoing climate warming. We employ the new GIA capability (Ivins and James, 1999) of

ISSM (Larour et al., 2012b), hereinafter referred to as the ISSM/GIA model, to combine these data of past and future ice loadings and calculate the change in bedrock topography over the same timescale projections as in the SeaRISE studies. Using appropriate analytical and numerical models, we then evaluate the stabilizing or destabilizing influence of predicted changes in bed topography on the ice sheet dynamics.

While the SeaRISE experiments employed state-of-the-art numerical treatments of ice flow, it should be noted that the majority of these models were not coupled to a comprehensive solid Earth model. Furthermore, the SeaRISE experiments do not capture the paleo-evolution of the AIS since the LGM, thus limiting the possibility for the participating ice sheet models (with GIA capability) to perform similar analysis presented in this study.

2.1   The ISSM/GIA model

ISSM is a continental-scale, high-resolution, multi-model simulation code developed for understanding the dynamic behavior of large ice sheets (Larour et al., 2012b). This open source finite-element software is capable of simulating ice-flow mechanics of varying degrees of complexity (Seroussi et al., 2012), performing sensitivity analysis (e.g., Larour et al., 2012a), inverting unknown field parameters (e.g., Morlighem et al., 2010), and assessing mechanics of rift propagation and eventual collapse of ice shelves (e.g., Borstad et al., 2012). The semi-analytical GIA solution of

Ivins and James (1999) is one of several new features being actively implemented in ISSM. Here we briefly summarize the key elements of the model.

We assume that the ice sheet rests on top of the solid non-rigid Earth, which is considered to be a simple two-layered incompressible continuum with an upper elastic lithosphere floating on the viscoelastic (Maxwell material) mantle half-space. The theory governing the deformation of pre-stressed solid Earth subject to a normal surface traction of ice sheet relies upon the fundamental equations of motion and is discussed elsewhere (e.g., McConnell, 1965; Wolf, 1985; Ivins and James, 1999). For axisymmetric loading problems, it is possible to obtain the semi-analytical solution of vertical displacement at the lithosphere surface (i.e., both the ice–bed and ocean–bed interfaces). This is the essence of the solution for GIA assessment. For the equilibrium sea level hypothesis, however, the GIA solutions perturb the ice sheet only within the area of grounded ice. For the AIS that is surrounded on most of its periphery by floating ice, the extent of the grounded ice may evolve, as we assume in this study, according to the hydrostatic balance between the ice shelf and ocean water.

Given the appropriate ice loading history and choice of the model/material parameters (cf. Sect. 2.3), the GIA solution depends on the size of the ice load itself and the radial distance of the evaluation point from the load center. Assumed axisymmetry implies that the shape of the ice load be essentially cylindrical (e.g., Ivins and James, 1999). In the Cartesian framework of the ISSM/GIA model, we treat the size of the ice load as the property of a mesh element and compute the GIA solution at each node of the element. Individual 2-D (xy plane) mesh elements are defined as the equivalence of a footprint (i.e., projection onto the xy plane) of cylindrical disc loads, ensuring that the corresponding element and disc both share the same origin and plan-form area (cf. inset of Supplement Fig. S1a). The height of the ice load is then assigned to each element such that the average normal tractional force on the corresponding area of ice–bedrock contact is conserved. At each node within the domain, the final GIA solutions are computed by integrating the solutions due to individual disc loads, defined as the property of mesh elements.

The ISSM/GIA model is tested against the benchmark experiments (Wolf, 1985; Ivins and James, 1999) and found to be sensitive to the mesh resolution. For reasonably fine resolution (element size typically on the same order of magnitude as ice thickness, or two orders of magnitude smaller than the characteristic size of an ice sheet), however, the model performs well within the acceptable accuracy. Sample results are provided in Supplement Fig. S1.

2.2 Differential ice height

Prior to describing the ISSM/GIA model applied to the AIS, we briefly discuss how the change in Antarctic ice thickness is translated into the height of ice load. In this study, we as-

sume that the sea level remains in its present-day state. We define the height of ice load at any time, t, as the differential ice height (DIH), $\Delta h(x, y, t)$, with reference to the present-day configuration of the AIS (James and Ivins, 1998).

In the regions where ice is grounded both presently and at the given time t, DIH is simply the change in ice thickness over the corresponding time period. Similarly, in the regions where ice is floating both at present and at time t, DIH is zero as we assume that ice shelves are freely floating over the ocean, respecting the hydrostatic equilibrium. Defining DIH is a bit complicated in the areas over which the GL migrates over the course of time. If ice thickness at time t is smaller than the present-day value (i.e., ice floats at time t, but is now grounded), the DIH is defined as follows:

$$\Delta h(x, y, t) = \frac{\rho_{\mathrm{w}}}{\rho} b(x, y, t) - h(x, y, 0), \qquad (1)$$

where ρ_w is the ocean water density, ρ is the ice density, $b(x, y, t) < 0$ is the isostatically corrected bedrock elevation (with respect to the present-day sea level) of the marine portions at time t (James and Ivins, 1998), and $h(x, y, 0)$ is the present-day ice thickness. DIH in such a case would be negative. Similarly, in the areas where the ice sheet holds thicker ice at time t than at present (i.e., ice now floating is grounded at time t), DIH is defined as follows:

$$\Delta h(x, y, t) = h(x, y, t) - \frac{\rho_{\mathrm{w}}}{\rho} b(x, y, 0), \qquad (2)$$

where $h(x, y, t)$ is the ice thickness at time t and $b(x, y, 0) < 0$ is the present-day bedrock elevation of the marine portions of the ice sheet. DIH in such a case would be positive.

As a general convention in this study, we use $t < 0$ to denote the past, i.e., before present, and $t > 0$ for the future unless stated otherwise.

2.3 Model tuning

We apply the ISSM/GIA model to the AIS. We mesh the footprint of present-day AIS (Bamber et al., 2009) into triangular elements. In order to capture the potentially interesting features, the domain is discretized to consist of a high-resolution mesh around the coast (typical element size of 10 km) than in the interior of the ice sheet (element size of 25 km). This unstructured mesh captures the model inputs (e.g., past or future ice loads) in sufficient detail and provides a reasonable compromise between solution accuracy and computational efficiency. Doubling the mesh density, for example, improves the GIA solution (under present-day ice loading) only slightly ($< 0.1\%$), but it increases the high computational cost by one order of magnitude.

A comprehensive list of modern global positioning system (GPS) measurements is presented by Thomas et al. (2011). However, we tune our model by testing it against 18 high-precision data. Following Ivins et al. (2013), we first eliminate records from the Antarctic Peninsula north of $72°$ S due to the associated difficulty of dealing with large elastic and transitional viscoelastic signals present there. We then average the values from stations located within 100 km of one another and eliminate some stations with reported errors greater than the signal amplitude.

In order to make reasonable predictions of the present-day GIA uplift, the slow response of highly viscous solid Earth demands that the evolution of AIS during the past several thousand years be considered in the ISSM/GIA model. There are a few reliable GIA ice loading histories for Antarctica (e.g., Peltier, 2004; Ivins and James, 2005; Whitehouse et al., 2012). These generally describe the timing and magnitude of deglaciation since the LGM based on geological and ice core data. In this study, we employ a much improved loading history discussed in Ivins et al. (2013). By upgrading the loading history of Ivins and James (2005) with recently available geochronological constraint data, the later model was able to provide a more accurate GIA correction to GRACE (Gravity Recovery and Climate Experiment) data measured over the period AD 2003–2012.

Based on Ivins et al. (2013), we have isostatically corrected DIHs available for 11 time stamps in the past (at -1, -2.2, -3.2, -6.8, -7.6, -11.5, -15, -17, -19, -21 and -102 kyr; see Supplement Fig. S2). Note that $t = -21$ kyr roughly corresponds to the LGM of the AIS, while -15 kyr marks the onset of deglaciation of the WAIS (e.g., Clark et al., 2009). For $t = -1$ kyr, we consider zero DIH that could be constrained using the recently available surface mass balance data (e.g., Verfaillie et al., 2012; Favier et al., 2013). However, this process is not straightforward because the magnitude and spatial distribution of ice flux during the periods of inferred mass balance are vastly unknown. Furthermore, we consider $t = -102$ kyr to mark the initial configuration for AIS as being identical to the present-day configuration. This implicitly assumes that the DIHs before the LGM have a minimal impact on the current and future response of the solid Earth. (We demonstrate in Sect. 3.2 that this is indeed a valid assumption.) Note also that the ice loading on the ISSM/GIA model is assumed to vary in a piece-wise linear fashion between the adjacent time stamps.

The model and material parameters considered in this study approximate the preliminary reference Earth model (Dziewonski and Anderson, 1981) and are taken from Ivins et al. (2013) unless otherwise specified. For several values for lithosphere thickness, Ivins et al. (2013) performed a parameter-space study in their two-layer mantle model. Noting that 65 and 115 km may represent, respectively, the average thickness of the lithosphere for the WAIS and EAIS, Ivins et al. (2013) provided appropriate combinations of the upper and lower mantle viscosity. Because the past (see Supplement Fig. S2) and future DIHs (cf. Sect. 2.4 and Supplement Fig. S3) indicate that the majority of changes occur in the WAIS, we consider a 65 km thick lithosphere in our model. We cannot pick the corresponding mantle viscosity from Ivins et al. (2013), as our model does not have two

mantle layers. We therefore consider the mantle viscosity as a tuning parameter, such that the difference in the mean between the measured modern GPS data (Thomas et al., 2011) and modeled current GIA uplift at 18 data points is minimized (Fig. 2b). The optimized solutions for current uplift rate are shown in Fig. 2a. Key characteristic features of our predictions include greater uplift rate around the Mt. Ellsworth territory and a mild rate of bed subsidence in the interior of EAIS. Such spatial patterns of uplift rate essentially reflect signatures of the employed ice loading history (cf. Supplement Fig. S2). Note that the optimal predictions of uplift rate (Fig. 2a) correspond to a mantle viscosity of 7×10^{20} Pa s. As expected, this magnitude falls in between the upper (2×10^{20} Pa s) and lower mantle viscosity (1.5×10^{21} Pa s) recommended for the chosen lithosphere thickness (Ivins et al., 2013). While the architecture of the ISSM/GIA model can capture high-resolution spatial variability in solid Earth material parameters, we do not experiment with lateral inhomogeneities in this study.

2.4 Future ice loading

The AIS mass change may become more dynamic in the future due to ice shelf melting (e.g., Pritchard et al., 2012; Depoorter et al., 2013; Rignot et al., 2013). The SeaRISE participating ice sheet models, primarily driven by melt-dominated forcing, quantified the future evolution of AIS under the so-called "R8 scenario", which is the proxy of representative concentration pathway emission scenario 8.5 (RCP 8.5) (Bindschadler et al., 2013; Nowicki et al., 2013). The RCP 8.5 scenario represents an ongoing rise in emissions throughout the century, reaching 8.5 W m^{-2} at AD 2100 (e.g., van Vuuren et al., 2011). The radiative forcing associated with the RCP 8.5 scenario is loosely related to the R8 scenario via all three components of the SeaRISE model forcing, namely the surface climate, basal sliding and ice shelf melting. As these forcings are the ones that govern the future evolution of AIS, it is relevant in the present context to provide a brief account of them.

Firstly, the SeaRISE surface climate forcing follows a $1.5 \times$ A1B scenario (IPCC-AR4, 2007) until AD 2200. (The A1B scenario generally describes a future world of rapid growth in economy and population that peaks in mid-century, and technologies that rely equally on both fossil and non-fossil sources of energy.) A mild increase in surface temperature, a total of 0.5 °C, is assumed during the period AD 2200–2500. Secondly, no sliding amplification is considered until AD 2100 assuming that the Antarctic surface temperature will remain below zero, thus ignoring the potential for surface melt-induced basal sliding prior to this time. Thereafter, sliding increases linearly at a rate of 20 % of its original value per century, but only in coastal regions. Inland, the sliding amplification factor decreases linearly as a function of surface elevation such that no sliding enhancement is applied above 1200 m a.s.l. Thirdly, ice shelf melting is assumed to

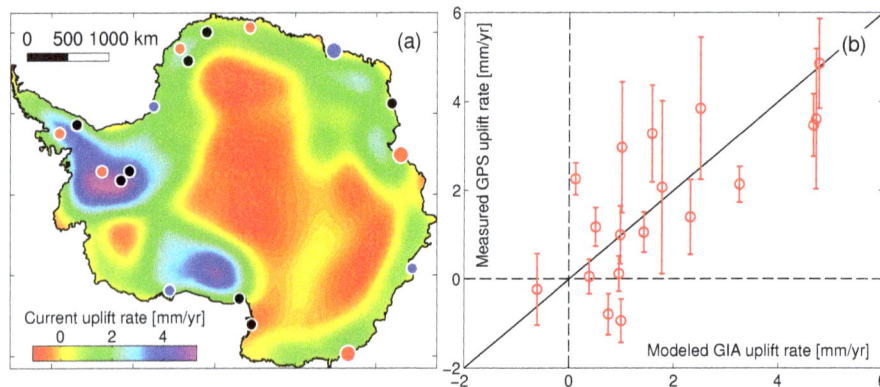

Figure 2. Model tuning and predictions of current uplift rate. (**a**) Modeled GIA uplift rate at the present day. Calculations are made by forcing the ISSM/GIA model by ice loading history over the past 21 kyr (Supplement Fig. S2) (Ivins et al., 2013). Black circles locate the position where model results are within the 1-σ uncertainty range of GPS measurements (Fig. 2b). Red circles represent cases where the model overestimates the measurements, whereas blue circles indicate locations where the model underperforms. Big circles are to denote the absolute misfits that are > 0.75 mm yr^{-1}. (**b**) Validation of the model against 18 high-precision GPS uplift data (Thomas et al., 2011). Error bars depict 1-σ uncertainties associated with the GPS measurement.

increase linearly from its present-day value to 60 m yr^{-1} at AD 2200. Additional 10 m yr^{-1} melt extends linearly over the next 300 yr. Changes in basal melting conditions are only applied to the Amundsen Sea Sector (90–120° W) and the Amery Ice Shelf (60–75° E), not to the Weddell and Ross seas. Such a restriction of ice shelf melting considered in the SeaRISE experiment seems reasonable with reference to current observations (Depoorter et al., 2013; Rignot et al., 2013) that reveal that ice shelves around the Amundsen Sea are the most susceptible to melting. Furthermore, spatial distribution of ocean temperature anomalies (e.g., Pritchard et al., 2012) also supports this hypothesis.

Under the R8 scenario, a total of four ice sheet models simulates the future evolution of the AIS through AD 2500. These models are the Anisotropic Ice Flow Model (AIF) (Wang et al., 2012), the Penn State Ice Sheet Model (PennState) (Pollard and DeConto, 2009), the Potsdam Parallel Ice Sheet Model (PISM-PIK) (Winkelmann et al., 2011), and the Simulation Code for Polythermal Ice Sheets (SiCoPolIS) (Greve, 1997). These ice sheet models employ different assumptions and methods for solving the full physics involved in simulating ice flow (e.g., shallow-ice vs. first-order flow mechanics; different treatments for basal sliding, subglacial hydrology, and GL migration), they employ different numerics (e.g., different spatial and temporal resolutions), they have unique techniques for dealing with the prescribed model forcing, and so forth (cf. Bindschadler et al., 2013; Nowicki et al., 2013). Consequently, each ice sheet model produces a unique evolution of the AIS. We extract the ice thickness from each model prediction for five time stamps (at $t = 100, 200, 300, 400$ and 500 yr) and compute DIHs using Eqs. (1) and (2) as appropriate. For simplicity, we assume $b(x, y, t) \approx b(x, y, 0)$ in Eq. (1) (i.e., future bedrock elevation is not isostatically corrected while com-

puting future DIHs). We anticipate that the associated errors translated into the GIA solutions would be minimal. Examples of future DIHs are provided in Supplement Fig. S3. Again, ice loads in the ISSM/GIA model are assumed to vary linearly between the adjacent time stamps.

3   Future bed topography

In order to predict the future bed topography for AIS, the calibrated ISSM/GIA model (cf. Sect. 2.3) is forced by an appropriate sequence of ice load changes into the future. As we have four independent sets of future ice loading (cf. Sect. 2.4), we may compute four unique GIA solutions at any evaluation time in the future. Based on these solutions, here we present the estimates of future bed uplift, isolate the role of past and future ice loading, and also evaluate how predicted change in bed topography alters the bedrock slope. Note that in this section we deviate from the general time convention and present our results (predictions) in "AD".

3.1   Vertical bed displacement

The GIA solutions for individual future ice loading, combined with the consideration of a lone spin-up loading history, are computed at AD 2100 and AD 2500 and shown in Supplement Figs. S4 and S5, respectively. Although these solutions differ from each other in both magnitude and their spatial distribution, some common features are noteworthy: (i) all models predict minor subsidence in a few places, particularly along the Wilkes Land coast; (ii) the topography of the interior of the EAIS is likely to remain unchanged; and (iii) a pervasive and large uplift is predicted in the WAIS (except for the AIF simulations) and around the Amery Ice Shelf (except for the PennState and PISM-PIK simulations).

We can, of course, offer no one GIA solution as being more reliable than any other. We therefore calculate the average of model solutions (hereafter termed "model-average solutions") (Fig. 3a and b) and consider these as reasonable estimates of the Antarctic bed uplift. It is possible that our ensemble approach for these predictions is insufficient. Nonetheless, we assert that they provide the correct order-of-magnitude estimates and the likely spatial patterns of the future bed uplift, which are sufficient to evaluate some of the IS/SE interactions outlined in Sect. 1.3.

By AD 2100, the Amundsen Sea Sector and the Amery Ice Shelf may rise by about four and three meters, respectively (Fig. 3a). The rest of the WAIS is likely to rise by up to two meters. The interior of the EAIS is predicted to remain unchanged. The Adelie and Wilkes lands, where all ice sheet models predict large snow accumulation (Nowicki et al., 2013), are likely to subside by about less than one meter. It should be noted that, for the chosen climate change scenario, all ice sheet models but SiCoPolIS predict moderate snow accumulation in the Queen Maud, Ross and Weddell basins, and minimal accumulation in the Amundsen and Amery basins (cf. Nowicki et al., 2013). Roughly similar spatial patterns of GIA uplift are predicted for AD 2500 as well (Fig. 3b). In this case, the bed may uplift by about 25 m in the Amundsen Sector, and by about 10–15 m in the rest of WAIS and the Amery Ice Shelf. The interior of EAIS may remain mostly unperturbed. Bed subsidence of about four meters is likely to occur along the Adelie and Wilkes Land coasts, as well as along the coast north of the Amery Ice Shelf.

We also find similar spatial patterns for the model-average bed uplift rates (cf. Supplement Fig. S6). At AD 2100, the Amundsen Sea Sector and Wilkes Land are predicted to rise and subside at the highest rates of about 45 and -5 mm yr^{-1}, respectively. The interior of Marie Byrd Land is predicted to rise at the highest rate of 75 mm yr^{-1} at AD 2500; The Amundsen Sea Sector also rises at a large rate of about 40 mm yr^{-1}. The greatest rate of subsidence (of about -15 mm yr^{-1}) is predicted along the east coast except around the Amery Ice Shelf.

At both evaluation times, as noted earlier, we obtain different GIA solutions associated with individual future ice loading. Here we briefly outline how well these predictions for the Antarctic bed uplift match one another. The standard deviation shown in Fig. 3c illustrates that the model predictions are generally in good agreement with each other at AD 2100, except around the Amundsen Sector and the Amery Ice Shelf. As depicted in Fig. 3d, similar agreement is found amongst the model predictions at AD 2500. Large deviations are predicted once again around the Amundsen Sea Sector. Moderate deviations can be seen along the east coast including the Amery Ice Shelf. Such deviations amongst model predictions for the Antarctic bed topography predicted both at AD 2100 and AD 2500 can generally be attributed to the limiting values of GIA solutions predicted by the PennState and AIF models (cf. Supplement Figs. S4 and S5). In both

years, PennState predicts large uplift in the Amundsen Sector and the least uplift (in fact, subsidence) around the Amery Ice Shelf; the opposite is true in the case of AIF predictions.

Our predictions might slightly overestimate the GIA solutions in the EAIS, as the modeled lithosphere thickness (i.e., 65 km) is much thinner than the more common value (115 km, as reported in Sect. 2.3). Nevertheless, the following general findings should remain unaltered. With reference to the WAIS GIA solutions, (i) Amery Ice Shelf may rise moderately; (ii) the interior of the EAIS may remain mostly unperturbed; and (iii) minor subsidence may occur along the coastal EAIS. It also needs to be mentioned here that we do not model the subglacial erosion in this study, although quite rapid evacuation of soft sediments is now occurring at the bed of the Pine Island Glacier. Erosion has been roughly estimated to cause the topography to lower at a rate of 0.6 ± 3 m yr^{-1} as ice flow exceeding 1 km yr^{-1} erodes material in deep longitudinal fjord valleys (Smith et al., 2012). This erosive action typically takes place in 20 km wide valleys of approximately 200 km length. While an important consideration in ice sheet modeling (Kessler et al., 2008), GIA topographic responses occur over much broader length scales exceeding the areal dimensions of fast erosion by nearly two orders of magnitude, thus having an impact on the evolution of the ice sheet on the scale of the drainage basin itself.

3.2 Role of past and future loading

Our predictions of bed uplift reflect the combined effects of long-term viscous creep of solid Earth driven by the ice loading history and its short-term (centennial timescale) viscoelastic response to the future ice load change. It might be useful to isolate the contribution of past and future ice loading to the evolution of future bed topography. First, we let the calibrated ISSM/GIA model (driven by the past loading alone) run for the next 500 yr into the future under the idealized condition that the current distribution of ice thickness prevails as is, thus imposing $\Delta h(x, y, t) = 0$ m for all $t \in [0, 500]$ yr. We find similar spatial patterns of bed uplift, as shown in Fig. 2a for the current uplift rate, in the future (cf. Supplement Fig. S7b); the total amount of GIA uplift at years AD 2100 and AD 2500 is in the respective ranges of about $[-0.1, 0.5]$ (Fig. 4a) and $[-0.5, 2.5]$ m (Fig. 4b). In such an idealized scenario of the unchanged future AIS, the notable features are that the peninsula, the whole of the WAIS, and coastal regions of the EAIS are likely to uplift, with the highest uplift occurring around the Mt. Ellsworth territory, and that the interior of the EAIS may remain unaffected or subside minimally (cf. Supplement Fig. S7b). These are consistent with characteristic features of the employed ice loading history (cf. Supplement Fig. S2).

Next, we compute model-average GIA solutions at years AD 2100 (Fig. 4c) and AD 2500 (Fig. 4d) by forcing the ISSM/GIA model by the future ice load changes alone. We

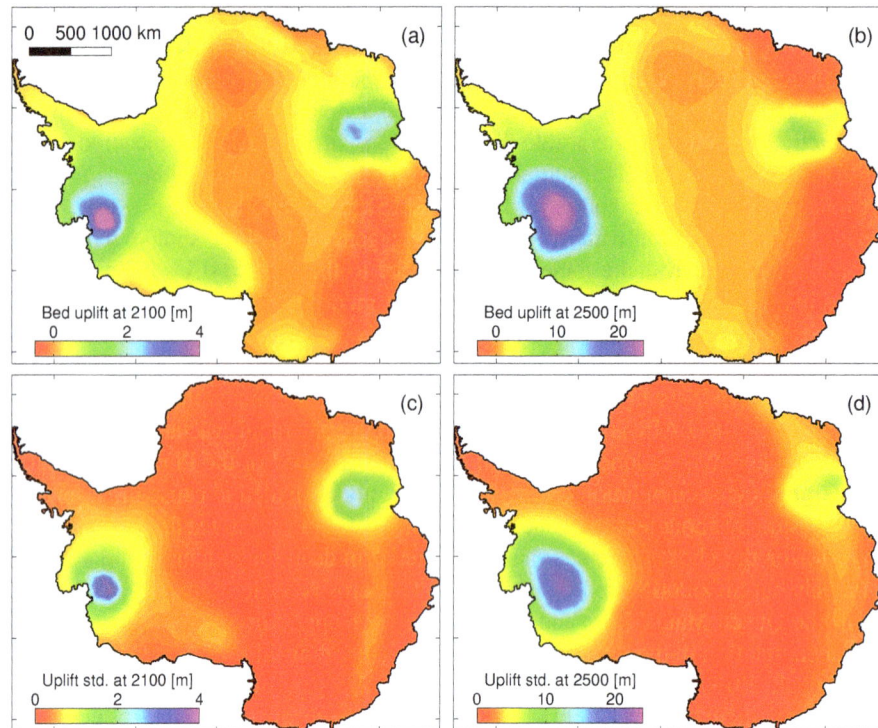

Figure 3. Estimates of the future Antarctic bed uplift. Model-average predictions for bed uplift at (**a**) AD 2100 and (**b**) AD 2500 under the R8 scenario. Associated standard deviations are shown in subplots (**c**) and (**d**), respectively. Calculations are made by forcing the calibrated ISSM/GIA model (Fig. 2) by the future ice loading (Supplement Fig. S3) obtained from the SeaRISE project (Bindschadler et al., 2013; Nowicki et al., 2013). See Supplement Fig. S6 for corresponding solutions for the bed uplift rate. Roughly similar spatial patterns are obtained, with the respective range of values $[-5, 45]$ and $[-15, 75]$ mm yr^{-1} at AD 2100 and AD 2500.

impose $\Delta h(x, y, t) = 0$ m for all $t \in [-120, 0]$ kyr to ensure that the model is properly spun up. Alternative solutions are found by subtracting the GIA solutions associated with the past loading alone (Fig. 4a and b) from the corresponding final predictions (Fig. 3a and b). The resulting solutions are essentially identical to those shown in Fig. 4c and d, implying that the principle of linear superposition holds (cf. Supplement Fig. S7).

Comparing the estimates of GIA uplift associated with the past (Fig. 4a and b) and future ice loading alone (Fig. 4c and d) with those depicted in Fig. 3a and b, we find that the future ice loading dominates and that the contribution of long-term viscous creep associated with the past is only about one tenth the magnitude of the predicted GIA uplift. This suggests that the errors associated with the ice loading history may have minor consequences for the future predictions of the Antarctic bed topography, provided the differing scenarios properly sample the possible amplitude of the ice sheet loading. Significant changes in magnitude and timing of the loading history may, however, require that the different mantle viscosity be employed in the ISSM/GIA model to predict the current uplift rates correctly (cf. Sect. 2.3). This, in turn, may yield the different contribution (not necessarily higher) of long-term viscous creep of the solid Earth to predictions

of the future bed uplift. Nonetheless, it is important here to highlight the need for better constraining the DIHs, particularly in the recent past (during the past 1 kyr).

3.3 Change in bed slope

Strong spatial variability in the future GIA uplift (cf. Sect. 3.1) implies that the Antarctic bed slope will be modulated in the future (e.g., Gomez et al., 2013; Konrad et al., 2013). We compute current and future bedrock slopes (associated with the model-average GIA solutions) following $\alpha_b(x, y, t) = \sqrt{\alpha_{bx}^2(x, y, t) + \alpha_{by}^2(x, y, t)}$, where $\alpha_{bx}(x, y, t)$ and $\alpha_{by}(x, y, t)$ are the x and y components of the bed slope, respectively. Figure 5b, for example, depicts the present-day bedrock slope of Antarctica. While this plot reveals the degree of bed steepness, it does not provide the information regarding the aspect of slope. It is important to identify whether the bedrock has forward or reverse slope, particularly while evaluating the role of GIA uplift in the marine ice sheet instability (to be discussed later). We therefore plot the current bathymetry of the AIS in Fig. 5a; in order to facilitate the discussion, we only consider the areas with bedrock below sea level. Notice in the figure, for example, the blue color around

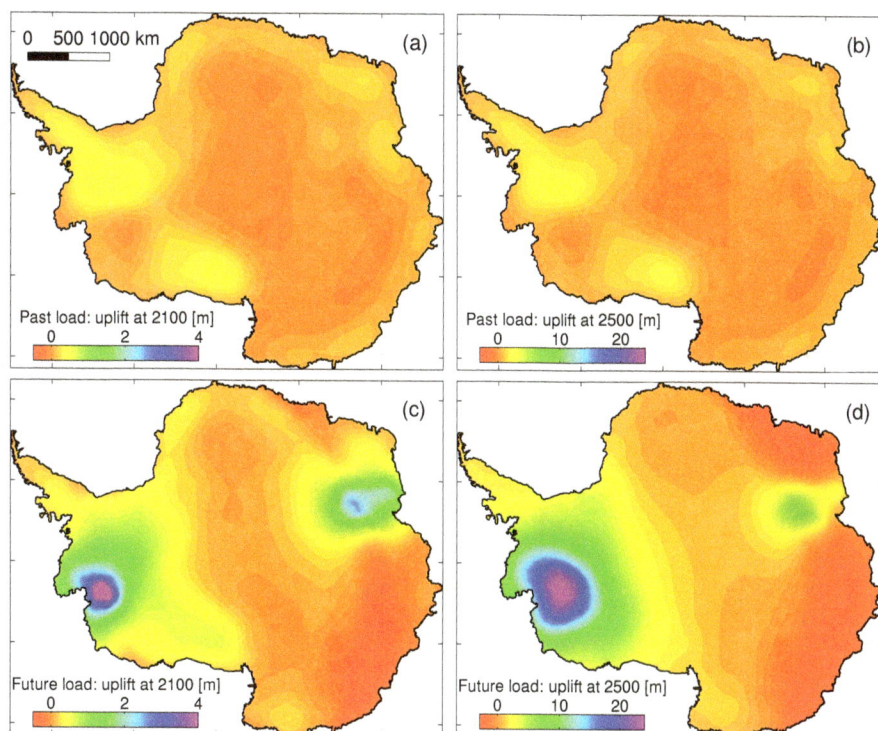

Figure 4. Role of past and future ice loading in the future bed topography. GIA uplift at **(a)** AD 2100 and **(b)** AD 2500 obtained by forcing the calibrated ISSM/GIA model (cf. Fig. 2.) by the past loading alone. Future DIHs are assumed to be zero. Corresponding solutions (model-average) at **(c)** AD 2100 and **(d)** AD 2500 associated with future ice loading alone. Past DIHs are assumed to be zero. For ease of comparison, we use the same color scale for each year. (Results in Figs. 4a and b are in the respective ranges of $[-0.1, 0.5]$ and $[-0.5, 2.5]$ m.) Also, see Supplement Fig. S7 where we replot some of these figures with a different color scale that better illustrates the spatial distribution.

the interior of WAIS that illustrates the existence of reverse bed slope in those regions.

We obtain the future bedrock topography by adding the current bedrock topography and the future GIA solution (Sect. 3.1). From the corresponding bedrock topography, we calculate the bed slope at AD 2100 and AD 2500. The changes in bedrock slope are then computed by subtracting the present-day slope (Fig. 5b) from the future bed slope. Sample results are shown in Fig. 5c for AD 2500. In the figure, we generally notice the reduced bed slope (apart from a few localized regions with enhanced slope) around the Amundsen Sea Sector and the Amery Ice Shelf where large uplift is predicted. The reverse bed in these regions will thus have generally less steep slope in the future. Similar spatial patterns (but small magnitudes) of change in bed slope are obtained for AD 2100 as well (results not shown).

Although the magnitudes of change are larger in the regions where the bed has experienced large uplift, such as in the Amundsen Sea Sector and in the Amery Ice Shelf, percent changes in bed slope are also significant around the Ronne and Ross ice shelves (Fig. 5d). The bedrock slopes beneath these ice shelves, for example, reduce by more than one percent at AD 2500. Under the circumstances when GL is to advance (this actually is the general case in the present

context; cf. Sect. 4.2), change in bed slopes around the GL beneath the ice shelves may impact the magnitude of GL migration (cf. Sect. 1.1 and Fig. 1a) and thus ice dynamics.

4 Implications for ice sheet dynamics

Our predictions for the future evolution of Antarctic bed topography may influence the future dynamics of AIS. In this section, we specifically quantify the potential effect of the predicted change in bed topography on the gravitational driving stress (Sect. 4.1), GL migration (Sect. 4.2), and ice surface velocities (Sect. 4.3). Because our model is not yet capable of computing IS/SE interactions with full dynamic feedbacks, it should be noted that some of the results presented below are obtained by bootstrapping the relevant future bedrock topography. The general procedure includes the following. First, we consider the present-day settings (geometric setting and boundary conditions) of the AIS for our calculations. Next, we upgrade the geometry and relevant boundary conditions (e.g., basal friction while calculating ice surface velocities in Sect. 4.3) to account for the future GIA uplift and perform recalculations. Comparing corresponding results, we finally isolate the influence of GIA uplift.

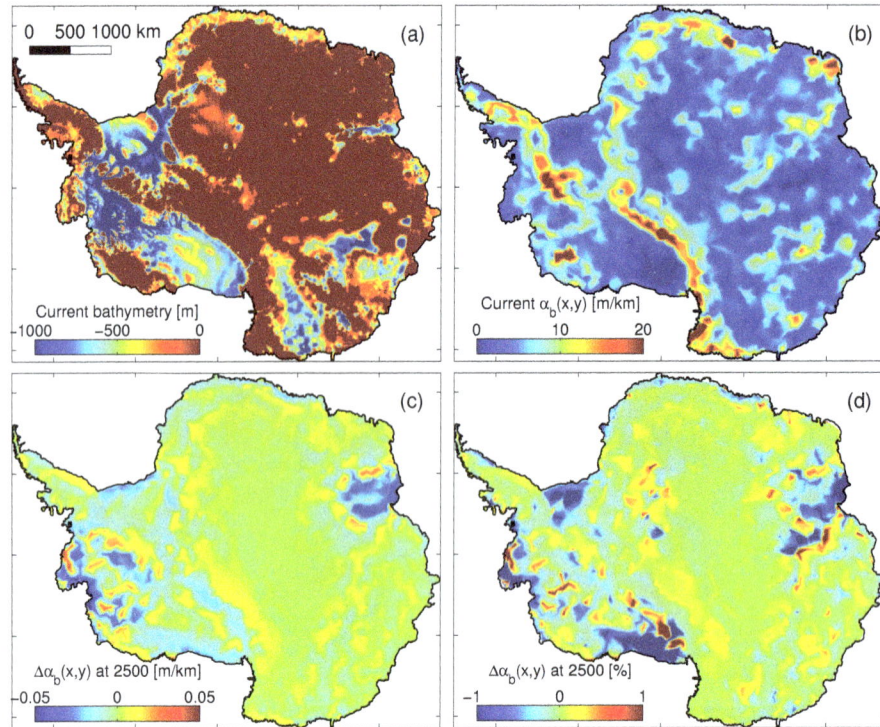

Figure 5. Estimated change in Antarctic bed slope in the future. Present-day (**a**) bathymetry and (**b**) bed slope, $\alpha_b(x, y)$, of Antarctica. To facilitate discussions, only data in the range $[-1000, 0]$ m are shown in Fig. 5a. (**c–d**) Model-average change in bed slope, $\Delta\alpha_b(x, y)$, at AD 2500. Negative magnitudes of $\Delta\alpha_b(x, y)$ imply that bedrock will have less steep slopes in the future.

In order to minimize the potential compounding effects on the ice sheet dynamics, we keep the present-day settings (particularly, ice thickness and thermo-mechanical boundary conditions) fixed for the further calculations. We therefore advise caution in overinterpreting any individual results, obtained from the present-day settings of AIS perturbed by the predicted GIA uplift, for AD 2100 and 2500. For clarity, we revert to the general time convention (cf. Sect. 2.2) in order to stress that the following results illustrate the potential effect of the predicted bed uplift, not the predictions of the total GIA effect, on the future dynamics of AIS.

4.1   Gravitational driving stress

Here, we discuss the GIA effects on the gravitational driving stress. We compute driving stress following $\tau_d(x, y, t) = \sqrt{\tau_{dx}^2(x, y, t) + \tau_{dy}^2(x, y, t)}$, where $\tau_{dx}(x, y, t)$ and $\tau_{dy}(x, y, t)$ are the x and y components of the driving stress, respectively. For $i = x, y$, we define $\tau_{di}(x, y, t) = -\rho\, g\, h(x, y, t)\alpha_{si}(x, y, t)$ as the ith component of driving stress (e.g., Cuffey and Paterson, 2010), where g is the gravitational acceleration, $\alpha_{si}(= \partial_i h + \alpha_{bi})$ is the ith component of surface slope, and $\partial_i h$ is the thickness gradient in the ith direction. In order to isolate the GIA effect, as noted earlier, we use the present-day ice thickness, i.e., $h(x, y, t) = h(x, y, 0)$, and its gradients in all calculations. Hence, the change in bed

slope due to the GIA uplift (Sect. 3.3) is responsible for modulating the driving stress.

Figure 6 shows the change in driving stress associated with the bed uplift predicted at years AD 2100 and AD 2500. In both cases (Fig. 6a and b), we notice small but similar trends of change in driving stress. Relatively large changes are predicted at positions of larger bed uplift. Reduction in driving stress is evident around the Amery Ice Shelf, implying that the local surface slopes are likely to flatten in this region. Minor increments in driving stress are predicted in the area around Dome C. Complex patterns are predicted around the Amundsen Sea Sector; an extensive zone of reduced driving stress is surrounded by zones with enhanced driving stress (see particularly Fig. 6b).

Theoretically, for the given distribution of ice thickness, changes in bed slope and surface slope (and, hence, the driving stress) should generally be in phase for the topography with forward slope, but out of phase for cases with reverse bed slope. Due to the complex nature of the Antarctic bathymetry, it is an arduously difficult task to find a robust and consistent relationship between changes in bed slope and driving stress, particularly around the Amundsen Sector (compare, for example, Fig. 5c vs. Fig. 6b). Nonetheless, we generally find the reduction in local driving stress associated with the GIA uplift (compare, for example, Fig. 3b vs. Fig. 6b). Given the small order-of-magnitude predictions

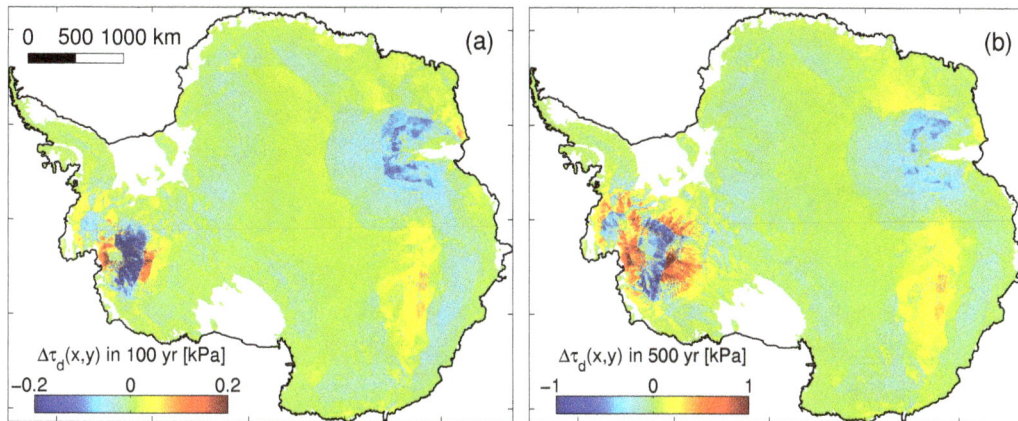

Figure 6. Influence of GIA uplift on the gravitational driving stress. Change in driving stress, $\Delta\tau_d(x,y)$, due to the predicted bed uplift at (a) AD 2100 and (b) AD 2500. Calculations are made for the current distribution of ice thickness. Negative magnitudes imply that surface slopes flatten in the future. As the GIA solutions only perturb the ice–bedrock contact area, we mask out the ice shelves in the figures.

for change in driving stress (i.e., several hundreds of Pascal only), it is important here to note that, as will also be shown in Sect. 4.3, minor changes in driving stress may affect the ice sheet dynamics substantially, as ice velocities are directly proportional to the driving stress by about a power of three (e.g., Cuffey and Paterson, 2010).

4.2 GL migration

In this section, we evaluate the effects of GIA uplift on the GL. We employ the simple hydrostatic equilibrium criterion to identify the transition points by seeking the floating ice thickness, $h_f(x,y,t)$, such that

$$h_f(x,y,t) = -\frac{\rho_w}{\rho}\, b(x,y,t). \tag{3}$$

Regions with ice thickness $h(x,y,t) > h_f(x,y,t)$ are assumed to be grounded and the rest to be floating. Because we use the present-day ice thickness, i.e., $h(x,y,t) = h(x,y,0)$, in all calculations, the extent of the current and future grounded ice (i.e., GL) is determined by the corresponding bathymetry, thus highlighting the influence of predicted GIA solutions.

Using Eq. (3), we compute the extent of the grounded ice for $t = 0$, 100, and 500 yr. Changes in GL are then identified by subtracting, in turn, the first solution from the latter two. Figure 7a, for example, shows the GL migration associated with the GIA uplift predicted for AD 2500. The mask shown in the figure primarily represents GL advance, implying that more ice will be grounded in the future due to the GIA effects. However, we also predict the minor GL retreat in a few scattered locations, particularly along the Wilkes Land coast, where the bedrock is generally predicted to subside partly due to the large snow accumulation simulated under the chosen climate change scenario (Nowicki et al., 2013). Figure 7b, for example, depicts the mask of GL retreat

around the Shackleton Ice Shelf. Note that we also obtain similar but less extensive migration in GL associated with the GIA uplift predicted for AD 2100 (results not shown); there is however no evidence of GL retreat in this case.

Based on the measured ice surface velocities (Rignot et al., 2011), we locate more than 2700 ice flowlines (cf. Supplement Fig. S8) to quantify the magnitude of GL migration (mostly advance) associated with the predictions of GIA uplift at AD 2500. Results are shown for two important regions, namely the Ronne (Fig. 7c) and Ross (Fig. 7d) ice shelves. Figures reveal that the GL may advance by more than 100 km in these two ice shelves. Significant GL advance (by tens of km) is also predicted in the Amery Ice Shelf, Amundsen Sector, Larsen Ice Shelf, Brunt Ice Shelf, and in other regions. In a few locations, e.g., the Shackleton Ice Shelf (Fig. 7b), as noted earlier, we predict a minor retreat (by ≤ 10 km) of the GL.

Although the primary control of GL migration in our calculations is bedrock elevation (Eq. 3), it should be noted that the bedrock slope plays an equally important role (e.g., Lingle and Clark, 1985; Gomez et al., 2010), as summarized in Sect. 1.1. Extensive advancement in GL associated with the GIA uplift is generally consistent with what is expected when slope reduces in the reverse bedslope topography (compare Fig. 5a vs. c, for example, around the Ronne Ice Shelf). For bedrock with forward slopes, however, the advancement in GL can be explained by the GIA-induced increment in bed slopes (compare Fig. 5a vs. c, for example, around the Getz Ice Shelf). The minor GL retreat predicted, for example, in the Shackleton Ice Shelf is associated with the flattening (Fig. 5c) of the forward bed slope (Fig. 5a). Although we are able to show a systematic relationship between the bathymetry, change in bed slope, and the direction of the GL migration in a few cases, it is complicated to provide such one-to-one relationships consistently over the entire ice sheet. Nonetheless, the results indicate that the GIA effects

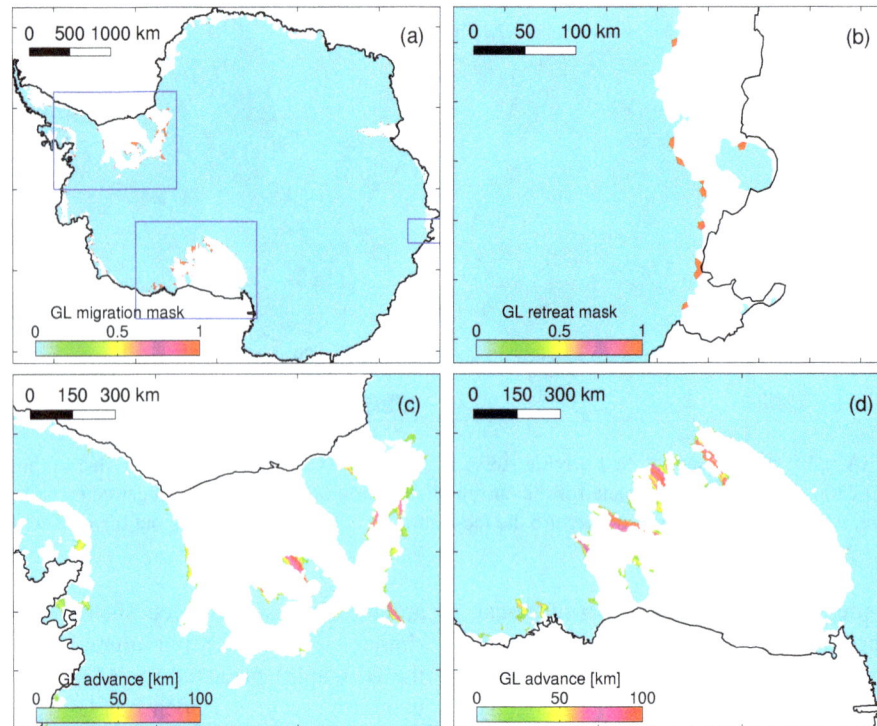

Figure 7. Influence of GIA uplift on the GL. (**a**) The mask of GL migration associated with the GIA solution at AD 2500. Calculations are based on the hydrostatic equilibrium criterion for the current distribution of ice thickness. Cyan depicts the extent of present-day grounded ice. Red shows the GL migration (mostly advance) due to the GIA uplift. Minor retreats in GL are predicted in a few areas along Wilkes Land in the EAIS; their extents are limited to one or two mesh elements, i.e., ≤ 10 km. (**b**) Mask of example retreat around the Shackleton Ice Shelf, as seen by zooming into the small blue box in Fig. 7a. Other large boxes in Fig. 7a enclose two important regions that are magnified: (**c**) the Ronne Ice Shelf and the Bellingshausen Sea Sector; and (**d**) the Ross Ice Shelf. Color codes illustrate the magnitude of GL advance measured along the ice flowlines.

generally support the extension of grounded ice (i.e., GL advance) in the future, thereby promoting the stability in the marine portions of the ice sheet that rest on a reverse bed slope.

4.3   Ice surface velocities

Finally, we analyze the influence of GIA uplift on the ice surface velocities. By solving the quasi-static thermomechanical problem of ice flow, we calculate the englacial velocity field of the AIS with and without GIA effects. We assume that higher-order physics based on the equations governing mass and momentum conservation (Pattyn, 2003) together with the constitutive relations for isotropic ice (Glen, 1955) describe the internal creep deformation of ice, and that a viscous law of friction (e.g., MacAyeal, 1993) governs basal sliding. We rely on a steady-state thermal problem identical for all simulations, based on present-day conditions. For simplicity, we do not consider the possibility of till deformation underneath the ice sheet. A description of ice rheology (Glen, 1955) and other common assumptions related to ice flow modeling can be found elsewhere (e.g., Cuffey and Paterson, 2010).

For the present-day setting of the AIS, we solve this problem of ice dynamics through diagnostic simulation of the 3-D ice flow capability of ISSM, satisfying a number of boundary conditions (e.g., Larour et al., 2012b). We impose (i) a stress-free condition at the ice–atmosphere interface, (ii) water pressure directing towards the ice sheet at the ice–ocean (peripheral) interface, and (iii) a sliding condition governed by the basal friction at the ice–bed interface. Zero friction is applied at the base of the ice shelf (i.e., free-floating condition), while basal friction under the grounded ice is inferred from InSAR (Interferometric Synthetic Aperture Radar) based surface velocities (Rignot et al., 2011) using a data assimilation technique (e.g., Morlighem et al., 2010). The basal friction pattern is similar to the one described in Morlighem et al. (2013).

We re-run the simulation for two additional cases, associated with the predictions of GIA uplift at years AD 2100 and AD 2500. In each case, we upgrade the bedrock and surface elevations of the grounded ice; the extent of the grounded ice (GLs) is also updated (cf. Sect. 4.2). For floating ice as well, we upgrade both bed and surface elevations so that all floating ice is in hydrostatic equilibrium, which

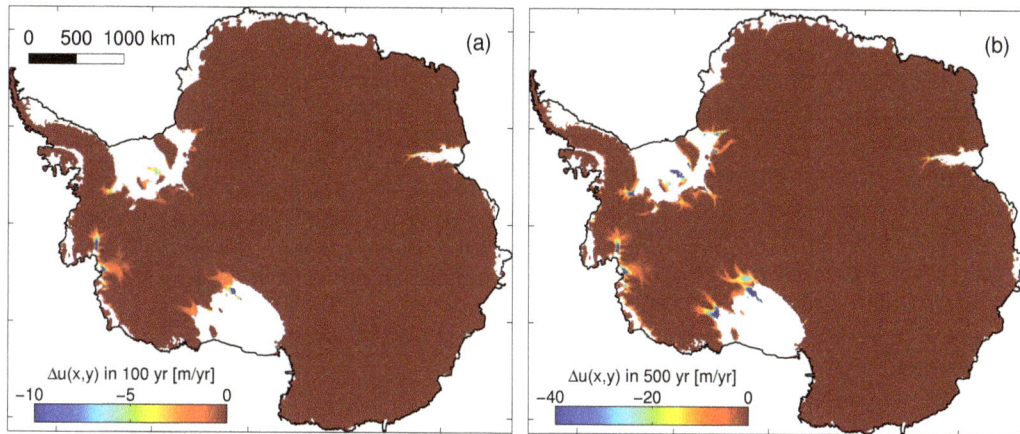

Figure 8. Influence of GIA uplift on the ice surface velocities. Change in ice surface velocity, $\Delta u(x, y)$, due to the predicted bed uplift at (**a**) AD 2100 and (**b**) AD 2500. Calculations are made by running the diagnostic simulation of the 3-D ice flow (higher-order) capability of ISSM. Other model setup and boundary conditions are consistent with those of the SeaRISE control experiment (cf. Bindschadler et al., 2013). A systematic reduction in velocity, which can be attributed partly to the reduction in $\tau_d(x, y)$ (around the Amundsen Sector and the Amery Ice Shelf; cf. Fig. 6) and partly to the GL advance (particularly in the large ice shelves; cf. Fig. 7), indicates the stabilizing effects of GIA uplift on the future dynamics of AIS.

ensures continuity of the driving stress at the GL. Applying the same boundary conditions discussed above (except in the newly grounded or floating areas), we compute the englacial velocity for both cases associated with the future GIA uplift. In areas presently floating that become grounded at $t = 100$ and 500 yr (cf. Fig. 7a), we update basal friction assuming that it is roughly equal to the gravitational driving stress (Morlighem et al., 2013). Similarly, we impose a free-floating condition in areas presently grounded that float in the future (cf. Fig. 7b).

For a given vertical profile of an ice sheet, the maximum velocity is always observed at the surface. We therefore place our emphasis upon the GIA influence on the ice surface velocities. Using the simulation results discussed above, we compute the GIA-induced change in surface velocity associated with the predictions of GIA uplift at years AD 2100 (Fig. 8a) and AD 2500 (Fig. 8b). In both cases, we find similar patterns of change in ice surface velocity. Although the predicted changes are small, about two to three orders of magnitude smaller than the surface velocities themselves (Rignot et al., 2011), a systematic reduction in velocity is evident around the sheet–shelf margins. This suggests that the GIA effects generally contribute to decelerating the flow speed across the GL, and hence promote stability in the marine portions of the ice sheet. Note that we mask out the ice shelves in our figures, for we have no intention of making predictions in the ice shelves.

The predicted changes in surface velocity for the grounded ice can be interpreted as the combined effects of changes in driving stress (cf. Sect. 4.1 and Fig. 6) and the GL (cf. Sect. 4.2 and Fig. 7) associated with the GIA uplift. Around the Ross and Ronne ice shelves, the GIA-induced reduction in surface velocity is consistent with the GL advance. In other

regions, e.g., the Amundsen Sea Sector and the Amery Ice Shelf, predicted reduction in ice velocity can be attributed partly to the GL advance and partly to the reduced driving stress. All in all, the effects of GIA on several aspects of ice dynamics (e.g., driving stress, GL, and ice surface velocities) are consistent in that the GIA promotes systematic stability in marine portions of the AIS in the future.

5 Conclusions

This study has examined the interplay between the ice sheet evolution and the solid Earth responses for the AIS. First, we compute the future uplift of the Antarctic bedrock using the calibrated ISSM/GIA model driven by the inferred and predicted evolution of the ice sheet. Next, we evaluate how such a response of the solid Earth impacts AIS dynamics.

Our calculations are based on several approximations of model physics and numerics; it is important to highlight some of these here. The GIA model describes a simple two-layer representation of the solid Earth; the model and material viscoelastic parameters are kept constant spatially and the viscosity is Newtonian. Our ice sheet model solves the quasi-static thermomechanical flow problem for higher-order mechanics; the GLs are determined by the hydrostatic equilibrium criterion. A more comprehensive exploration of the positive or negative IS/SE feedbacks is warranted in the future. There is much to be learned from GIA models that employ additional GPS data, possibly right in the heart of the Amundsen Sea Sector, where viscoelastic uplift rates may approach 40 mm yr^{-1} (Groh et al., 2012). Our computations of ice loading after the present-day rely on several ice sheet models driven by the melt-dominated forcing under the R8 scenario. The computed model-average GIA response provides our assessment of its impact on the ice sheet dynamics. While there are limitations to the data and methods employed, our research brings us to the following two important conclusions of broader interest.

First, the short-term viscoelastic response of the solid Earth to the future ice load change, rather than its long-term viscous response to the past loading, governs the future evolution of the Antarctic bed topography. The magnitude and spatial variability in the future bed uplift are therefore determined by the nature of the future evolution of the AIS. A larger uplift is expected, for example, where the ice sheet loses more mass, while its far-field consequences seem to involve a relatively small amplitude subsidence. Our calculations suggest that the Antarctic bed may rise by a few meters and a few tens of meters around the WAIS, particularly the Amundsen Sea Sector and the Amery Ice Shelf, at years AD 2100 and AD 2500, respectively. Minor subsidences of about one meter and a few meters are predicted along the Wilkes Land coast at the respective times, partly caused by the net accumulation in the climate scenario runs (Nowicki et al., 2013). The interior of the EAIS is likely to remain unchanged.

Second, a pervasive and large uplift predicted in the interior of the WAIS, a substantially marine-based ice sheet, has particular significance because it generally corresponds to the flattening of the reverse bed slope. This drives the GL forward and consequently promotes the stability of the ice sheet. Our calculations, based on the present-day setting of the AIS perturbed by the predicted GIA uplift, reveal that the GL may advance by more than 100 km in the Ross and Ronne ice shelves due to the predicted GIA uplift for AD 2500. This may reduce the future ice surface velocities across the GLs by several tens of meter per annum.

The conclusions summarized above indicate a negative feedback between the ice sheet evolution and the solid Earth response for the marine ice sheet. For areas with reverse bed slope, for example, loss in ice mass flattens the bed and drives the GL forward and hence decelerates the rate of mass loss. Although our model is capable of illustrating this mechanism systematically, accurate quantification of its significance requires a dynamically coupled IS/SE model including the intricate details of ice sheet stability to breakup. This negative feedback is consistent with the ice sheet and sea level simulations computed by Gomez et al. (2010, 2013) wherein loss in ice mass reduces the local sea level due to self-gravitation and hence decelerates the rate of mass loss. For accurate simulations of the AIS on centennial timescales under the reasonable climate change scenarios, both the solid Earth and sea level changes proximal to the GL may need to be properly accounted for.

Acknowledgements. This study was performed at the Caltech Jet Propulsion Laboratory under a contract with the NASA's Cryosphere Science and the Solid Earth Surface and Interior Focus Area programs. The authors are indebted to the SeaRISE participants for providing them with the valuable data. Conversations with Volker Klemann and Pippa Whitehouse are acknowledged. Constructive comments by two anonymous reviewers greatly improved this manuscript. Surendra Adhikari is thankful to Victor Tsai for hosting him as a postdoctoral scholar at Caltech, which made the completion of this writeup possible.

Edited by: M. Poutanen

References

Bamber, J. L. and Aspinall, W. P.: An expert judgement assessment of future sea level rise from the ice sheets, Nature Clim. Change, 3, 424–427, 2013.

Bamber, J. L., Gomez-Dans, J. L., and Griggs, J. A.: A new 1 km digital elevation model of the Antarctic derived from combined satellite radar and laser data - Part 1: Data and methods, The Cryosphere, 3, 101–111, http://www.the-cryosphere.net/3/101/2009/, 2009.

Bindschadler, R., Nowicki, S., Abe-Ouchi, A., Aschwanden, A., Choi, H., Fastook, J., Granzow, G., Greve, R., Gutowski, G., Herzfeld, U., Jackson, C., Johnson, J., Khroulev, C., Levermann, A., Lipscomb, W., Martin, M., Morlighem, M., Parizek, B., Pollard, D., Price, S., Ren, D., Saito, F., Sato, T., Seddik, H., Seroussi, H., Takahashi, F., Walker, R., and Wang, W.: Ice-Sheet Model Sensitivities to Environmental Forcing and Their Use in Projecting Future Sea-Level (The SeaRISE Project), J. Glaciol., 59, 195–224, doi:10.3189/2013JoG12J125, 2013.

Borstad, C. P., Khazendar, A., Larour, E., Morlighem, M., Rignot, E., Schodlok, M. P., and Seroussi, H.: A damage mechanics assessment of the Larsen B ice shelf prior to collapse: Toward a physically-based calving law, Geophys. Res. Lett., 39, 1–5, doi:10.1029/2012GL053317, 2012.

Clark, P. U., Dyke, A. S., Shakun, J. D., Carlson, A. E., Clark, J., Wohlfarth, B., Mitrovica, J. X., Hostetler, S. W., and Marshall McCabe, A.: The last glacial maximum, Science, 5941, 710–714, doi:10.1126/science.1172873, 2009.

Conway, H., Hall, B. L., Denton, G. H., Gades, A. M., and Waddington, E. D.: Past and future grounding-line retreat of the West Antarctic Ice Sheet, Science, 286, 280–283, doi:10.1126/science.286.5438.280, 1999.

Cook, C. P., van de Flierdt, T., Williams, T., Hemming, S. R., Iwai, M., Kobayashi, M., Jimenez-Espejo, F. J., Escutia, C., González, J. J., Khim, B.-K., McKay, R. M., Passchier, S., Bohaty, S. M., Riesselman, C. R., Tauxe, L., Sugisaki, S., Lopez Galindo, A., Patterson, M. O., Sangiorgi, F., Pierce, E. L., Brinkhuis, H., Klaus, A., Fehr, A., Bendle, J. A. P., Bijl, P. K., A. Carr, S., Dunbar, R. B., Flores, J. A., Hayden, T. G., Katsuki, K., Kong, G. S., Nakai, M., Olney, M. P., Pekar, S. F., Pross, J., Röhl, U., Sakai, T., Shrivastava, P. K., Stickley, C. E., Tuo, S., Welsh, K., and Yamane, M.: Dynamic behaviour of the East Antarctic ice sheet during Pliocene warmth, Nat. Geosci., 6, 765–769, doi:10.1038/ngeo1889, 2013.

Cuffey, K. and Paterson, W. S. B.: The Physics of Glaciers, 4th Edition, Elsevier, 2010.

Depoorter, M. A., Bamber, J. L., Griggs, J. A., Lenaerts, J. T. M., Ligtenberg, S. R. M., van den Broeke, M. R., and Moholdt, G.: Calving fluxes and basal melt rates of Antarctic ice shelves, Nature, 502, 89–92, doi:10.1038/nature12567, 2013.

Dziewonski, A. M. and Anderson, D. L.: Preliminary reference Earth model, Phys. Earth Planet. Inter., 25, 297–356, 1981.

Favier, L., Gagliardini, O., Durand, G., and Zwinger, T.: A three-dimensional full Stokes model of the grounding line dynamics: effect of a pinning point beneath the ice shelf, The Cryosphere, 6, 101–112, doi:10.5194/tc-6-101-2012, 2012.

Favier, V., Agosta, C., Parouty, S., Durand, G., Delaygue, G., Gallee, H., Drouet, A., Trouvilliez, A., and Krinner, G.: An updated and quality controlled surface mass balance dataset for Antarctica, The Cryosphere, 7, 583–597, doi:10.5194/tc-7-583-2013, 2013.

Fjeldskaar, W.: The amplitude and decay of the glacial forebulge in Fennoscandia, Norsk Geologisk Tidsskrift, 74, 2–8, 1994.

Glen, J.: The creep of polycrystalline ice, Proc. R. Soc. A, 228, 519–538, 1955.

Gomez, N., Mitrovica, J. X., Huybers, P., and Clark, P. U.: Sea level as a stabilizing factor for marine-ice-sheet grounding lines, Nat. Geosci., 3, 850–853, doi:10.1038/ngeo1012, 2010.

Gomez, N., Pollard, D., and Mitrovica, J.: A 3-D coupled ice sheet - sea level model applied to Antarctica through the last 40 ky, Earth Planet. Sci. Lett., 384, 88–99, 2013.

Greve, R.: A continuum-mechanical formulation for shallow polythermal ice sheets, Phil. Trans R. Soc. A, 355, 921–974, 1997.

Groh, A., Ewert, H., Scheinert, M., Fritsche, M., Rülke, A., Richter, A., Rosenau, R., and Dietrich, R.: An investigation of glacial isostatic adjustment over the Amundsen Sea Sector, West Antarctica, Glob. Planet. Change, 98–99, 45–53, 2012.

Hughes, T., Sargent, A., and Fastook, J.: Ice-bed coupling beneath and beyond ice streams: Byrd Glacier, Antarctica, J. Geophys. Res., 116, 1–17, doi:10.1029/2010JF001896, 2011.

IPCC-AR4: Fourth Assessment Report: Climate Change 2007: The AR4 Synthesis Report, Geneva: IPCC, http://www.ipcc.ch/ipccreports/ar4-wg1.htm, 2007.

Ivins, E. R. and James, T. S.: Simple models for late Holocene and present-day Patagonian glacier fluctuations and predictions of a geodetically detectable isostatic response, Geophys. J. Int., 138, 601–624, doi:10.1046/j.1365-246x.1999.00899.x, 1999.

Ivins, E. R. and James, T. S.: Antarctic glacial isostatic adjustment: a new assessment, Antarct. Sci., 17, 541–553, 2005.

Ivins, E. R., James, T. S., Wahr, J., O. Schrama, E. J., Landerer, F. W., and Simon, K. M.: Antarctic contribution to sea level rise observed by GRACE with improved GIA correction, J. Geophys. Res., 118, 3126–3141, doi:10.1002/jgrb.50208, 2013.

James, T. S. and Ivins, E. R.: Predictions of Antarctic crustal motions driven by present-day ice sheet evolution and by isostatic memory of the Last Glacial Maximum, J. Geophys. Res., 103, 4993–5017, doi:10.1029/97JB03539, 1998.

Katz, R. and Worster, M.: Stability of ice-sheet grounding lines, Proc. R. Soc. A, 466, 1597–1620, doi:10.1098/rspa.2009.0434, 2010.

Kessler, M. A., Anderson, R. S., and Briner, J. P.: Fjord insertion into continental margins driven by topographic steering of ice, Nat. Geosci., 1, 365–369, doi:10.1038/ngeo201, 2008.

Konrad, H., Thoma, M., Sasgen, I., Klemann, V., Grosfeld, K., Barbi, D., and Martinec, Z.: The deformational response of a viscoelastic solid Earth model coupled to a thermomechanical ice sheet model, Survey in Geophysics, pp. 1–18, doi:10.1007/s10712-013-9257-8, 2013.

Larour, E., Schiermeier, J., Rignot, E., Seroussi, H., Morlighem, M., and Paden, J.: Sensitivity Analysis of Pine Island Glacier ice flow using ISSM and DAKOTA, J. Geophys. Res., 117, F02009, 1–16, doi:10.1029/2011JF002146, 2012a.

Larour, E., Seroussi, H., Morlighem, M., and Rignot, E.: Continental scale, high order, high spatial resolution, ice sheet modeling using the Ice Sheet System Model (ISSM), J. Geophys. Res., 117, F01022, 1–20, doi:10.1029/2011JF002140, 2012b.

Le Meur, E. and Huybrechts, P.: A comparison of different ways of dealing with isostasy: examples from modelling the Antarctic ice sheet during the last glacial cycle, Ann. Glaciol., 23, 309–317, 1996.

Le Meur, E. and Huybrechts, P.: A model computation of the temporal changes of surface gravity and geoidal signal induced by the evolving Greenland ice sheet, Geophys. J. Int., 145, 835–849, 2001.

Lingle, C. and Clark, J.: A numerical model of interactions between a marine ice sheet and the solid earth: application to a west Antarctic ice stream, J. Geophys. Res., 90, 1100–1114, 1985.

MacAyeal, D.: A tutorial on the use of control methods in ice-sheet modeling, J. Glaciol., 39, 91–98, 1993.

McConnell, R. K. J.: Isostatic adjustment in a layered Earth, J. Geophys. Res., 70, 5171–5188, doi:10.1029/JZ070i020p05171, 1965.

Morlighem, M., Rignot, E., Seroussi, H., Larour, E., Ben Dhia, H., and Aubry, D.: Spatial patterns of basal drag inferred using control methods from a full-Stokes and simpler models for Pine Island Glacier, West Antarctica, Geophys. Res. Lett., 37, L14502, 1–6, doi:10.1029/2010GL043853, 2010.

Morlighem, M., Seroussi, H., Larour, E., and Rignot, E.: Inversion of basal friction in Antarctica using exact and incomplete adjoints of a higher-order model, J. Geophys. Res., 118, 1746–1753, doi:10.1002/jgrf.20125, 2013.

Naish, T., Powell, R., Levy, R., Wilson, G., Scherer, R., Talarico, F., Krissek, L., Niessen, F., Pompilio, M., Wilson, T., Carter, L., DeConto, R., Huybers, P., McKay, R., Pollard, D., Ross, J., Winter, D., Barrett, P., Browne, G., Cody, R., Cowan, E., Crampton, J., Dunbar, G., Dunbar, N., Florindo, F., Gebhardt, C., Graham, I., Hannah, M., Hansaraj, D., Harwood, D., Helling, D., Henrys, S., Hinnov, L., Kuhn, G., Kyle, P., Läufer, A., Maffioli, P., Magens, D., Mandernack, K., McIntosh, W., Millan, C., Morin, C., Ohneiser, C., Paulsen, T., Persico, D., Raine, I., Reed, J., Riesselmann, C., Sagnotti, L., Schmitt, D., Sjunneskog, C., Strong, P., Taviani, M., Vogel, S., Wilch, T., and Williams, T.: Obliquity-paced pliocene West Antarctic ice sheet oscillations, Nature, 458, 322–329, doi:10.1038/nature07867, 2009.

Nowicki, S., Bindschadler, R., Abe-Ouchi, A., Aschwanden, A., Bueler, E., Choi, H., Fastook, J., Granzow, G., Greve, R., Gutowski, G., Herzfeld, U., Jackson, C., Johnson, J., Khroulev, C., Larour, E., Levermann, A., Lipscomb, W., Martin, M., Morlighem, M., Parizek, B., Pollard, D., Price, S., Ren, D., Rignot, E., Saito, F., Sato, T., Seddik, H., Seroussi, H., Takahashi, K., Walker, R., and Wang, W.: Insights into spatial sensitivies of ice mass response to environmental change from the SeaRISE ice sheet modeling project I: Antarctica, J. Geophys. Res., 118, 1–23, doi:10.1002/jgrf.20081, 2013.

Pattyn, F.: A new three-dimensional higher-order thermomechanical ice sheet model: Basic sensitivity, ice stream development, and ice flow across subglacial lakes, J. Geophys. Res., 108, 1–15, doi:10.1029/2002JB002329, 2003.

Peltier, W.: Global glacial isostasy and the surface of the ice-age Earth: The 699 ICE-5G (VM2) model and GRACE, Annu. Rev. Earth Planet. Sci., 32, 111–149, 2004.

Pollard, D. and DeConto, R.: Modelling West Antarctica ice sheet growth and collapse through the past five million years, Nature, Letters 458, 329–332, http://www.nature.com/nature/journal/v458/n7236/abs/nature07809.html, 2009.

Pritchard, H. D., Ligtenberg, S. R. M., Fricker, H. A., Vaughan, D. G., van den Broeke, M. R., and Padman, L.: Antarctic ice-sheet loss driven by basal melting of ice shelves, Nature, 484, 502–505, doi:10.1038/nature10968, 2012.

Raymo, M., Mitrovica, J. X., O'Leary, M., DeConto, R., and Hearty, P.: Departures from eustasy in Pliocene sea-level records, Nature Geosci., 4, 328 – 332, doi:http://dx.doi.org/10.1038/ngeo1118, 2011.

Rignot, E., Mouginot, J., and Scheuchl, B.: Ice Flow of the Antarctic Ice Sheet, Science, 333, 1427–1430, doi:10.1126/science.1208336, 2011.

Rignot, E., Jacobs, S., Mouginot, J., and Scheuchl, B.: Ice shelf melting around Antarctica, Science, doi:10.1126/science.1235798, 2013.

Schoof, C.: Ice sheet grounding line dynamics: Steady states, stability, and hysteresis, J. Geophys. Res., 112, 1–19, doi:10.1029/2006JF000664, 2007.

Seroussi, H., Ben Dhia, H., Morlighem, M., Rignot, E., Larour, E., and Aubry, D.: Coupling ice flow models of varying order of complexity with the Tiling Method, J. Glaciol., 58 (210), 776–786, doi:10.3189/2012JoG11J195, 2012.

Siegert, M., Ross, N., Corr, H., Kingslake, J., and Hindmarsh, R.: Late Holocene ice-flow reconfiguration in the Weddell Sea sector of West Antarctica, Quaternary Science Rev., 78, 98–107, 2013.

Smith, A. M., Bentley, C. R., Bingham, R. G., and Jordan, T. A.: Rapid subglacial erosion beneath Pine Island Glacier, West Antarctica, Geophys. Res. Lett., 39, L12501, doi:10.1029/2012GL051651, 2012.

Thomas, I. D., King, M. A., Bentley, M. J., Whitehouse, P. L., Penna, N. T., Williams, S. D. P., Riva, R. E. M., Lavallée, D. A., Clarke, P. J., King, E. C., Hindmarsh, R. C. A., and Koivula, H.: Widespread low rates of Antarctic glacial isostatic adjustment revealed by GPS observations, Geophys. Res. Lett., 38, L22302, doi:10.1029/2011GL049277, 2011.

Thomas, R.: Thickening of the Ross Ice Shelf and equilibrium stat of the West Antarctic Ice Sheet, Nature, 259, 180–183, 1976.

van Vuuren, D., Edmonds, J., Kainuma, M., Riahi, K., Thomson, A., Hibbard, K., Hurtt, G., Kram, T., Krey, V., Lamarque, J., Masui, T., Meinshausen, M., Nakicenovic, N., Smith, S., and Rose, S.: The representative concentration pathways: an overview, Clim. Dynam., 109, 5–31, 2011.

Vaughan, D. and Arthern, R.: Why is it hard to predict the future of ice sheets?, Science, 315, 1503–1504, 2007.

Verfaillie, D., Fily, M., Le Meur, E., Magand, O., Jourdain, B., Arnaud, L., and Favier, V.: Snow accumulation variability derived from radar and firn core data along a 600 km transect in Adelie Land, East Antarctic plateau, The Cryosphere, 6, 1345–1358, doi:10.5194/tc-6-1345-2012, 2012.

Walker, R. T., Holland, D. M., Parizek, B. R., Alley, R. B., Nowicki, S. M. J., and Jenkins, A.: Efficient flowline simulations of ice shelf-ocean interactions: sensitivity studies with a fully coupled model, J. Phys. Oceanogr., 43, 2200–2210, doi:10.1175/JPO-D-13-037.1, 2013.

Wang, W., Li, J., and Zwally, J.: Dynamic inland propagation of thinning due to ice loss at the margins of the Greenland ice sheet, J. Glaciol., 58, 734–740, 2012.

Weertman, J.: Stability of the junction of an ice sheet and an ice shelf, J. Glaciol., 13(67), 3–11, 1974.

Whitehouse, P., Bentley, M. J., and Brocq, A. M. L.: A deglacial model for Antarctica: geological constraints and glaciological modelling as a basis for a new model of Antarctic glacial isostatic adjustment, Quaternary Sci. Rev., 32, 1–24, 2012.

Winkelmann, R., Martin, M. A., Haseloff, M., Albrecht, T., Bueler, E., Khroulev, C., and Levermann, A.: The Potsdam Parallel Ice Sheet Model (PISM-PIK) - Part 1: Model description, Cryosphere, 5, 715–726, doi:10.5194/tc-5-715-2011, 2011.

Winkelmann, R., Levermann, A., Martin, M. A., and Frieler, K.: Increased future ice discharge from Antarctica owing to higher snowfall, Nature, 492, 239–242, doi:10.1038/nature11616, 2012.

Wolf, D.: The normal modes of a uniform, incompressible Maxwell half-space, J. Geophys., 56, 106–117, http://elib.uni-stuttgart.de/opus/volltexte/2010/5078, 1985.

Permissions

The contributors of this book come from diverse backgrounds, making this book a truly international effort. This book will bring forth new frontiers with its revolutionizing research information and detailed analysis of the nascent developments around the world.

We would like to thank all the contributing authors for lending their expertise to make the book truly unique. They have played a crucial role in the development of this book. Without their invaluable contributions this book wouldn't have been possible. They have made vital efforts to compile up to date information on the varied aspects of this subject to make this book a valuable addition to the collection of many professionals and students.

This book was conceptualized with the vision of imparting up-to-date information and advanced data in this field. To ensure the same, a matchless editorial board was set up. Every individual on the board went through rigorous rounds of assessment to prove their worth. After which they invested a large part of their time researching and compiling the most relevant data for our readers.

The editorial board has been involved in producing this book since its inception. They have spent rigorous hours researching and exploring the diverse topics which have resulted in the successful publishing of this book. They have passed on their knowledge of decades through this book. To expedite this challenging task, the publisher supported the team at every step. A small team of assistant editors was also appointed to further simplify the editing procedure and attain best results for the readers.

Apart from the editorial board, the designing team has also invested a significant amount of their time in understanding the subject and creating the most relevant covers. They scrutinized every image to scout for the most suitable representation of the subject and create an appropriate cover for the book.

The publishing team has been an ardent support to the editorial, designing and production team. Their endless efforts to recruit the best for this project, has resulted in the accomplishment of this book. They are a veteran in the field of academics and their pool of knowledge is as vast as their experience in printing. Their expertise and guidance has proved useful at every step. Their uncompromising quality standards have made this book an exceptional effort. Their encouragement from time to time has been an inspiration for everyone.

The publisher and the editorial board hope that this book will prove to be a valuable piece of knowledge for researchers, students, practitioners and scholars across the globe.

List of Contributors

I. Janutyte
NORSAR, Kjeller, Norway
Vilnius University, Vilnius, Lithuania
Lithuanian Geological Survey, Vilnius, Lithuania

M. Majdanski
Institute of Geophysics Polish Academy of Sciences, Warsaw, Poland

P. H. Voss
Geological Survey of Denmark and Greenland – GEUS, Copenhagen, Denmark

E. Kozlovskaya
Sodankylä Geophysical Observatory/Oulu Unit, University of Oulu, Oulu, Finland

D. L. de Castro
Programa de Pós-Graduação em Geodinâmica e Geofísica, Universidade Federal do Rio Grande do Norte, Natal, Brazil

F. H. R. Bezerra
Programa de Pós-Graduação em Geodinâmica e Geofísica, Universidade Federal do Rio Grande do Norte, Natal, Brazil

B. Gaite
Institute of Earth Sciences Jaume Almera, ICTJA-CSIC, Lluis Sole i Sabaris s/n, 08028 Barcelona, Spain

A. Villaseñor
Institute of Earth Sciences Jaume Almera, ICTJA-CSIC, Lluis Sole i Sabaris s/n, 08028 Barcelona, Spain

A. Iglesias
Institute of Geophysics, Universidad Nacional Autónoma de México, Mexico City, Mexico

M. Herraiz
Department of Geophysics and Meteorology, Universidad Complutense de Madrid, Madrid, Spain
Institute of Geosciences (UCM, CSIC), Madrid, Spain

I. Jiménez-Munt
Institute of Earth Sciences Jaume Almera, ICTJA-CSIC, Lluis Sole i Sabaris s/n, 08028 Barcelona, Spain

L. T. White
Southeast Asia Research Group, Department of Earth Sciences, Royal Holloway University of London, Egham, Surrey, UK

Research School of Earth Sciences, The Australian National University, Canberra, ACT, Australia
Geoscience Australia, Canberra, Australia

M. P. Morse
Research School of Earth Sciences, The Australian National University, Canberra, ACT, Australia
Geoscience Australia, Canberra, Australia

G. S. Lister
Research School of Earth Sciences, The Australian National University, Canberra, ACT, Australia
Geoscience Australia, Canberra, Australia

M. Pantaleo
Department 2, Physics of the Earth, Helmholtz Centre Potsdam, GFZ German Research Centre for Geoscience, Potsdam 14473, Germany

T. R. Walter
Department 2, Physics of the Earth, Helmholtz Centre Potsdam, GFZ German Research Centre for Geoscience, Potsdam 14473, Germany

L. Parras-Alcántara
Department of Agricultural Chemistry and Soil Science, Faculty of Science, Agrifood Campus of International Excellence –ceiA3, University of Cordoba, 14071 Cordoba, Spain

B. Lozano-García
Department of Agricultural Chemistry and Soil Science, Faculty of Science, Agrifood Campus of International Excellence –ceiA3, University of Cordoba, 14071 Cordoba, Spain

T. J. Jones
Department of Earth, Ocean and Atmospheric Sciences, University of British Columbia, Vancouver, V6T 1 Z4, Canada
School of Earth Sciences, University of Bristol, Wills Memorial Building, Bristol, BS8 1RJ, UK

J. K. Russell
Department of Earth, Ocean and Atmospheric Sciences, University of British Columbia, Vancouver, V6T 1 Z4, Canada

L. A. Porritt
Department of Earth, Ocean and Atmospheric Sciences, University of British Columbia, Vancouver, V6T 1 Z4, Canada

School of Earth Sciences, University of Bristol, Wills Memorial Building, Bristol, BS8 1RJ, UK

R. J. Brown
Department of Earth Sciences, Science Labs, Durham University, Durham, DH1 3LE, UK

H. Steffen
Lantmäteriet, Lantmäterigatan 2c, 80182 Gävle, Sweden

G. Kaufmann
Freie Universität Berlin, Institut für Geologische Wissenschaften, Fachrichtung Geophysik, Malteserstr. 74–100, Haus D, 12249 Berlin, Germany

R. Lampe
Ernst-Moritz-Arndt-Universität Greifswald, Institut für Geographie und Geologie, F.-L.-Jahn-Str. 16, 17487 Greifswald, Germany

P. Wu
Department of Geoscience, University of Calgary, 2500 University Drive NW, Calgary, AB, T2N 1N4, Canada
Department of Earth Sciences, The University of Hong Kong, Pokfulam Road, Hong Kong

H. Wang
State Key Laboratory of Geodesy and Earth's Dynamics, Institute of Geodesy and Geophysics, Chinese Academy of Sciences, Wuhan 430077, China

M. Dec
Institute of Geophysics, Polish Academy of Sciences, Ks. Janusza 64, 01-452 Warsaw, Poland

M. Malinowski
Institute of Geophysics, Polish Academy of Sciences, Ks. Janusza 64, 01-452 Warsaw, Poland

E. Perchuc
Bernardynska 21/57, 02-904 Warsaw, Poland

M. E. T. Quinquis
Geodynamics Team, Geological Survey of Norway (NGU), Trondheim, Norway

S. J. H. Buiter
Geodynamics Team, Geological Survey of Norway (NGU), Trondheim, Norway
Centre for Earth Evolution and Dynamics, University of Oslo, Oslo, Norway

S. Adhikari
Jet Propulsion Laboratory, California Institute of Technology, 4800 Oak Grove Dr., Pasadena, CA 91109, USA

E. R. Ivins
Jet Propulsion Laboratory, California Institute of Technology, 4800 Oak Grove Dr., Pasadena, CA 91109, USA
Division of Geological and Planetary Sciences, California Institute of Technology, 1200 E. California Blvd., Pasadena, CA 91125, USA

E. Larour
Jet Propulsion Laboratory, California Institute of Technology, 4800 Oak Grove Dr., Pasadena, CA 91109, USA

H. Seroussi
Jet Propulsion Laboratory, California Institute of Technology, 4800 Oak Grove Dr., Pasadena, CA 91109, USA

M. Morlighem
Department of Earth System Science, University of California – Irvine, 3200 Croul Hall, Irvine, CA 92697, USA

S. Nowicki
Code 615, NASA Goddard Space Flight Center, Greenbelt, Maryland, USA

www.ingramcontent.com/pod-product-compliance
Lightning Source LLC
Chambersburg PA
CBHW050455200326
41458CB00014B/5189